AF794884

Federated Learning for Smart Agriculture and Food Quality Enhancement

Scrivener Publishing
100 Cummings Center, Suite 541J
Beverly, MA 01915-6106

Artificial Intelligence and Soft Computing for Industrial Transformation

Series Editor: Dr. S. Balamurugan (sbnbala@gmail)

The "Artificial Intelligence and Soft Computing for Industrial Transformation" book series was launched in 2020 and covers various subjects under promising application areas including healthcare, automotive industry, robotics, manufacturing, electronics, marketing, gaming, transportation, energy management, human resource management, power system optimization, social media analytics, environmental engineering, agriculture, entertainment, civil structures, fintech, education, chemical engineering, biotechnology, physics, smart textiles, astronomy, autonomous vehicles, food processing, data security, e-commerce, law, medicine, materials science, and computer vision.

Publishers at Scrivener
Martin Scrivener (martin@scrivenerpublishing.com)
Phillip Carmical (pcarmical@scrivenerpublishing.com)

Federated Learning for Smart Agriculture and Food Quality Enhancement

Edited by

Padmesh Tripathi
Department of Computer Science and Engineering, Delhi Technical Campus, Greater Noida, India

Bhanumati Panda
Department of Applied Sciences, ABES Engineering College, Ghaziabad, UP, India

Shanthi Makka
Department of Computer Science and Engineering (Data Science), Vardhaman College of Engineering, Hyderabad, India

Reeta Mishra
School of Computer Science and Engineering, IILM University, Greater Noida, India

S. Balamurugan
Intelligent Research Consultancy Services, Coimbatore, India

and

Sheng-Lung Peng
Department of Creative Technologies and Product Design, National Taipei University of Business, Taiwan

Scrivener Publishing

WILEY

This edition first published 2026 by John Wiley & Sons, Inc., 111 River Street, Hoboken, NJ 07030, USA and Scrivener Publishing LLC, 100 Cummings Center, Suite 541J, Beverly, MA 01915, USA
© 2026 Scrivener Publishing LLC
For more information about Scrivener publications please visit www.scrivenerpublishing.com.

All rights reserved. No part of this publication may be reproduced, stored in a retrieval system, or transmitted, in any form or by any means, electronic, mechanical, photocopying, recording, or otherwise, except as permitted by law. Advice on how to obtain permission to reuse material from this title is available at http://www.wiley.com/go/permissions.

Wiley Global Headquarters
111 River Street, Hoboken, NJ 07030, USA

For details of our global editorial offices, customer services, and more information about Wiley products visit us at www.wiley.com.

The manufacturer's authorized representative according to the EU General Product Safety Regulation is Wiley-VCH GmbH, Boschstr. 12, 69469 Weinheim, Germany, e-mail: Product_Safety@wiley.com.

Limit of Liability/Disclaimer of Warranty
While the publisher and authors have used their best efforts in preparing this work, they make no representations or warranties with respect to the accuracy or completeness of the contents of this work and specifically disclaim all warranties, including without limitation any implied warranties of merchantability or fitness for a particular purpose. No warranty may be created or extended by sales representatives, written sales materials, or promotional statements for this work. The fact that an organization, website, or product is referred to in this work as a citation and/or potential source of further information does not mean that the publisher and authors endorse the information or services the organization, website, or product may provide or recommendations it may make. This work is sold with the understanding that the publisher is not engaged in rendering professional services. The advice and strategies contained herein may not be suitable for your situation. You should consult with a specialist where appropriate. Neither the publisher nor authors shall be liable for any loss of profit or any other commercial damages, including but not limited to special, incidental, consequential, or other damages. Further, readers should be aware that websites listed in this work may have changed or disappeared between when this work was written and when it is read.

Library of Congress Cataloging-in-Publication Data

ISBN 978-1-394-33869-6

Cover image: Generated with AI using Adobe Firefly
Cover design by Russell Richardson

Set in size of 11pt and Minion Pro by Manila Typesetting Company, Makati, Philippines

Printed in the USA

10 9 8 7 6 5 4 3 2 1

Contents

Preface

Recent years have witnessed the convergence of artificial intelligence (AI), machine learning (ML), and data science with agriculture and food, providing remarkable opportunities to improve quality, sustainability, and productivity in the agricultural sector. Federated Learning (FL) is a promising technology that has emerged at this intersection. This pioneering approach enables decentralized training of machine learning models while keeping sensitive data localized—offering matchless advantages to an industry often challenged by data privacy concerns, resource constraints, and geographically dispersed farming systems. Agricultural researchers, farmers, and food quality specialists can collaborate more meritoriously without sharing proprietary or sensitive data directly, nurturing innovation while preserving privacy and security. This technology can enhance various aspects of agriculture, including climate resilience, pest detection, crop management, and the optimization of food quality standards.

This volume brings together contributions from experts, practitioners, and researchers across agriculture and federated learning. It explores potential challenges, applications, and future directions of FL in agriculture and food quality enhancement through chapters focused on precision farming, soil monitoring, post-harvest processing, and food safety management. The aim is to familiarize the agriculture community with Federated Learning and highlight its potential to enhance food quality and safety in a privacy-preserving and sustainable manner. By integrating domain expertise with advanced computational methods, this book seeks to spark further collaboration and innovation toward more efficient, smarter, and more resilient global food and agriculture systems.

This book contains 14 chapters. A brief description of each chapter follows:

Chapter 1 presents the applications of federated learning in precision agriculture for optimizing crop management practices, pest detection systems, improving yield forecasting accuracy, and enhancing soil health monitoring. Chapter 2 discusses the use of federated learning in food calorie estimation, emphasizing its benefits, methods, and guiding principles.

Chapter 3 explores the application of federated learning, a decentralized technology, to improve food safety, quality standards, and product compliance management. Chapter 4 examines the challenges and factors that must be considered when implementing federated learning in agriculture, such as model fusion, communication limitations, and disparate data.

Chapter 5 analyzes the use of federated learning in food inspection and grading, highlighting its potential to improve accuracy, consistency, and scalability through real-world case studies. Chapter 6 presents a federated learning-based system designed to optimize crop production and enhance market stability by offering personalized, data-driven recommendations while maintaining farmer data privacy.

Chapter 7 explores the use of federated learning for plant disease detection, a vital responsibility for maintaining agricultural output and ensuring food security. Chapter 8 investigates how federated learning enables the development of robust crop classification models without compromising data security, protecting sensitive agricultural information such as crop health, yield predictions, and weather patterns.

Chapter 9 elaborates on the potential of federated learning to transform crop yield prediction, including cause-and-effect relationships and the influence of soil, water, and pest management. Chapter 10 explores the integration of federated learning into various agricultural applications, including precision farming, pest and disease detection, crop yield prediction, soil health monitoring, and climate impact analysis.

Chapter 11 outlines how federated learning supports robust decision-making frameworks to improve food quality, safety, and agricultural efficiency, fostering a more sustainable and resilient food system. Chapter 12 presents a federated differential privacy method using Pyramid Residual Networks (Pyramid ResNet) within a decentralized architecture for estimating crop yields, benchmarking performance against other machine learning models.

Chapter 13 reviews the use of federated learning to detect food adulteration by integrating data from food manufacturers, quality control laboratories, and regulatory bodies to build strong and generalizable models. Chapter 14 demonstrates the potential of federated learning to develop efficient, environmentally friendly models that support the agri-food sector's shift toward sustainability.

We extend our gratitude to all contributors who shared their knowledge, expertise, and insights in making this book a valuable resource. We hope it serves as a stepping stone toward a more data-driven, transparent, and equitable future for agriculture and food quality enhancement. Finally, we offer our thanks to Martin Scrivener and the team at Scrivener Publishing for their effort and guidance in this publication.

Padmesh Tripathi
Department of Computer Science and Engineering, Delhi Technical Campus, Greater Noida, India
Bhanumati Panda
Department of Applied Sciences, ABES Engineering College, Ghaziabad, UP, India
Shanthi Makka
Department of Computer Science and Engineering (Data Science), Vardhaman College of Engineering, Hyderabad, India
Reeta Mishra
School of Computer Science and Engineering, IILM University, Greater Noida, India
S. Balamurugan
Intelligent Research Consultancy Services, Coimbatore, India
Sheng-Lung Peng
Department of Creative Technologies and Product Design, National Taipei University of Business, Taiwan
August 2025

1

Harnessing the Power of Federated Learning for Agricultural Innovation

Abhishek[1*], Mritunjay Rai[1], Anand Prakash Singh[2] and Vishwanath Jha[3]

[1]*Department of Electrical and Electronics Engineering, Shri Ramswaroop Memorial University Barabanki, (U.P.), India*
[2]*IIMT College of Engineering, Greater Noida, UP, India*
[3]*Department of Computer Science Engineering (DS), Indrerprastha Engineering College, Ghaziabad, UP, India*

Abstract

Federated learning (FL) represents a paradigm shift in agricultural technology, offering novel solutions to pressing challenges in food production and quality management. This chapter provides an in-depth examination of FL's transformative application in agriculture, underscoring its pivotal role in advancing data driven decision-making processes while safeguarding data privacy. By enabling decentralized model training across a diverse array of stakeholders including farmers, agronomists, and agricultural researchers, FL facilitates the seamless integration of heterogeneous data sources without necessitating centralization. This decentralized approach not only enhances data security but also addresses privacy concerns associated with traditional data sharing methods. The chapter delves into several key applications of FL within the agricultural domain. In precision agriculture, FL is instrumental in optimizing crop management practices, improving yield forecasting accuracy, and enhancing soil health monitoring. Additionally, FL plays a crucial role in pest detection systems, where it enables the aggregation of data from multiple sensors and sources to improve detection capabilities and response strategies. The chapter also thoroughly addresses the operational and technical challenges associated with deploying FL in diverse agricultural settings. The discussion highlights how these challenges can be mitigated through tailored FL strategies and innovative solutions. The chapter presents a comprehensive analysis of current FL implementations and explores its potential future applications

**Corresponding author*: abverma1991@gmail.com

Padmesh Tripathi, Bhanumati Panda, Shanthi Makka, Reeta Mishra, S. Balamurugan and Sheng-Lung Peng (eds.) Federated Learning for Smart Agriculture and Food Quality Enhancement, (1–26) © 2026 Scrivener Publishing LLC

within agriculture. By incorporating decentralized, privacy-preserving data-processing techniques, FL offers a novel approach to optimizing agricultural practices. This includes improving resource management, increasing efficiency, and facilitating better decision-making in food production processes. This chapter underscores the transformative impact of FL on the agricultural sector. It illustrates how FL can be harnessed to achieve substantial improvements in agricultural practices, address critical challenges, and contribute to the development of robust food management systems. By bridging the gap between advanced data analytics and practical agricultural applications, FL paves the way for a more innovative and secure approach to managing global food resources. This chapter offers a comprehensive analysis of FL's application in agriculture, emphasizing its role in enhancing data driven decision-making while preserving data privacy. FL allows diverse stakeholders including farmers, agronomists, and researchers to collaboratively develop machine learning models using localized data, thereby avoiding central data aggregation and enhancing data security. The chapter meticulously explores FL's application in precision agriculture, where it significantly improves crop management and yield forecasting by integrating diverse data sources. FL's collaborative learning approach enhances predictive accuracy and supports advanced soil health monitoring. Additionally, in pest detection, FL aggregates data from various sensors to refine detection algorithms, thereby improving pest identification and management strategies. Operational and technical challenges of deploying FL in heterogeneous agricultural environments are also critically examined. These challenges include managing data diversity, addressing infrastructure constraints, and integrating FL with existing agricultural systems. The chapter discusses strategies to overcome these issues and optimize FL implementation. Furthermore, the chapter provides a detailed review of current FL applications and investigates potential future directions.

***Keywords*:** Deep learning, FL, precision agriculture, yield forecasting, pest detection, CNN, ANN

1.1 Introduction

The agricultural sector is facing new challenges in 21st century, as the global population predicted to exceed 9.6 billion, as it faces the challenge associated with feeding a global population that is predicted to become 10 billion by the year 2050, which is a huge increase from the 7.7 billion in 2019. To meet this demand, global food production must increase by approximately 60% from 2007 levels, a daunting task considering the simultaneous pressures of climate change, land degradation, and resource scarcity. Climate change, in particular, poses severe threats to agricultural productivity, with altered weather patterns, increased frequency of extreme events, and shifting pest populations, all contributing to reduced yields and greater uncertainty in food production. In this context, the role of digital technologies in transforming agricultural practices cannot be overstated. Among these technologies, precision agriculture has emerged as a crucial approach to optimizing resource use, increasing crop yields, and minimizing environmental impact. Precision agriculture uses advanced technologies like remote sensing, Global Positioning System (GPS)-mounted machinery, and Internet of Things (IoT) sensors to collect and use data on soil condition, crop health, and weather patterns. This data enables farmers to make more precise decisions, applying water, fertilizers, and pesticides exactly where they are needed, in that way improve efficiency and sustainability [1]. Figure 1.1 shows the application of FL in precision farming.

Traditional centralized data-processing models, where data is collected from various sources and analyzed in a central location, face significant barriers, including concerns about data privacy, security, and interoperability. In agriculture, data is often sensitive and proprietary, with farmers and agricultural businesses reluctant to share information due to fears of data breaches, misuse, or loss of competitive advantage. Additionally, the heterogeneous nature of agricultural data, which can vary widely in terms of format, quality, and source, complicates the development of comprehensive models that can be applied across different regions and farming

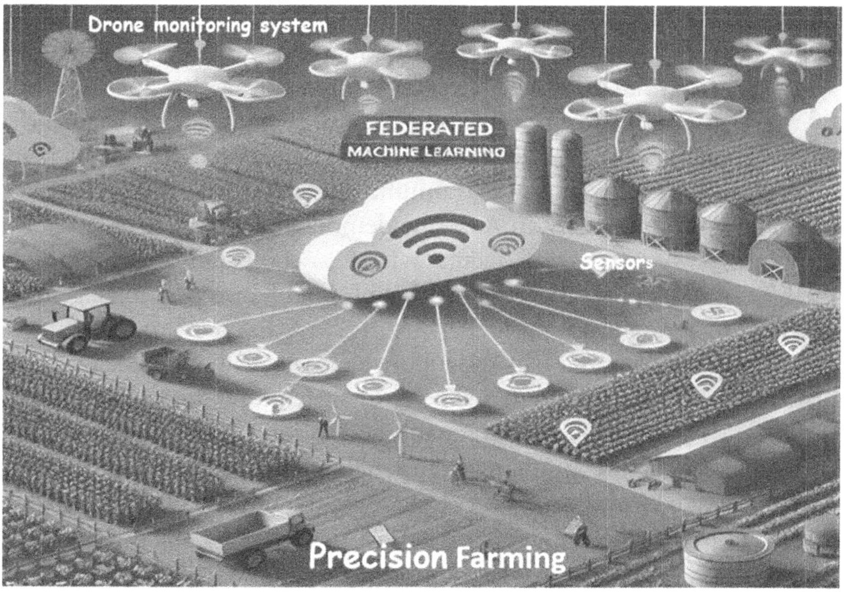

Figure 1.1 FL and agriculture.

systems [2]. The FL offers a transformative solution to these challenges, enabling decentralized data processing while preserving privacy and security. FL is a machine learning (ML) technique that is used to train models across multiple decentralized devices without privacy concerns. In FL, only the model updates and gradients are shared, which ensures that the data will remain on the local device. This approach not only addresses privacy issues but also mitigates the risks associated with data-related breaches and enhances the scalability of data driven solutions across diverse agricultural landscapes.

The application of FL in agriculture represents a paradigm shift in how data is leveraged to drive innovation and improve food production systems. By enabling the use of data from a wide range of sources like sensors, satellite imagery, and historical farm records without compromising privacy, FL helps in the development of more robust and accurate predictive models. For instance, in precision agriculture, FL can enhance crop management practices by providing tailored recommendations based on localized data, improving both yield forecasts and resource efficiency. A study found that FL-based models could improve the accuracy of crop yield predictions by up to 15% compared to traditional centralized models, underscoring the potential of FL to revolutionize decision-making in agriculture. Another critical application of FL in agriculture is in pest and disease management. Effective pest control is essential for maintaining crop health and ensuring food security, but it requires timely and accurate detection of pest outbreaks [3]. Traditional pest detection methods, which rely on manual scouting or centralized data analysis, are often slow and prone to errors. By aggregating data from multiple farms and regions, FL can enable the development of more precise and timely pest detection models, which are crucial for early intervention and minimizing crop losses. For example, a case study on wheat crops in Australia demonstrated that FL-based models improved pest detection accuracy by 12% compared to conventional methods, highlighting the potential for FL to enhance pest management practices in agriculture.

The successful use of FL in agriculture has come with its challenges. The primary challenge is the heterogeneity in agricultural data, which varies significantly in terms of format, quality, and origin. Agricultural data can be highly unstructured, ranging from satellite images and sensor readings to handwritten farm records and observational notes. This diversity makes it difficult to develop standardized models that can be applied universally across different farming systems and regions. Additionally, the quality of agricultural data can be inconsistent, with missing or inaccurate information further complicating the model training process. Infrastructure limitations also pose a significant challenge to the widespread adoption of FL in agriculture [4]. Many rural and remote agricultural regions, particularly in developing countries, lack reliable internet connectivity and the computational resources necessary to support FL. These regions often rely on outdated technology and have limited access to advanced digital tools, making it difficult to implement sophisticated data driven solutions. Addressing these infrastructure gaps will be critical to ensuring that the benefits of FL can be realized across the global agricultural sector.

The adoption of FL in agriculture is important in addressing the growing concerns about data privacy and security. Recently, there has been increasing awareness of the risks associated with sharing agricultural data, particularly as it relates to sensitive information about farm operations, crop yields, and market strategies. Farmers and agricultural organizations are often hesitant to share their data due to fears of misuse, loss of competitive advantage, and potential breaches of sensitive information. By ensuring that only model parameters are shared, instead of the raw data itself, FL reduces the risk of data breach and enhances the trustworthiness of data driven solutions in agriculture. In addition to improving privacy and security, FL also offers a scalable solution for integrating data from diverse sources across the agricultural value chain. This is particularly important for the global food systems, where data from different regions and stakeholders must be combined to develop comprehensive models that can address complex challenges such as climate change, resource food scarcity, and food security. By enabling the collaborative analysis of data from multiple sources, FL can facilitate the development of more accurate and context specific models, leading to better decision-making and improved outcomes for farmers and food producers.

The forthcoming of agriculture lies in the successful combination of advanced digital technologies that can address the complex challenges of food production and sustainability. FL offers a powerful tool for achieving this integration by enabling collaborative data analysis while preserving privacy and security. As FL evolves, its application in agriculture is going to increase, which will offer innovation and efficiency. Researchers are already exploring the use of FL in a variety of agricultural domains, including soil health monitoring, livestock management, and supply chain optimization, highlighting the broad potential of this technology to transform the agricultural sector. The integration of FL with other emerging technologies, such as blockchain and IoT, could unlock even greater opportunities for innovation in agriculture. For example, the combination of FL and blockchain could provide a decentralized and transparent platform for sharing agricultural data while ensuring that data ownership and control remain with the individual farmers. This could enable the development of more resilient and adaptive agricultural systems, capable of responding to the dynamic challenges of global food production. Similarly, the combination of FL with IoT will enable the data (real-time) collection that will lead to more accurate and timely interventions in crop and livestock management. FL represents a critical innovation in the ongoing transformation of agriculture. By the use of FL, agriculture can achieve large advancements in production, sustainability, and food security. Continued research by policymakers and industry will be necessary to realize full potential of FL and ensure that its benefits are accessible to farmers and food producers around the world.

1.2 Various Methods for Providing Solutions to Challenges in Agriculture

FL is a cutting-edge method that enables multiple farms to collaborate in ML models while keeping the data decentralized, offering significant advantages for agriculture. This approach enables farmers to share crucial insights about crop conditions, soil health, and pest issues without compromising sensitive information [5]. Such collaborative efforts improve the accuracy of crop monitoring and yield predictions by combining diverse experiences. FL delivers personalized recommendations for precision agriculture, optimizing resource use based on local data. It enhances data privacy and security, lowers transfer costs by reducing the need for raw data sharing, and accelerates model training through distributed computation. By utilizing FL, the agricultural sector can tackle challenges effectively, promoting collaboration and ultimately boosting productivity and sustainability.

1.2.1 Artificial Intelligence in Agriculture

New technologies and data collection and analysis technologies lead to the advancement in agriculture. These sensors, capable of capturing spatial, temporal, and image-based data, are instrumental in enhancing predictive accuracy in agricultural systems. The deployment of sensors that monitor agricultural parameters like crop health, soil moisture, weather patterns, and nutrient composition across varying temporal and spatial scales contributes to optimizing ML algorithms, improving management of agriculture, increasing production, and fostering sustainable farming practices. Smart sensors are pivotal in precision agriculture, where the objective is to apply data-driven methodologies to manage crops and livestock with heightened accuracy [6]. The sensors' capacity to collect granular data from diverse agricultural environments facilitates the development of precise, real-time insights. This helps farmers to target interventions that optimize allocated resources, enhance crop yields, and reduce environmental impact. For instance, sensors (soil moisture) allow for the exact timing of irrigation, thereby minimizing water usage while ensuring optimal plant health.

The integration of smart sensors into nutrient and pesticide management strategies further exemplifies their role in advancing agricultural efficiency. By continuously monitoring soil nutrient levels and the health of plants, these sensors give critical data that can inform the judicious application of fertilizers and pesticides. Such targeted applications mitigate the risks associated with overuse, which can lead to environmental degradation, and concurrently reduce the cost burden on farmers by enhancing the efficiency of input use. The efficacy of ML models in agriculture is markedly improved through the incorporation of high-resolution data derived from smart sensors [7]. These models, which require substantial datasets for training and prediction, benefit from the precision and accuracy of sensor data. Enhanced ML models can generate more reliable forecasts related to crop yields, pest infestations, and optimal harvest timings, thereby supporting decision-making processes in agriculture. In addition to bolstering predictive capabilities, the integration of agriculture machinery with smart sensors has catalyzed the automation of various farming operations. Automated systems equipped with sensors are capable of executing tasks such as spraying, planting and harvesting with superior precision and consistency compared to manual labor. This technological evolution not only augments productivity but also alleviates the physical demands placed on farmers, allowing them to focus on higher-order strategic decisions that contribute to farm optimization.

The contribution of smart sensors to the sustainability of agricultural systems is also noteworthy. The longitudinal data collected by these sensors is invaluable for monitoring the long-term effects of agricultural interventions on soil health, biodiversity, and ecosystem stability. Such data-driven insights are essential for developing and implementing sustainable agricultural practices that are resilient to environmental challenges. In conclusion, the influence of smart sensor technology on agricultural development is profound, underpinned by significant technological advancements that have enhanced the precision, efficiency, and sustainability of agricultural practices [8]. The ability of these sensors to capture and analyze diverse agricultural data has not only improved the performance of ML models but has also transformed agricultural management strategies, leading to increased productivity and environmental stewardship. As sensor technologies continue to evolve, their role in driving innovation and progress within the agricultural sector is expected to expand, offering new avenues for enhancing the sustainability and resilience of global food systems.

1.2.2 Application Machine Learning in Agriculture

ML uses data and algorithms to iteratively refine its predictive capabilities, thereby enhancing its accuracy over time. ML algorithms are capable of processing extensive agricultural datasets, including those related to cash-crop yields, conditions soil, and weather conditions, to discern underlying patterns and generate actionable insights. These insights are integral to knowledge-based systems that assist farmers in making informed decisions. The incorporation of ML into agricultural practices unlocks new opportunities for the prediction and comprehension of complex data and processes across various domains within the field. The application of ML in several key areas in agriculture like water and soil, livestock management, and crop is explored. Within crop management, ML has been particularly effective in yield prediction, disease, and weed identification as well as species recognition. These advancements of ML in optimizing agricultural productivity and sustainability.

1.2.3 Deep Learning Application in Agriculture

Deep learning (DL) represents an advanced subset of ML that enhances traditional ML by focusing on greater depth and complexity in its models, thereby enabling the learning of multiple stages of abstraction. Unlike classical ML, DL utilizes multiple processing layers, which facilitates hierarchical learning. This hierarchical structure allows for a non-linear approach to learning, with every layer successively adding more intricate information. There are several different types of DL architectures that have been created. These include FL, recurrent neural networks (RNNs), convolutional neural networks (CNNs), and deep neural networks (DNNs). These architectures have been successfully applied across multiple domains, including agriculture. The CNNs are particularly against object detection and classification of image. CNNs achieve this by utilizing hidden layers composed of convolutional layers that extract features from data, followed by fully connected layers that further refine the output. RNNs, on the other hand, are designed to model temporal dynamic behavior. The study demonstrates that DL significantly enhances the accuracy of classification and prediction tasks. The researchers discovered 14 agriculture-specific uses of DL, including weed identification, land cover classification, plant recognition, fruit counting, and crop type classification.

1.2.4 Application of Federated Learning in Agriculture

Many ML and DL techniques traditionally rely on centralized data repositories, where large datasets are stored and utilized during the training process. This centralized approach often compromises data privacy and necessitates significant amounts of data. These distributed datasets, which include weather, drone imagery, satellite, and other temporal and spatial inputs, are crucial for training DL algorithms in agricultural applications. The agricultural sector generates vast amounts of real-time data through cheap, energy-efficient devices, and sensors. While agricultural surveys are costly and only engage limited areas, remote sensing provides a more cost-effective solution for monitoring agricultural activities. Given the reliance on internet communication for these devices, minimizing the risks related to data privacy is essential, particularly for individual farmers and not just large organizations.

FL, a new ML technique, offers a viable way to solve data security issues by enabling several data farmers to work together to train and use models while maintaining local data. Federated averaging (FedAvg), the first aggregation method, was developed in 2016 [7], marking the beginning of FL, which was originally presented by the Google team in October of that year. FL, a decentralized type of ML, involves various sources contributing to the model's training without exchanging data with a central repository. In FL, only local data is used to train ML models across numerous decentralized sources. This approach improves data security by doing away with the need to send raw data to a central server or to other farmers.

FL is being adopted in various domains due to its effectiveness in handling distributed, decentralized data, particularly when privacy concerns are paramount. As DL methods transition toward FL techniques, the focus shifts from centralized data processing to a more distributed model, where data remains on individual devices or sources. These sources train their local models and share only model parameters, like the biases and weights of DL networks, rather than the data itself. In this process, multiple sources independently train their models using data (local) and then send their model updates. Many of these updates refine the global model, facilitating a collaborative learning process without direct data exchange [9]. New sources can also join the FL environment, although this requires additional training iterations to integrate their contributions.

The first step in FL is for a central server to generate a randomized global model, like a CNN. Here are the main components of the training:

- For each iteration, a subset of potential sources is selected.
- Selected sources' local devices get the global model.

- Usually using gradient descent, each participant trains the model with its own data, making local updates better.
- To improve communication efficiency, the central server merges these local model changes, which may include lossy compression.
- The server incorporates the aggregated data into the global model and returns the updated model parameters to the sources for subsequent learning iterations.

FL differs from centralized learning (CL) in several fundamental ways. Data in CL is collected and processed on a central server, whereas, in FL, data remains on the sources, with learning distributed across edge devices. While CL involves direct data submission to the server, FL focuses on sharing model updates iteratively, leading to a more efficient and secure training process, particularly when dealing with sensitive or distributed data. FL is particularly advantageous when training on data from scattered devices, as it distributes the computational load among sources and reduces the need for centralized resources. This technology has numerous applications in agriculture, where it can play a critical role.

1.3 Rice Leaf Disease Classification

Agriculture is vital for human survival, serving as the foundation of the global food system and guaranteeing the sustainability of the food chain. Historically, farmers depended on regular physical field inspections to assess crop progress, frequently resulting in food shortages caused by natural disasters and human errors. Conventional agricultural methods are often labor-intensive and inefficient, leading to diminished profitability. The food manufacturing business not only meets consumer demand but also significantly contributes to alleviating poverty and conserving water resources [10]. Wheat, rice, and maize are the predominant grains in global production, with wheat occupying the biggest harvested area of 214 million hectares, followed by rice at 154 million hectares and maize at 140 million hectares [11]. Human consumption constitutes 85% of total rice output, in contrast to 72% for wheat and 19% for maize. Rice is an essential global food supply, with production predominantly localized in domestic areas. It is a widely consumed and economical staple throughout Asia, functioning as a crucial source of nourishment. FAOSTAT indicates that rice is cultivated across five continents, including Asia, America, Africa, and Europe, underscoring its worldwide importance.

The manual identification of these diseases (bacterial, cultivated, and pathogen) is labor intensive and poses considerable challenges, particularly in regions lacking sufficient crop protection expertise. Therefore, the development of automated disease identification systems, coupled with advanced decision support tools, is crucial to facilitate rice leaf disease management and mitigate crop losses. Although various classification systems for rice leaf disease diagnosis have been proposed, there remains a critical gap in reliable and secure methodologies that can address both diagnostic accuracy and data privacy concerns. In order to preserve strict data privacy, this study presents a revolutionary lightweight FL architecture for the categorization of rice leaf diseases. Effective model training is made possible by the framework's decentralized client-server design, which also protects client data. The validity of the FL models was thoroughly assessed using both independent and identically distributed (IID) and non-IID data. The results highlight the potential of this federated architecture to offer safe and effective disease categorization, tackling the twin issues of diagnostic performance and privacy protection in rice crop management. Figure 1.1 illustrates the steps that used in FL for crop disease classification.

- Each client independently trains a model at their local area by using their own dataset, consisting of paddy leaf disease images, and subsequently uploads to the local model to the central server.
- The central-server aggregates the locally trained models, updates the parameters, and trains a global model. This updated global model is then made accessible to all sources.
- Each client utilizes the parameters from the globally trained model provided by the central server to enhance their local decision-making processes and participate in the subsequent round of model updates.

The dataset comprises a total of 5,932 images representing various rice diseases, including bacterial leaf blight, blast, brown spot, and tungro. A portion of the images was sourced from rice fields in western Odisha, whereas others were obtained from an agricultural pest and insect pest image database [12]. Figures 1.2 to 1.5 illustrate the four distinct types of rice diseases captured in the dataset. Each image is appropriately labeled and stored in JPG format. As summarized in Table 1.2, the dataset includes 1,584 images of bacterial leaf blight, 1,440 images of blast, 1,600 images of brown spot, and 1,308 images of tungro. To simulate a FL environment, the dataset was partitioned based on the number of sources involved in

Figure 1.2 Bacterial blight disease in paddy crop.

Figure 1.3 Blast disease in paddy crop.

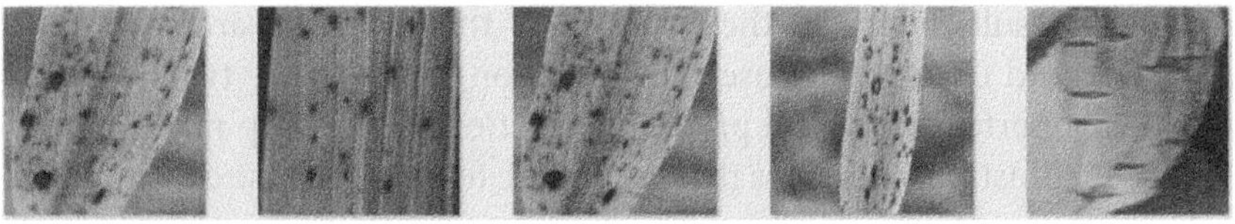

Figure 1.4 Brown spot disease in paddy crop.

Figure 1.5 Bacterial blight disease in paddy crop.

the simulation, with the total number of images divided equally among sources. Each client was then allocated a randomly selected portion of the dataset. This distribution strategy aimed to replicate a real world scenario where each client possesses a unique and diverse dataset.

The process of transforming raw data into a set of meaningful features that can be utilized for analysis or ML tasks is known as feature extraction. This technique is widely employed in fields such as artificial intelligence, natural language processing, and signal processing to identify significant patterns, characteristics, or representations within input data. By simplifying and enhancing the subsequent analysis or modeling tasks, feature extraction is pivotal in improving the performance of ML systems. In DL networks, this process is particularly important within convolutional and pooling layers, which are designed to extract relevant

Table 1.2 Data categorization and description.

S no.	Disease	Sample-images	Number of images
1.	Bacterial Blight Disease	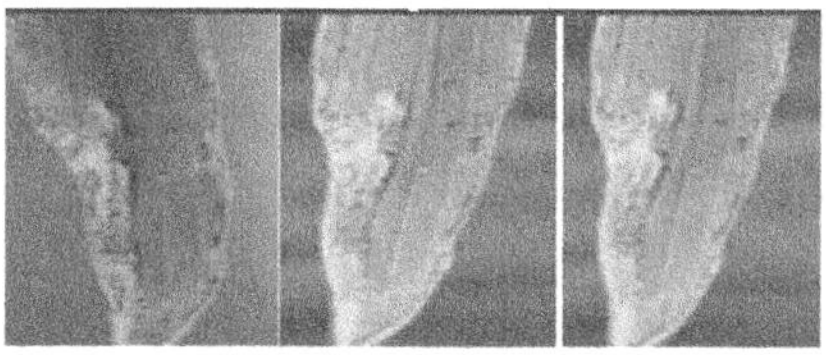	476 (Mild Blight) + 1108 (Sever Blight)
2.	Brown Spot	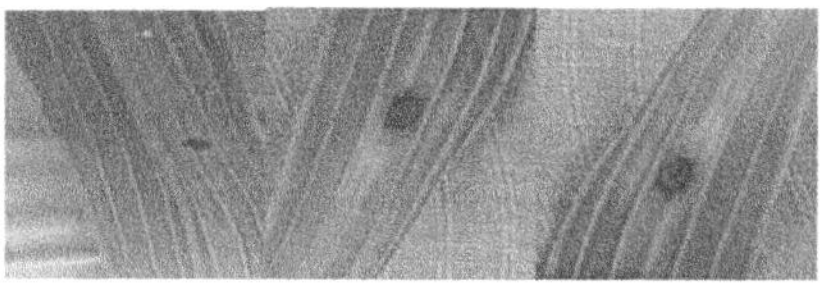	900 (Mild Brown Spot) + 700 (Sever Brown Spot)
3.	Blast Disease		1000 (Mild Blast) + 440 (Sever Blast)
4.	Tungro		450 (Mild-Tungro) + 314 (Severe-Tungro)

image features for tasks such as object detection and localization [13]. In the case of DL, feature extraction involves obtaining deep features from the fully connected layers, which are then used as input for classifier training.

To optimize model performance in rice leaf disease detection, various configurations can be tested, including adjusting the number of layers, modifying the learning rate, or altering the number of neurons per layer. The use of pre-trained models further accelerates this process by providing significant time and computational savings [14]. Pre-trained models such as VGG, ResNet, and Inception, which have been trained on large datasets like ImageNet, are particularly useful in extracting meaningful features from images. These features are then used for tasks such as object recognition, segmentation, and classification, facilitating the accurate identification of rice leaf diseases.

By modifying the feature extraction process, even with limited data, the accuracy of these extracted features can be improved. This method also significantly accelerates training by leveraging specific model architectures that incorporate layers such as reshape, flatten, dense, dropout, and activation functions [15].

1.3.1 A Federated Learning Framework for Classifying Rice Leaf Disease Images

The classification of rice leaf disease images is crucial for safeguarding rice crops, enhancing agricultural productivity, and ensuring global food security. By leveraging advanced technology to develop accurate and

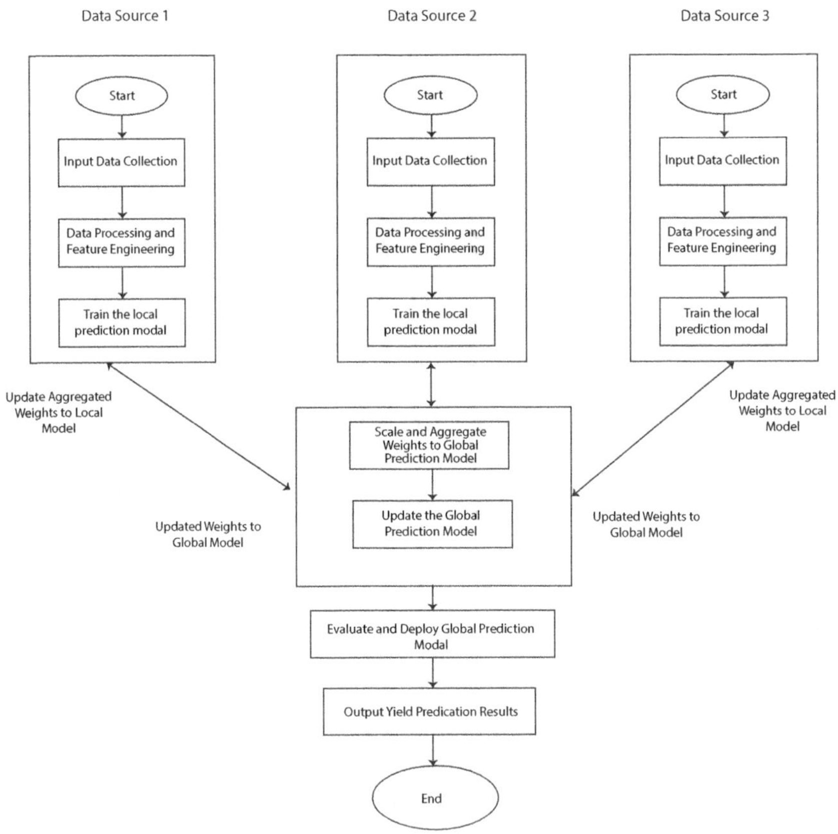

Figure 1.6 A federated deep learning framework for classifying rice leaf diseases.

efficient detection systems, farmers can be equipped with tools to effectively manage and mitigate the effects of rice leaf diseases. In the process of collecting data, farmers and researchers gather sensitive information related to their crops, such as images of diseased rice leaves, geographic locations, and specific farming practices. This data often includes proprietary information or personally identifiable information, which requires careful protection.

Figure 1.6 focuses on the implementation and performance analysis of FL for rice leaf disease classification, offering a solution that preserves data privacy while facilitating accurate disease detection.

1.4 Federated Learning-Based CNNs for Sunflower Leaf Disease Detection

Sunflower is an important crop in India and globally, yet leaf diseases significantly impact its quality and productivity [16]. These diseases vary in severity and are categorized into four levels: light (1%–25%), medium (26%–50%), severe (71%–83%), and profound, with the grading system based on the percentage of leaf area affected. This study introduces a novel FL framework, enabling collaborative but private machine model training across six distinct sources, each located in diverse geographic and climatic regions. Each source contributes disease related data, ensuring the model's adaptability to various conditions, which is vital for its resilience. A key innovation of this approach lies in the advanced averaging techniques used in FL, specifically the aggregation of local model updates. This addresses the challenge of central data storage, which is often a barrier to large-scale agricultural data processing while maintaining data privacy. Additionally, the framework leverages CNNs [17] for their superior image recognition capabilities, crucial for diagnosing leaf diseases from images [18]. This approach holds significant promise for sunflower agriculture in regions like India, where disease outbreaks severely limit productivity.

FL-based CNN approach aims to improve decision-making in crop management by offering a more accurate and efficient system for early disease detection and severity assessment. This technology enhances agricultural sustainability [19], productivity, and crop quality. It offers a fresh perspective on the potential of CNNs and FL in agricultural applications, contributing to the growing body of knowledge in agricultural technology [20].

Figure 1.7 Fresh leaves in sunflower disease.

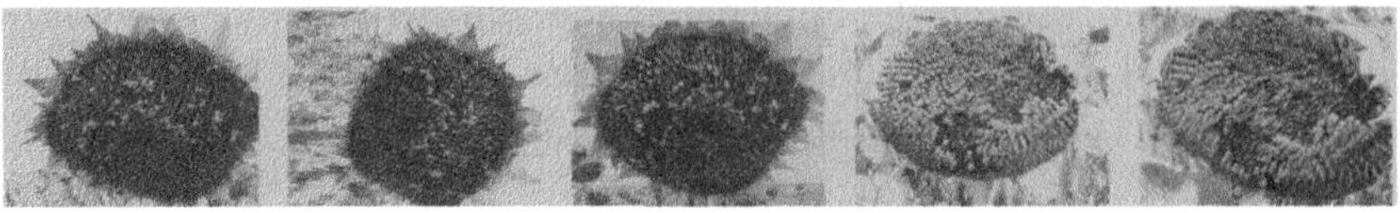

Figure 1.8 Gray mold in sunflower disease.

Figure 1.9 Downy mildew in sunflower disease.

Figure 1.10 Leaf scars in sunflower disease.

The images underwent standardization, with filters applied to enhance contrast and reduce noise [21]. Image augmentation techniques, such as rotation, flipping, and scaling, were employed to improve dataset resilience (Figures 1.6 to 1.10). A decision tree classifier categorizes diseases based on severity levels using features extracted by the CNN. Classification Criteria: Based on CNN processed images, the decision tree classifies disease severity using features such as texture, edge sharpness, and changes in color intensity [22].

1.4.1 Sunflower Disease

- Gray mold in sunflowers is caused by the fungus *Botrytis cinerea*. This disease can result in wilting, rot, and decay, especially in damp or humid environments.

- Downy mildew in sunflowers is caused by the pathogen *Bremia lactucae.* It usually manifests as yellow or light green spots on the upper side of the leaves, accompanied by a fuzzy, grayish mold on the underside.
- Leaf scars in sunflowers result from leaf drop or damage, often caused by environmental stressors, diseases, or pests.

1.4.2 Model Aggregation Decision Tree Classifier with CNN Integration

Evaluation metrics, precision, recall, and F1 score metrics evaluate the decision tree's performance. The classification is based on leaf [23] area damage, categorized into four severity levels; mild (1%–25%), moderate (26%–50%), significant (51%–75%), and severe (76%–100%). Labeling: Expert agronomists and plant pathologists label the dataset for accurate algorithm training.

- Training and Evaluation: Iterative training is conducted, where local models are trained on client data and then aggregated on a central server. Testing: The aggregated model is tested on unseen data to assess its generalization capabilities [24].
- Performance Evaluation and Optimization: Cross-validation methods assess the model's resilience.
- Hyperparameter Tuning: The number of layers and neurons is adjusted according to performance metrics, such as the learning rate. Client Input: Continuous feedback from sources helps refine and adapt the model for real world applications.

In the FL model, a client server structure is used, where each client independently trains its model using their local data. The server then collects these individual models and combines them, a process known as the FL framework [25]. Model aggregation, a widely used method for this aggregation, is the FedAvg algorithm, which works by averaging the weights from all the sources' models to create a globally updated model. To maintain data privacy [26], all data remains stored on the client side, ensuring that it is not shared with the central server. To further enhance security, special protocols are used during the data aggregation process to prevent any potential data leakage while the models are being transmitted. The CNN is structured with multiple layers, including convolutional layers for feature extraction, pooling layers for down sampling, and fully connected layers for decision-making [27]. The Rectified Linear Unit (ReLU) activation function introduces non-linearity to the network, whereas dropout layers help prevent the model from overfitting during training. In local training, each client utilizes their local data to train the CNN. The model is fine-tuned by optimizing its parameters through backpropagation using an optimizer like Adam, which ensures that the model efficiently learns from the data and improves in performance over time [28].

The input for the CNN consists of images measuring 324 by 324 pixels. This dimension was selected to balance computational efficiency while allowing for comprehensive feature extraction [29]. The convolutional layers: The first convolutional layer utilizes a kernel size of 9 × 9 with 32 filters, enabling the capture of larger patterns in the images. The second and third convolutional layers employ smaller kernel sizes of 5 × 5 but increase the number of filters to 128 and 64, respectively, allowing the network to capture more intricate details. Each convolutional layer is succeeded by a maximum pooling layer with a 2 × 2 kernel and a stride of 2, which reduces the vertical and horizontal dimensions of the input, thereby lowering computational demands and mitigating overfitting in subsequent layers. Following the convolutional and pooling layers, a fully connected layer flattens the output. The first of these linked layers contains 1,024 neurons, whereas the second has 512 neurons, and they classify the extracted features into 1 of the 12 categories. To prevent overfitting, a dropout layer is incorporated. The output layer employs a softmax function, producing 12 outputs corresponding to the severity levels of the diseases [30]. The model was trained on a dataset comprising 4,758 images, showcasing various severity levels and disease types [21]. The number of epochs and batch size were determined on the basis of the desired accuracy and available computational resources. This CNN architecture is specifically designed to manage the complexities of the large dataset, classify sunflower leaf diseases according to severity, and capture various attributes effectively [31]. It efficiently employs convolutional layers for feature extraction and fully connected layers for classification.

1.5 Federated Learning–Based CNNs for Banana Leaf Disease Detection

In agriculture, particularly in banana plantations, leaf [32] diseases present a common challenge. These diseases significantly threaten plant production, adversely impacting the financial stability of farmers and, consequently, the economies of various regions. The severity of leaf damage is categorized into four levels: mild (1%–25%), moderate (26%–50%), severe (51%–75%), and critical (76%–100%). This issue is complex, requiring attention from senior officials and effective mitigation strategies. Traditional methods of disease detection and classification often struggle to provide reliable, rapid, and high throughput results. Therefore, it is essential to adopt more advanced technologies that can address emerging challenges. Notably, FL and CNNs are being developed to transform plant protection. This research seeks to create a model leveraging the capabilities of FL and CNNs in a scalable, robust manner while ensuring data privacy to assess the severity of bunt.

Our approach begins with an experimental framework that employs FL, facilitating distributed learning among six different sources. These sources, representing various geographical areas or independent agricultural companies, contribute parts of their learning models, which are then integrated. This collaborative effort between commercial and governmental entities allows for data sharing without compromising sensitive information. One key advantage of this technique is its focus on data privacy and the authentic collection of datasets, which helps to understand the unique characteristics of local disease presentations. The effectiveness of CNNs [33], a type of DL algorithm adept at identifying distinct features in images of banana leaves at various damage stages, is enhanced through a novel multifractal method.

1.5.1 Banana Leaf Diseases

- Banana Yellow Sigatoka, caused by the fungus *Mycosphaerella fijensis*, is a major disease that affects banana crops. It primarily targets the leaves, resulting in yellowing and the development of dark streaks, which can ultimately diminish photosynthesis and weaken the plant.
- Banana Panama Disease, caused by the soil-borne fungus *Fusarium oxysporum* f. sp. *cubense*, is a serious threat to banana plants, especially the Cavendish variety: Yellowing of Leaves, Wilting, Dark Streaks in the Vascular Tissue, Splitting of the Pseudostem.
- Banana Moko Disease, caused by the bacterium *Ralstonia solanacearum*, poses a serious threat to banana and plantain crops. This disease primarily targets the vascular system of the plants, resulting in considerable yield losses: Wilting and Yellowing of Leaves, Dark Streaks in Vascular Tissues, Oozing of Bacterial Exudate, Premature Fruit Ripening, Necrosis of Flower Bracts.
- Banana Bract Mosaic Virus (BBrMV) Disease is a viral infection that impacts banana and plantain plants. Caused by the BBrMV, this disease can result in considerable economic

Figure 1.11 Healthy banana leaf.

Figure 1.12 Banana yellow sigatoka disease of banana leaf.

Figure 1.13 Banana panama disease of banana leaf.

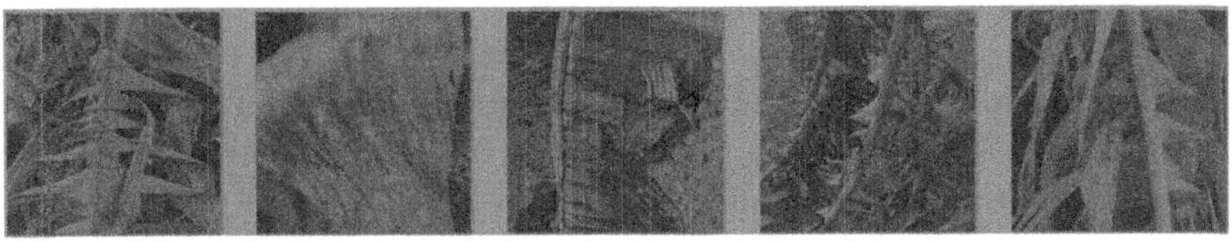

Figure 1.14 Banana moko disease of banana leaf.

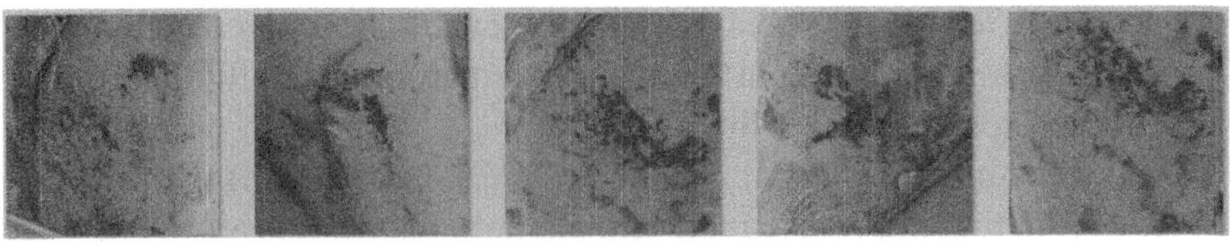

Figure 1.15 Banana bract mosaic virus of banana leaf.

losses in banana production: Mosaic Patterns on Bracts and Leaves, Streaks and Stripes on Leaf Sheaths, Discoloration of Pseudostems, Distorted and Wrinkled Leaves, Reduced Plant Vigor.

The following sections outline the algorithm pipeline for developing and testing a FL framework integrated with a CNN model to classify the severity of banana leaf diseases [34] in Figures 1.11 to 1.15. The proposed methodology employs a decentralized learning approach using six separate clients to accurately categorize disease severity into four levels: mild (1% to 25%), moderate (26% to 50%), severe (51% to 75%), and critical (76% to 100%). This technique is thoughtfully designed to enhance model functionality while ensuring data privacy and accommodating large datasets across various farming scenarios. The core of our approach relies on a large dataset consisting of images of diseased banana leaf parts, each rated according to the severity of the disease. Data collection is conducted through various channels, including agricultural organizations and contributions from small-scale farmers, to thoroughly capture disease susceptibility. Hotspots are identified on the basis of specific severity thresholds and mapped onto the corresponding images, serving as a basis for classification procedures.

Before training the model, the data undergoes several pre-processing steps to enhance performance accuracy. These steps include the following:

- **Resizing:** All images are automatically resized to a uniform standard size, regardless of their original aspect ratios (e.g., 4:3, 16:9, and 1:1), ensuring consistency across the dataset.
- **Normalization:** Pixel values are scaled from 0 to 1 to facilitate faster training convergence.
- **Data Augmentation:** Techniques such as rotation, flipping, and zooming are employed to make the model less prone to overfitting and to enhance its generalization capabilities.

1.5.2 Model Aggregation

Evaluating a model's performance involves various metrics, including accuracy and precision. Metrics such as recall, F1 score, sensitivity, specificity, accuracy, F-measure, receiver operating characteristic (ROC) curve, and confusion matrix provide a thorough assessment of the model's ability to distinguish between different severity levels. Precision indicates the model's overall accuracy, while recall emphasizes its sensitivity and the impact of false positives, which is vital for evaluating the model's effectiveness in classifying severity levels. The F1 score, representing the harmonic mean of precision and recall, serves as a unified measure to assess the balance between these two metrics. The implementation of the FL-CNN [35] model will leverage established DL frameworks like TensorFlow or PyTorch, which facilitate FL operations. Initial experiments will follow a trial and error methodology to identify the optimal hyperparameters such as learning rate, batch size, and number of training epochs to maximize performance. This approach aims to utilize CNNs and FL to accurately classify disease categories in banana leaves. The study also addresses critical challenges, including data privacy and scalability, through the use of FL.

One of the key components of the classifier is a CNN, recognized as a powerful tool for image recognition. The CNN [26] architecture consists of multiple convolutional layers, pooling layers, and fully connected layers. These convolutional layers are combined with the ReLU activation function, which introduces non-linearity, enabling the model to process data in a detailed manner and uncover intricate patterns within the input.

The building block layers include downsampling, which reduces the dimensionality of the data, enhancing efficiency and improving feature extraction. The final layers of the CNN are fully connected layers, which apply a softmax activation function to generate probabilities across the different categories.

The FAVORISE model employs a ML function that provides a statistical distribution across four categories of disease severity. A significant aspect of this approach is the FL setup, which facilitates the training of multiple models across six clients. Each client uses data from a distinct area or farm, reflecting the reality that data is often sourced from various locations rather than a single entity.

The FL process follows these steps:

- **Initialization:** A generalized version of the model is provided to all participants.
- **Local Training:** Each client receives a dataset to adjust the weights and biases of its model, customizing it for its specific data.
- **Weight Aggregation:** After local training, clients send the updated weights and biases to a shared server. The server aggregates these updates using a FedAvg method to create an updated global model.
- **Model Redistribution:** The global model is then distributed back to all clients, and this cycle is repeated for several rounds until convergence is achieved or a specific performance standard is met.

The input images have dimensions of 454 × 454 pixels. They are preprocessed to align with the network's input size, facilitating effective feature extraction. The convolutional layers play a key role in identifying specific features within the banana leaf images. The first convolutional layer is responsible for initial feature extraction, whereas subsequent layers deepen the network's ability to capture more complex patterns associated with various disease severity levels. After each convolutional layer, max pooling layers reduce the spatial dimensions of the output feature maps, optimizing computational resource usage and ensuring that the model focuses on the presence of features rather than their precise location in the image. Following the convolutional and pooling layers, the architecture consists solely of fully connected layers. This structure, combined with advanced reasoning derived from the retiree analogy, ultimately classifies the leaf images into one of 12 severity categories. The final layer employs a softmax activation function to generate a probability distribution across these classes, enabling the model to confidently identify the most likely category for each input.

1.6 Conclusion

This chapter has explored the combination of FL and CNNs to classify banana leaf diseases, tackling key challenges in agricultural disease management. By utilizing a decentralized learning framework, we facilitate collaborative model training across various clients while ensuring data privacy. This method promotes effective learning from diverse disease symptoms and environmental factors, resulting in more reliable and generalizable predictions. Our detailed algorithm pipeline starts with meticulous data collection from different agricultural sources, creating a diverse dataset that enhances classification accuracy. Key pre-processing steps such as resizing, normalization, and data augmentation boost input quality, which is essential for optimal CNN performance. We assess the FL-CNN model using metrics like accuracy, precision, recall, and F1 score, highlighting the necessity of reducing false positives in disease classification. The CNN architecture is tailored to effectively extract features from banana leaf images, employing multiple convolutional and pooling layers to recognize both simple and intricate disease patterns. Ultimately, this research not only improves disease detection but also provides farmers with timely insights for better crop management. The synergy of FL and CNNs offers a scalable, privacy-conscious solution that supports agricultural sustainability. This work establishes a foundation for future innovations in smart agriculture, underscoring the potential of technology to enhance food security and economic resilience in farming communities.

AI disclosure statement: The author acknowledges the use of AI-based tools for minor language editing or grammatical correction. However, all intellectual contributions, analyses, and conclusions presented in this chapter are entirely the author's own.

References

1. Alpaydin, E., *Introduction to Machine Learning*, MIT Press, 2020.
2. Samuel, A.L., Some studies in machine learning using the game of checkers. *IBM J. Res. Dev.*, *3*, 3, 210–229, 1959, https://doi.org/10.1147/rd.33.0210.
3. Benos, L., Tagarakis, A.C., Dolias, G., Berruto, R., Kateris, D., Bochtis, D., Machine learning in agriculture: A comprehensive updated review. *Sensors*, *21*, 11, Article 3758, 2021, https://doi.org/10.3390/s21113758.
4. Lee, S., Kim, H., Lieu, Q.X., Lee, J., CNN-based image recognition for topology optimization. *Knowl.-Based Syst.*, *198*, 105887, 2020, https://doi.org/10.1016/j.knosys.2020.105887.
5. Kamilaris, A. and Prenafeta-Boldú, F.X., Deep learning in agriculture: A survey. *Comput. Electron. Agric.*, *147*, 70–90, 2018, https://doi.org/10.1016/j.compag.2018.02.016.
6. Cao, K., Liu, Y., Meng, G., Sun, Q., An overview on edge computing research. *IEEE Access*, 8, 85714–85728, 2020, https://doi.org/10.1109/ACCESS.2020.2991734.
7. Ghosh, A.M. and Grolinger, K., Edge-cloud computing for Internet of Things data analytics: Embedding intelligence in the edge with deep learning. *IEEE Trans. Ind. Inf.*, *17*, 3, 2191–2200, 2021, https://doi.org/10.1109/TII.2020.3008711.
8. Wang, X., Han, Y., Leung, V.C.M., Niyato, D., Yan, X., Chen, X., Convergence of edge computing and deep learning: A comprehensive survey. *IEEE Commun. Surv. Tutor.*, *22*, 2, 869–904, 2020, https://doi.org/10.1109/COMST.2020.2970550.
9. Morais, R., Mendes, J., Silva, R., Silva, N., Sousa, J.J., Peres, E., A versatile, low-power and low-cost IoT device for field data gathering in precision agriculture practices. *Agriculture*, *11*, 7, Article 619, 2021, https://doi.org/10.3390/agriculture11070619.
10. Wójtowicz, M., Wójtowicz, A., Piekarczyk, J., Application of remote sensing methods in agriculture. *Commun. Biometry Crop Sci.*, *11*, 1, 31–50, 2016.
11. Kaur, J., Hazrati Fard, S.M., Amiri-Zarandi, M., Dara, R., Protecting farmers' data privacy and confidentiality: Recommendations and considerations. *Front. Sustain. Food Syst.*, 6, Article 903230, 2022, https://doi.org/10.3389/fsufs.2022.903230.
12. Mammen, P.M., Federated learning: Opportunities and challenges. *arXiv:2101.05428*, https://doi.org/10.48550/arXiv.2101.05428, 2021.
13. Mothukuri, V., Parizi, R.M., Pouriyeh, S., Huang, Y., Dehghantanha, A., Srivastava, G., A survey on security and privacy of federated learning.

Future Gener. Comput. Syst., *115*, 619–640, 2021, https://doi.org/10.1016/j.future.2020.10.007.

14. Li, Z., Sharma, V., Mohanty, S.P., Preserving data privacy *via* federated learning: Challenges and solutions. *IEEE Consum. Electron. Mag.*, *9*, 3, 8–16, 2020, https://doi.org/10.1109/MCE.2019.2959108.
15. Wani, J.A., Sharma, S., Muzamil, M., Ahmed, S., Sharma, S., Singh, S., Machine learning and deep learning based computational techniques in automatic agricultural diseases detection: Methodologies, applications, and challenges. *Arch. Comput. Methods Eng.*, *29*, 1, 641–677, 2022, https://doi.org/10.1007/s11831-021-09588-5.
16. Kodama, T. and Hata, Y., Development of classification system of rice disease using artificial intelligence, in: *2018 IEEE International Conference on Systems, Man, and Cybernetics (SMC)*, pp. 3699–3702, 2018, https://doi.org/10.1109/SMC.2018.00626.
17. Kivrak, K.G., *et al.*, Integration of extra-large-seeded and double-podded traits in chickpea (*Cicer arietinum* L.). *Agronomy*, *10*, 6, Article 901, 2020, https://doi.org/10.3390/agronomy10060901.
18. Sethy, P.K., Barpanda, N.K., Rath, A.K., Behera, S.K., Deep feature based rice leaf disease identification using support vector machine. *Comput. Electron. Agric.*, *175*, 105527, 2020, https://doi.org/10.1016/j.compag.2020.105527.
19. Dara, S. and Tumma, P., Feature extraction by using deep learning: A survey, in: *2018 Second International Conference on Electronics, Communication and Aerospace Technology (ICECA)*, pp. 1795–1801, 2018, https://doi.org/10.1109/ICECA.2018.8474912.
20. Ghosal, S. and Sarkar, K., Rice leaf diseases classification using CNN with transfer learning, in: *2020 IEEE Calcutta Conference (CALCON)*, pp. 230–236, 2020, https://doi.org/10.1109/CALCON49167.2020.9106423.
21. Zhu, H., Xu, J., Liu, S., Jin, Y., Federated learning on non-IID data: A survey. *arXiv:2106.06843*, https://arxiv.org/abs/2106.06843, 2021.
22. Kumar, D. and Kukreja, V., Deep learning in wheat diseases classification: A systematic review. *Multimed. Tools Appl.*, *81*, 7, 10143–10187, 2022, https://doi.org/10.1007/s11042-022-12160-3.
23. Lamba, S., Baliyan, A., Kukreja, V., GAN based image augmentation for increased CNN performance in paddy leaf disease classification, in: *2022 2nd International Conference on Advance Computing and Innovative Technologies in Engineering (ICACITE)*, pp. 2054–2059, 2022, https://doi.org/10.1109/ICACITE53722.2022.9823799.
24. Baliyan, A., Kukreja, V., Salonki, V., Kaswan, K.S., Detection of corn gray leaf spot severity levels using deep learning approach, in: *2021 9th International Conference on Reliability, Infocom Technologies and Optimization (ICRITO)*, pp. 1–5, 2021, https://doi.org/10.1109/ICRITO51393.2021.9596540.
25. Mehta, S., Kukreja, V., Vats, S., Manwal, M., Scalable and privacy-severity analysis of pomegranate leaf diseases: Federated learning with CNNs, in: *2023 14th International Conference on Computing Communication and*

Networking Technologies (ICCCNT), pp. 1–6, 2023, https://doi.org/10.1109/ICCCNT56998.2023.10306771.
26. Islam, A., Sama Raisa, S.R., Khan, N.H., Rifat, A., II, A deep learning approach for classification and segmentation of leafy vegetables and diseases, in: *2023 International Conference on Next-Generation Computing, IoT and Machine Learning (NCIM)*, pp. 1–6, 2023, https://doi.org/10.1109/NCIM59001.2023.10212506.
27. Mehta, S., Kukreja, V., Gupta, R., Decentralized detection of cassava leaf diseases: A federated convolutional neural network solution, in: *2023 International Conference on Circuit Power and Computing Technologies (ICCPCT)*, pp. 381–386, 2023, https://doi.org/10.1109/ICCPCT58313.2023.10245357.
28. Sangeetha, R., Logeshwaran, J., Rocher, J., Lloret, J., An improved agro deep learning model for detection of Panama wilts disease in banana leaves. *AgriEngineering*, *5*, 2, Article 42, 2023, https://doi.org/10.3390/agriengineering5020042.
29. Russel, N.S. and Selvaraj, A., Leaf species and disease classification using multiscale parallel deep CNN architecture. *Neural Comput. Appl.*, *34*, 21, 19217–19237, 2022, https://doi.org/10.1007/s00521-022-07521-w.
30. Kukreja, V. and Kumar, D., Automatic classification of wheat rust diseases using deep convolutional neural networks, in: *2021 9th International Conference on Reliability, Infocom Technologies and Optimization (ICRITO)*, pp. 1–6, 2021, https://doi.org/10.1109/ICRITO51393.2021.9596133.
31. Mehta, S., Kukreja, V., Bansal, A., Kaur, K., Singh, A., Exploring the potential of convolutional neural networks in automatic diagnosis of dragon fruit diseases from plant photographs, in: *2023 7th International Conference on Intelligent Computing and Control Systems (ICICCS)*, pp. 536–540, 2023, https://doi.org/10.1109/ICICCS56967.2023.10142811.
32. Andrew, J., Eunice, J., Popescu, D.E., Chowdary, M.K., Hemanth, J., Deep learning-based leaf disease detection in crops using images for agricultural applications. *Agronomy*, *12*, 10, Article 2395, 2022, https://doi.org/10.3390/agronomy12102395.
33. Tripathi, P., Kumar, N., Rai, M., *et al.*, Ayoub Khan Applications of Deep Learning in Agriculture, in: *Artificial Intelligence Applications in Agriculture and Food Quality Improvement*, M.A. Khan and (Eds.), pp. 17–28, IGI Global, 2022, https://doi.org/10.4018/978-1-6684-5141-0.ch002.
34. Tripathi, P., Kumar, N., Rai, M., *et al.*, Applications of Machine Learning in Agriculture, in: *Smart Village Infrastructure and Sustainable Rural Communities*, M.A. Khan and (Eds.), pp. 99–118, IGI Global, 2023, https://doi.org/10.4018/978-1-6684-6418-2.ch006.
35. Rai, M., *et al.*, Advancement in Agricultural Techniques with the Introduction of Artificial Intelligence and Image Processing, in: *Smart Village Infrastructure and Sustainable Rural Communities*, M.A. Khan and (Eds.), pp. 47–68, IGI Global, 2023, https://doi.org/10.4018/978-1-6684-6418-2.ch004.

2

Federated Learning–Based Food Calorie Estimation

Lingam Sunitha[1]*, Shanthi Makka[2], Kumavat Prakash[1] and Vankadaru Charan[1]

[1]Department of IT, Vardhaman College of Engineering, Hyderabad, Telangana, India
[2]Department of Computer Science and Engineering (Data Science), Vardhaman College of Engineering, Hyderabad, India

Abstract
Maintaining a balanced diet and controlling medical conditions like obesity and diabetes require accurate food calorie estimation. Conventional techniques frequently call for manual input, which can be error-prone and time-consuming. Automated systems for estimating food calories have been made possible by recent developments in artificial intelligence and machine learning. But these systems usually need a lot of data, which makes privacy issues arise. One potential remedy is federated learning, which allows for cooperative model training without the need for centralized data collection. This chapter examines the use of federated learning in food calorie estimation, emphasizing its benefits, methods, and guiding principles.

***Keywords*:** Artificial intelligence, machine learning, deep learning, federated learning, food calories

2.1 Introduction

A machine learning (ML) method called federated learning (FL) enables several parties to train a model without exchanging data. It is a reaction to privacy and security issues that crop up when conventional methods

**Corresponding author*: sunithavvit@gmail.com

Padmesh Tripathi, Bhanumati Panda, Shanthi Makka, Reeta Mishra, S. Balamurugan and Sheng-Lung Peng (eds.) Federated Learning for Smart Agriculture and Food Quality Enhancement, (27–56) © 2026 Scrivener Publishing LLC

are used to train artificial intelligence (AI) models. Many advances in the field of digital healthcare can be attributed to ML and deep learning in particular. Because access to data that approximates the genuine worldwide distribution is beneficial to all ML techniques, FL is a promising way to acquire strong. FL elegantly tackles difficulties associated to egress of sensitive medical data by allowing many parties to train jointly without requiring data exchange or centralization. As a result, it might pave the way for innovative corporate ventures and research directions, potentially enhancing medical care everywhere. But even now, FL affects almost every stakeholder and step of the treatment process. From enhanced medical image analysis that gives doctors more accurate diagnostic tools to true precision medicine that finds patients who are similar to one another, to cooperative and expedited drug discovery that lowers costs and shortens time to market for pharmaceutical companies, has an impact on all of these aspects of healthcare.

2.2 Foundations of Federated Learning

2.2.1 Overview of Federated Learning

ML is becoming more and more popular in domains like computer vision and natural language processing as a result of the growth of big data and Internet of Things (IoT). However, getting enough training data is challenging due to worries about concept drift and data privacy. In order to overcome these constraints, Google introduced FL in 2016, which involves three parties: the user client, the contributor client, and the learning coordinator.

2.2.2 Key Concepts and Architecture

FL is a system of collaborative learning that includes the ML model. Figure 2.1 depicts clients and servers. Data sources with their own datasets, such as smart phones, IoT sensors, or distant servers, are referred to as clients. Serving as a central hub, the server coordinates the updates and information exchanged between clients and helps the ML model to get better. In order to protect data security and privacy, the ML model is continuously improved without centralizing or sharing raw data. Clients and servers exchange information in FL, where clients receive an initial model and update it with local datasets. By gathering and combining these updates, the server improves the performance of the model [5].

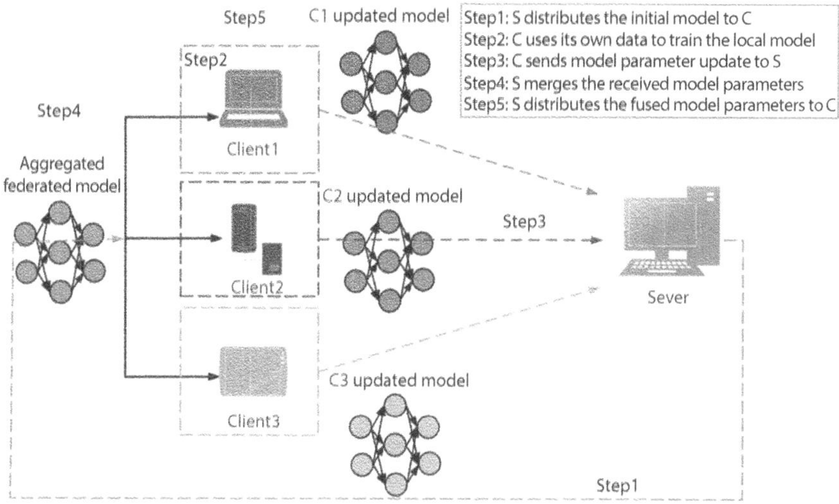

Figure 2.1 Client-server architecture.

By combining updates to the model's parameters, the federated averaging (FedAvg) algorithm enables the model to learn from a variety of data sources. On mobile devices, this can improve the predictive text suggestions.

FL is a cooperative ML technique that guarantees data stays on the devices, generating it during training by preventing the aggregation of all trained data at a particular point. This method lowers the possibility of unauthorized access while safeguarding training [1] data from different kinds of attacks.

2.2.3 Secure Data Communication for Protecting Training Data

FL creates a single global model without requiring access to the original data by training models on several tasks using local datasets. For this, the FedAvg algorithm is employed.

1. The server chooses n customers from the pool.
2. The server provides the initial parameters to these clients.
3. The server provides the initial parameters to each client.

In Figure 2.2, for the global model, the suggested algorithm employs a FedAvg strategy, taking into account a sample of 10% of clients that are available for the upcoming communication round. Before a training round

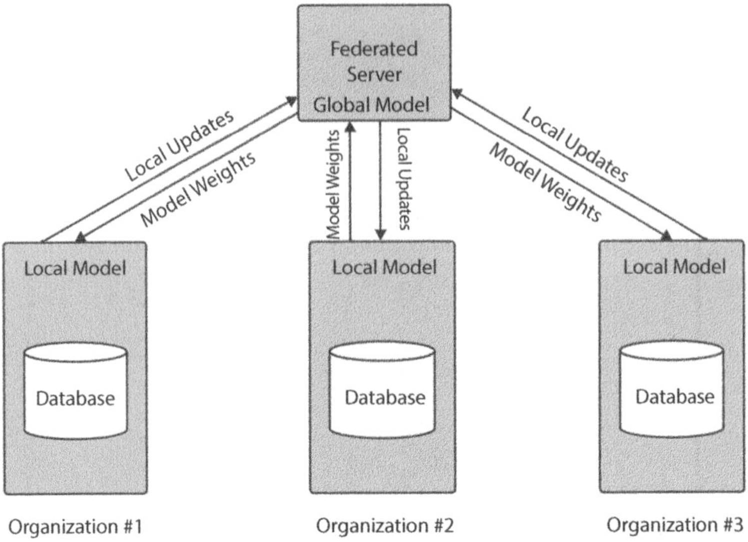

Figure 2.2 Federated learning strategies.

can begin, the strategy also establishes a minimum number of clients that must be connected to the server and sampled. The default batch size and learning rate are used to define the fitness configuration. A different approach is to use the FedAvg with momentum technique at the server side in conjunction with the LeNet convolutional neural network (CNN) model [25] at the client side. The purpose of the study is to determine if adding a momentum factor to the LeNet CNN model enhances its ability to dampen oscillations and speed up training.

2.2.4 Privacy Preserving Techniques

Data security and privacy are becoming more and more of a concern with the growth of ML applications. A possible solution is FL, but model parameters rob it of privacy guarantees. In order to comply with EU/UK General Data Protection Regulation (GDPR) regulations, this article reviews FL studies, talks about privacy-preserving techniques, and suggests future approaches [2]. Various techniques for privacy preservation in ML are as follows:

1. Data anonymization
2. Differential privacy
3. Secure multi-party computation (SMC)
4. Homomorphism encryption

Techniques for maintaining privacy that are used in distributed learning systems include homomorphic encryption, secure SMC, data anonymization, and differential privacy. These techniques guard against linkage attacks, but they are unable to protect sensitive attributes. SMC makes sure that only outputs are disclosed to third parties, but overheads make it difficult to implement in practice. Because of its computational overhead, homomorphic encryption is not as secure when used for large-scale data training.

2.2.5 The GDPR

All EU nations saw the GDPR, a significant revision to the 1995 Data Protection Directive, which goes into effect in May 2018. By mandating that personal data be processed lawfully, under stringent guidelines, and for a justifiable purpose, it seeks to protect it. The law contains 99 articles that address administrative, technical, and philosophical requirements for businesses that handle personal data. By requiring data controllers to obtain consent from data subjects, it protects data ownership and distinguishes between three participant roles: data subject, data controller, and data processor.

2.2.6 Challenges in Achieving GDPR Compliance

GDPR requires ML-based applications to safeguard personal data, but there are issues with data breaches, misuse, and transparency with centralized systems. The execution of ML algorithms can be challenging to interpret due to their black-box nature and data minimization. Redesigning ML algorithms internally will be necessary to comply with GDPR regulations, which will demand a large investment of time and money in technology.

2.3 Federated Learning: A Collaborative Learning Approach

Strict data protection laws make traditional cloud-centric ML techniques unsuitable for handling personal data. The increasing use of personal devices and growing privacy concerns have led to a rise in decentralized AI, which blends mobile edge computing with AI/ML techniques. An alternative that keeps original personal data on devices and enables collaborative training of ML models is FL [3]. In order to address vulnerabilities related to data privacy, FL is a cross-disciplinary technique that allows end-users' devices to locally train a shared ML model on local data. It is intended to function with identically distributed data that is not independent or balanced.

Figure 2.3 gives the four steps make up the workflow cycle in a centralized FL framework: local computation, local model aggregation, global model update, participant selection, and global model dissemination. For the upcoming training cycle, the server broadcasts a global ML model and chooses participants. Local nodes are used by participants to train their updated local ML models. In order to update the global model, the server aggregates locally trained ML models, utilizing methods such as secure aggregation and encryption that protect privacy. Participants in the following training session are then given access to the updated global model.

2.3.1 Ensuring GDPR Compliance in Centralized Federated Learning Systems

1. Roles and obligations
2. GDPR principles
3. Rights of data subject

FL is a distributed ML [26] technology that respects privacy and data protection regulations, but it incurs communication costs because devices constantly share updates. With an emphasis on security, privacy,

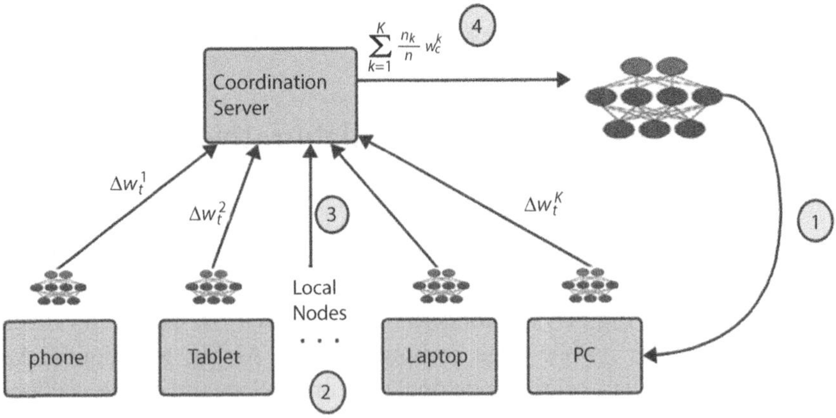

Figure 2.3 FL workflow.

and communication, this paper reviews research aimed at overcoming these limitations in FL settings. Research on FL is being conducted in the domains of autonomous vehicles, IoT applications, and transportation, with the goal of enhancing predictive healthcare ML by utilizing sensitive data from several hospitals [4].

2.3.2 The Different Types of FL Systems

1. Horizontal FL
2. Vertical FL
3. Federated transfer learning

Horizontal Federated Learning: This arises when datasets from various individuals have different samples but the same feature space. In essence, the datasets contain information about various entities yet share the same structure

Key Features:

- Despite having various samples, datasets share the same characteristics.
- Without disclosing their real data, participants can work together to train a model.
- Ideal for situations in which the data structures of various entities are comparable.

Applications: Without exchanging the actual transaction records, some banks use their transaction data to train a fraud detection algorithm.

Vertical Federated Learning: (feature-based FL) is the result of participants having datasets with different attributes but the same sample space. This indicates that every participant has unique opinions regarding the same group of entities.

Key Features:

- While datasets differ in their features, they share the same entities.
- Permits the training of a more complete model by combining features from many individuals.
- While participants only exchange intermediate computations, data privacy is preserved.

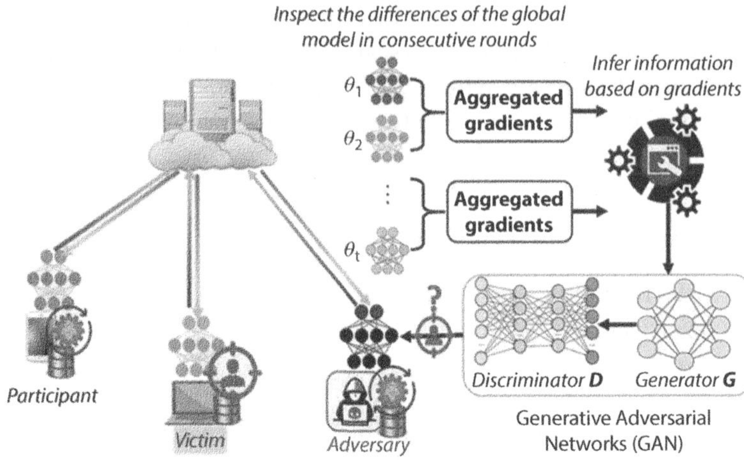

Figure 2.4 Encryption that protect privacy.

Applications: A partnership between a retail company and a bank where the retail company has purchasing data and the bank has financial data about the same customers.

Federated Transfer Learning: The novel technique of sharing information and understanding obtained from training ML models among several organizations without having to centralize the raw data is known as federated transfer learning. By addressing the issues of privacy and data silos, this method allows for decentralized learning across various datasets. Federated transfer learning is important because it may provide AI models collective intelligence while protecting the privacy of individual data sources. Figure 2.4 illustrates how an adversary in a FL system can exploit gradients and use GANs to infer private data from other users.

2.3.3 Mile Stones and Achievements in Federated Learning

1. **Privacy Preservation:**
   - **Data Confidentiality:** FL makes it possible to train models across devices without sending private information to a central server, which is essential for industries like finance, healthcare, and personalized services.
2. **Scalability:**
   - **Handling Large-Scale Data:** Massive volumes of data dispersed over numerous devices can be handled by FL;

centralized data canters would be challenging to maintain because of storage and bandwidth constraints.

3. **Resource Utilization:**
   - **Efficient Use of Local Resources:** FL reduces the need for potent centralized servers by utilizing the local computational resources on devices.
4. **Personalization:**
   - **Tailored Models:** FL makes it possible to create customized models that enhance both the user experience and model performance by adjusting to the unique data distribution on each device.
5. **Robustness to Data Breaches:**
   - **Minimized Risk of Centralized Data Breaches:** FL lowers the possibility of significant data breaches by maintaining data decentralized, which is less likely in centralized data storage systems. Figure 2.5 shows threats to FL, including insider and outsider attacks.

2.3.4 Challenges in Federated Learning

1. **Communication Overhead:**
   - **Bandwidth Consumption:** Significant bandwidth consumption may result from the necessity for devices to communicate model updates to the central server on a regular basis.

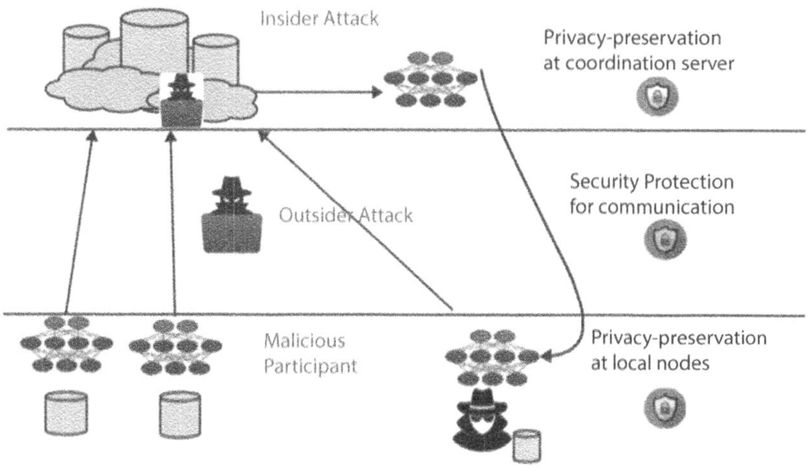

Figure 2.5 Session access global model.

- **Latency:** The entire training process can be slowed down by communication delays, especially if the devices are geographically separated or have erratic internet connections.

2. **System Heterogeneity:**
   - **Diverse Device Capabilities:** Variations may be seen in the processing power, storage capacity, and battery life of devices involved in FL.
3. **Data Heterogeneity:**
   - **Non-Independent and Identically Distributed (IID) Data:** Because data on various devices is frequently non-IID, training a global model with good cross-device generalization can be difficult.
4. **Security and Privacy Concerns:**
   - **Inference Attacks:** FL is susceptible to inference attacks, in which adversaries attempt to reconstruct data from model updates, even though it does not share raw data.
5. **Algorithmic Challenges:**
   - **Optimization Complexity:** Asynchronous updates, non-convex optimization landscapes, and different data distributions are some of the reasons why training a model in a federated environment is more difficult than in a centralized one. Figure 2.6 shows the FL architecture where we can train local models and send their updates to a central server for aggregation.

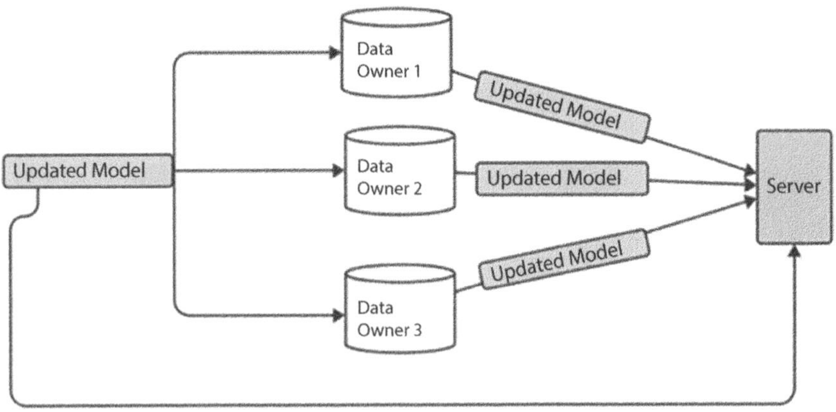

Figure 2.6 Federated learning architecture.

2.4 Machine Learning for Food Calorie Estimation

2.4.1 Traditional Methods

1) Image-Based Estimation

Recent developments in computer vision and deep learning have greatly increased the accuracy of food calorie estimations, which is an important task in dietary management and health monitoring. CNNs [24] are a widely used technique in this field that provide the fundamental architecture for identifying food items in images and estimating their caloric content.

We made use of the **ECUSTFD** (ECUST Food Dataset) in this chapter, which includes information on food masses and volumes in addition to labeled images of food. Food items like bread, bananas, and apples are among the 19 types of food included in the dataset. Bounding boxes are used to annotate the images, designating the reference coins and food items utilized in the volume estimation process. Without the need for manual data labeling, this annotation process streamlines the training of object detection models. We are able to precisely identify food items in images because we used the **TensorFlow Object Detection API** to build our CNN model. Pooling, convolutional, sequential, and fully connected layers make up the structure of the CNN model. Food detection accuracy depends on the network's ability to automatically learn features from the images, including shape, texture, and color. This is made possible by the architecture. We used **Random Forest and Support Vector Machine** (SVM) classifiers in addition to CNN to improve prediction accuracy.

The CNN model is constructed during the training phase using **Keras** and the **RMSprop optimizer**. The accuracy of the model is greatly impacted by the optimizer's crucial role in managing the algorithm's convergence. By adjusting the learning rate and neuron count, among other hyper-parameters, we were able to achieve a 92% accuracy rate in food detection. By directing the network's learning process, the activation function also helps the model learn from the training dataset.

We calculate the overall calorie content by multiplying the mass and volume of the identified food by its known calorie value after the food items have been identified in an image. The model counts the food items in an image and adds up their individual calorie values to get the estimated total number of calories. Because it provides a consistent scale for calculating food dimensions, the reference coin's inclusion in the dataset enables the model to estimate food volume accurately.

We created a dependable and effective system for estimating food calories by combining deep learning methods like CNN with additional classifiers like SVM and Random Forest. With its simple and automated calorie monitoring feature, this system has the potential to enhance dietary tracking [14].

A smart farming system can also be used to more easily gather data on crops grown, eliminating the need for manual data logging [15]. With the introduction of AI and image processing to the agricultural sector, many agricultural techniques have advanced. These methods boost input productivity and efficiency, supporting the sustainability of crop production [16]. A coin with a known volume will also be included as a probe object in the image. The faster R-CNN algorithm will be used to identify the food item in the picture. Because the faster Regions with Convolutional Neural Networks (R-CNN) algorithm is the fastest of all the object detection algorithms, we are using it. Instead of using a selective search algorithm on the feature map to find region proposals, faster R-CNN uses a different network to predict region proposals. Because of this, it is the fastest algorithm and can be applied to real-time object detection [17]. Additionally, various deep learning models, including CNN and RNN, are employed to improve the quantity and quality of agricultural output [18].

2) Manual Logging

A popular technique for estimating food calories is manual logging, in which people keep a journal of the foods they eat and use food labels or nutritional databases [27] to estimate the number of calories in each serving. With this method, users must record every meal and snack they have, including the kind and amount of food they eat. To determine the calorie content, they then cross-reference these entries with printed materials or online nutritional databases. Access to large databases is made possible by numerous well-known apps, which facilitate the search for calorie information for a variety of foods. Manual logging is accessible and customizable, but it has a number of drawbacks. The dependence on users' ability to precisely estimate portion sizes is one of the primary challenges. Inaccurate portion estimates can result in inaccurate calorie calculations even with meticulous logging. In addition to being time-consuming, this method requires constant effort to log every food item throughout the day. User compliance is also an important consideration; if a user fails to log certain meals or records them incorrectly, the total calorie estimate will be unreliable.

Nonetheless, the procedure does provide flexibility for people with particular dietary needs or preferences. Users can easily track not only calories but also specific nutrients, macros, and other dietary needs because they have complete control over what they log. The time and work involved in this method have, however, resulted in the development of more automated solutions, such as systems that use deep learning and image processing to recognize foods and estimate calories. Although CNNs are an emerging technology, manual logging is still a valuable tool for dietary management because it provides more precise, automated calorie estimation. These sophisticated systems can identify food items from photos [21], saving the user time and effort while producing more accurate results.

2.4.2 Limitations

2.4.2.1 Data Privacy Concerns

Users usually have to take pictures of their meals for ML-based food calorie estimation, after which the images are processed to determine the calorie content. This raises several privacy-related issues:

1. **Sensitive Personal Data:**
Users' uploaded photos can disclose a great deal about their cultural background, personal habits, and way of life. Food preferences frequently reveal personal details about a person's lifestyle, such as dietary limitations, health issues (like diabetes or allergies), customs from their culture or religion, and even their socioeconomic standing. Privacy violations may result from misuse of this data or from unauthorized access to it. For instance, if the data is obtained by a hostile party, then it might be used to deduce personal information about a user.

2. **Data Storage and Security:**
Cloud-based servers are used by many ML systems for data processing and storing. Concerns concerning the security of this data during transit and storage have been raised by the possibility of transmitting images and other sensitive information over the internet. Food photos and related user data are susceptible to hacking attempts and data breaches if they are kept on the cloud without

adequate encryption or security measures. Food data can occasionally be re-identified even after anonymization if it is connected to other personal data.

3. **Lack of Transparency:**
Once their data is uploaded, users frequently have no control over how it is handled. Users may not be able to tell if their data is being used only for calorie estimation or if it is also being used for other purposes, such as marketing, or if it is being sold to third parties, due to ambiguous or buried terms and conditions for data usage in complex legal language. The system's inability to be transparent may cause users to lose faith in it, which would prevent ML-based food calorie estimation apps from being widely used.

4. **Third-Party Data Sharing:**
Certain systems might exchange data for additional processing, data analytics, or product development with outside partners or vendors. Users may unintentionally give companies outside the service they originally intended to use access to their food data if they do not give their explicit consent or provide full disclosure. Significant ethical questions concerning user consent and the possibility of data misuse are brought up by this.

5. **Lack of Accuracy:**
A lot of food logging applications in naturalistic setting fails to provide accurate measurements of nutrients and poses usability challenges. Matching food sizes and estimating portions is very difficult and often predicted inaccurately by the applications [19].

2.4.2.2 Data Diversity Concerns

To ensure the effectiveness of ML-based food calorie estimation systems, a large and diverse dataset representing a range of foods from different cultures, regions, and preparation styles must be used for training. However, data diversity is frequently a major drawback that results in estimates that are skewed and incorrect.

1. **Limited Dataset Representation:**
A large number of ML models are trained on datasets that are limited in scope and mostly come from developed or Western

countries. While common foods like salads, pasta, and sandwiches may be included in these datasets, regional cuisines from South Asia, Africa, and Latin America, as well as dishes unique to those cultures, are not well-represented. As a result, users who eat these underrepresented foods might get inaccurate calorie estimates because the model might not identify them correctly. For instance, a dish like "dosas" or "baobab fruit" might not be included in the dataset, which could result in inadequate or unsuccessful recognition.

2. **Bias Toward Popular Food Items:**
Common food items that are well-represented in the dataset are frequently biased due to a lack of diversity in the training data. Due to this bias, the system may work better for users whose diets primarily consist of well-known or widely consumed foods, but it may work much less well for users whose diets consist of traditional, uncommon, or regional foods. The frequency with which users from different backgrounds discover that the system incorrectly categorizes their meals diminishes its accuracy and usefulness.

3. **Generalization Issues:**
ML models that are trained on small amounts of data find it difficult to generalize effectively in a variety of settings. For example, preparation, lighting, presentation, and even the kind of plate or utensils used can affect how a food item appears. In the event that the system is not exposed to a broad range of these factors, it may be unable to accurately identify a food item, particularly if it is prepared differently culturally or is home-cooked.

4. **Combination Dishes and Food Mixtures:**
A plate of rice, veggies, and meat is an example of a multi-component meal. Identifying different foods in an image can be a challenge for ML models, especially when the foods are mixed together or presented in an unusual way. These models are challenged by multi-ingredient dishes like stews, salads, and curries because they must identify and measure each ingredient separately in order to estimate the total caloric content. If the model never saw these kinds of meals in its training set, then this gets even more complicated.

2.5 Federated Learning in Food Calorie Estimation

Estimated calories in various items per 100 grams [6]: Table 2.1 shows the estimated calories in various cereals and grains per 100 grams, while Table 2.2 lists the estimated calories in various vegetables per 100 grams.

Table 2.1 Cereals and grains.

Rice	354
Wheat flour	320
Refined flour	352
Ragi	320
Rice flakes	354
Semolina	333
Vermicelli	332
Barley	315

Table 2.2 Vegetables.

Bottle gourd	11
Bitter gourd	21
Brinjal	25
Capsicum	16
Cauliflower	23
Cucumber	20
Tomato	21
Lady finger	27

2.5.1 Motivation and Benefits

Motivation:

In our daily lives, we can train a ML model and typically have unrestricted access to data. As long as data privacy does not get in the way, this method of working is acceptable. However, if we have to work toward COVID19 diagnoses, then a large amount of data from clinics or hospitals will be required. We are also stuck if they refuse to share due to data privacy laws.

We are unable to work with sensitive data, and this is precisely the issue with centralized learning. FL steps in to save the day. Through the use of this learning method, we can train a common model on all devices. First, using some initial data, some basic training is carried out on your local server or computer. This initiates the training process. Subsequently, every target device in the circuit downloads the model and uses the device's data (federated data) to enhance it. As a result, in this instance, we are sending the model to the data rather than the other way around, saving us the trouble of compiling the data on a single machine and ensuring data privacy [7].

Benefits:

FL serves as a liaison between the owners of the model and the data. Technically, the model owner does not see the data when training the model. The end users will have quicker predictions because the models are locally deployed, improving their overall user experience. We can train our model at scale to produce "intelligent models" because training is decentralized and privacy is ensured. While the model will be trained remotely across thousands of devices, we will require fewer processing resources locally.

Privacy Preservation: FL allows users to contribute to model training without sharing personal dietary data.

Enhanced Data Utilization: FL leverages data from a wide range of sources, improving model robustness and accuracy.

2.5.2 Federated Learning Framework for Calorie Estimation

2.5.2.1 Dataset Preparation: Fruits and Vegetables' Image Recognition

Objective

Prepare a dataset of fruits and vegetables images with corresponding calorie information for training a calorie estimation model.

Dataset Acquisition

- **Download:** Obtain the fruits and vegetables' image recognition dataset from Kaggle. This dataset contains images of various fruits and vegetables categorized into 36 classes.

Data Organization

- **Dataset Structure:** The dataset is organized into folders, each representing a different fruit or vegetable category, such as "apple," "banana," or "carrot."
- **Label Mapping:** Create a mapping between the image categories and their corresponding calorie values per 100 g. For instance, "apple" might be mapped to 52 calories/100 g.

Data Augmentation

- **Enhance Diversity:** Apply image augmentation techniques to increase data variability and improve model generalization. Common techniques include the following:
    - Rotation: Randomly rotate images by a specified angle.
    - Flipping: Horizontally or vertically flip images.
    - Zooming: Randomly zoom in or out on image regions.
    - Cropping: Randomly crop image sections.

Data Preprocessing

- **Standardization:**
    - **Resize:** Convert all images to a consistent size (e.g., 224 × 224 pixels) to accommodate the CNN architecture.
    - **Normalization:** Scale pixel values to a specific range (e.g., [0, 1] or [−1, 1]) for efficient computation.
- **Data Splitting:** Divide the dataset into three subsets:
    - **Training Set:** Approximately 70% of the data for model training.
    - **Validation Set:** Approximately 20% of the data for hyperparameter tuning and model evaluation.
    - **Testing Set:** Approximately 10% of the data for final model performance assessment.

2.5.3 Model Design: Convolutional Neural Network (CNN) for Image-Based Calorie Estimation

The main aim is to develop a CNN-based model to accurately estimate calorie content from food images [22, 23].

2.5.3.1 Model Architecture

A CNN is a suitable architecture for image-based tasks due to its ability to extract relevant features from images [9]. The proposed model consists of the following layers:

- **Input Layer:** Accepts image data with dimensions of 224 × 224 pixels and three-color channels red green blue (RGB).
- **Convolution Layers:** Extract features from the input image.
  - Conv Layer 1: Employs 32 filters with a 3 × 3 kernel size, followed by ReLU activation and max pooling to down sample the feature maps.
  - Conv Layer 2: Uses 64 filters with a 3 × 3 kernel size, followed by ReLU activation and max pooling.
- **Dense Layers:** Process the extracted features for classification or regression.
  - Flatten Layer: Converts the 2D feature maps into a 1D vector.
  - Dense Layer: Contains 128 neurons with ReLU activation for non-linearity.
- **Output Layer:** Produces the final prediction.
  - For multi-class classification (e.g., low, medium, and high calories): SoftMax activation is used to obtain class probabilities [9].
  - For regression (predicting exact calorie value): A single neuron with linear activation is used.

2.5.3.2 Loss Function and Optimizer

- **Classification:** Categorical cross-entropy loss is employed to measure the difference between predicted and true class probabilities.
- **Regression:** Mean squared error is used to quantify the difference between predicted and actual calorie values.

- The **Adam optimizer** is recommended due to its efficiency and adaptability.

Evaluation Metric

- **Classification:** Accuracy is used to assess the model's performance in correctly predicting calorie categories.
- **Regression:** Root Mean Square Error (RMSE) is employed to evaluate the model's prediction accuracy for calorie values.

2.5.4 Local Model Training for Fruits and Vegetables' Image Recognition

Challenges and Considerations
Before diving into the implementation, it is crucial to acknowledge the limitations of local training on user devices:

- Computational Resources: Mobile devices have significantly less computational power compared to servers, impacting training speed and model complexity.
- Data Storage: Storing the entire dataset on a device might be impractical due to storage limitations.
- Privacy Concerns: Users might be hesitant to share raw image data for local training due to privacy implications.

2.5.4.1 Potential Solutions and Approaches

i. Model Compression and Quantization

- Reduce Model Size: Employ techniques like pruning, quantization, and knowledge distillation to significantly reduce the model's size and computational requirements.
- Framework Support: Leverage frameworks like TensorFlow Lite or PyTorch Mobile, which offer tools for model optimization and deployment on mobile devices.

ii. Federated Learning

- Distributed Training: Train the model collaboratively across multiple devices without sharing raw data.
- Privacy Preservation: Aggregate model updates on a central server while keeping individual data local [10].
- Framework Support: Utilize TensorFlow Federated or PySyft for FL implementation.

iii. Transfer Learning

- Pre-Trained Models: Utilize pre-trained models like MobileNet or EfficientNet, which are designed for mobile devices.
- Fine-Tuning: Fine-tune the pre-trained model on a smaller, device-specific dataset to adapt it to the local environment.

iv. Model Selection and Architecture

- Lightweight Models: Choose CNN architectures optimized for mobile devices (e.g., MobileNet and ShuffleNet).
- Reduced Complexity: Consider smaller models with fewer parameters to accelerate training and inference.

2.5.4.2 *Implementation Steps*

1. Data Preparation:
   - Curate a smaller, representative subset of the dataset for local training.
   - Preprocess images to match the desired input size and format.
2. Model Selection and Optimization:
   - Choose an appropriate model architecture and apply compression techniques.
   - Consider transfer learning or incremental learning based on available data and computational resources [8].
3. Training Framework:
   - Select a suitable deep learning framework (TensorFlow Lite and PyTorch Mobile) with support for mobile deployment.
   - Configure training parameters, optimizer, and loss function.
4. Model Training:
   - Train the model on the device using the prepared dataset.
   - Monitor training progress and adjust hyperparameters as needed.
5. Model Evaluation:
   - Evaluate the model's performance on a validation set to assess accuracy and identify areas for improvement.
6. Deployment:
   - Integrate the trained model into a mobile application for real-time inference.
   - Optimize the inference process for low latency and power consumption.

2.5.5 Federated Averaging for Fruits and Vegetables' Image Recognition

Understanding Federated Averaging

FedAvg is a fundamental algorithm in FL. It allows multiple clients (in this case, user devices) to collaboratively train a model without sharing their raw data.

2.5.5.1 Key Steps Involved

1. Initialization: A central server initializes a global model.
2. Local Training: Each client downloads a copy of the global model and trains it on its local data for a specified number of epochs.
3. Model Update Sharing: Clients send the updated model weights to the central server.
4. Model Aggregation: The server averages the received model updates to create a new global model.
5. Iterative Process: Steps 2–4 are repeated for multiple rounds.

2.5.5.2 Application to Fruits and Vegetables' Image Recognition

For our fruits and vegetables' image recognition dataset, this process would look like:

1. Initialization: A central server initializes a CNN model for image classification.
2. Local Training: Users download the model to their devices. Each user trains the model on their local images of fruits and vegetables.
3. Model Update Sharing: Users send the updated model weights to the central server.
4. Model Aggregation: The server averages the received model updates to create a new, improved global model.
5. Iterative Process: Steps 2–4 are repeated multiple times.

While a full implementation would require a significant amount of code, here is a simplified outline using TensorFlow Federated:

2.5.5.3 Challenges and Considerations

- Communication Efficiency: Sending entire models can be inefficient. Consider techniques like quantization and sparsification.
- Heterogeneity: Different devices have varying computational power and data distribution. This can affect convergence.
- Privacy: While FL protects raw data, model updates might still reveal information. Differential privacy can be considered [11].
- System Complexity: Implementing a FL system requires careful engineering and management.

2.5.6 Model Evaluation: Evaluating the Global Model

Understanding the Challenge

Evaluating a federated model presents unique challenges compared to traditional centralized models. The model is trained across multiple devices, and accessing all raw data for a comprehensive evaluation might be impractical or impossible due to privacy concerns.

Evaluation Strategy

1. **Centralized Evaluation:**
    - **Data Collection:** If feasible, collect a representative subset of data from participating clients to form a centralized test set.
    - **Model Evaluation:** Evaluate the global model on this centralized test set using standard metrics (e.g., RMSE for calorie estimation and accuracy for classification).
2. **Federated Evaluation:**
    - **Local Evaluation:** Each client evaluates the global model on their local test data.
    - **Metric Aggregation:** The central server collects evaluation metrics from clients and computes an aggregate metric (e.g., average RMSE).
    - **Challenges:** Inconsistent data distributions across clients can impact the reliability of aggregated metrics [12].
3. **Hybrid Approach:**
    - Combine centralized and federated evaluation for a more robust assessment.

- Use centralized evaluation for initial model validation and federated evaluation for ongoing monitoring.

Metrics for Evaluation

- **RMSE:** For calorie estimation, RMSE measures the average magnitude of the error between predicted and actual calorie values.
- **Accuracy:** For classification tasks (e.g., calorie ranges), accuracy measures the proportion of correct predictions.
- **Other Metrics:** Consider additional metrics like mean absolute error, R-squared, or precision, recall, and F1-score based on the specific problem.

Considerations

- **Data Heterogeneity:** The distribution of data across clients can impact model performance. Consider techniques like weighted averaging of metrics to account for data imbalances.
- **Privacy:** Ensure that evaluation processes adhere to privacy regulations and protect user data.
- **Computational Efficiency:** Evaluate the computational cost of different evaluation methods, especially for resource-constrained devices.

2.6 Challenges and Future Directions

2.6.1 Challenges

i. Data Privacy and Security Concerns

FL was created to handle privacy concerns by storing user data locally on devices and only exchanging updates to the models. But vulnerabilities persist even with this strategy. The possibility that model updates will disclose sensitive information about the underlying data is one of the biggest threats to privacy. Reconstruction attacks or deducing personal information from updates shared with the central server could cause this. Although methods like secure SMC and differential privacy try to reduce these risks, they also come with additional costs like higher computational overhead and worse model performance. The risk is increased when food calorie estimation is involved because personal health information, such as dietary preferences or medical conditions, may be included in the data.

ii. Heterogeneity in Data Distribution

FL depends on training models over decentralized datasets, but the distribution of data in these datasets varies a lot. Because different users may have different dietary habits, cultural backgrounds, and food preferences, this heterogeneity poses a fundamental challenge for FL in food calorie estimation. Due to these variations, there may be model bias, meaning that users who consume less common food types may not see the full potential of the global model. Handling this non-IID data is necessary to make sure the federated model performs well across a variety of populations, which is a known problem in FL.

iii. Communication and Computational Overhead

In FL, the central server and edge devices—like smartphones—swap model updates. This procedure adds a significant amount of computational and communication overhead, which makes it difficult to use for tasks involving the estimation of food calories that call for regular updates and high model accuracy. For example, the model must adjust to new food items and changes in portion sizes, and users may need to continuously enter data about their meals. It can be challenging to maximize communication effectiveness while preserving model performance, particularly when managing a big user base.

iv. Labeling and Data Quality Issues

Data labeling in federated food calorie estimation is frequently difficult. Noisy data can result from users entering inaccurate information about the kinds or amounts of food they eat. Ensuring the quality and accuracy of the data becomes challenging because it is dispersed across various users' devices. The lack of a centralized data collection point to confirm the accuracy of the inputs exacerbates this problem.

v. Model Convergence and Personalization

Due to the differences in clients' access to data and computing power, FL can have trouble achieving model convergence. Some users may enter data for food calorie estimation on a regular basis, whereas others may do so infrequently. This unequal participation can hinder the training process and increase the difficulty of developing a global model of high quality. Furthermore, because dietary habits and preferences vary greatly from person to person, it is important to strike a balance between developing a universal model and making recommendations based on unique data.

vi. Regulatory and Ethical Issues

In food calorie estimation, FL must also negotiate a complicated regulatory landscape, particularly when it comes to handling sensitive health-related data. Laws such as Health Insurance Portability and Accountability Act (HIPAA) in the United States and GDPR in Europe must be followed. Legal and ethical ramifications could be severe for any infringement of privacy, no matter how small. Furthermore, in order to guarantee equity and reliability, ethical issues regarding the utilization of data and the possibility of prejudice in calorie estimation algorithms need to be addressed [13].

2.6.2 Future Directions

There are many prospects for improving food calorie estimation systems with FL. Future advancements in this area could boost the technology's efficiency, accuracy, and privacy:

1. **Enhanced Privacy Protections:** Advanced privacy-preserving techniques like homomorphic encryption and differential privacy will further protect against privacy breaches, even though FL already preserves data privacy by keeping information on users' devices. Future systems will be able to guarantee that private user information, like food preferences and medical histories, is secure during the model-training phase by implementing these techniques.
2. **Personalization of Models:** Personalization, or adjusting models to each individual user, is beneficial for FL. System customization to users' individual dietary requirements and preferences will be made possible by personalized FL. Models can become more adaptive and provide more accurate calorie estimates and recommendations by utilizing techniques such as transfer learning or multi-task learning.
3. **Handling Data Imbalance and Diversity:** Managing non-uniform, heterogeneous data is a common challenge in FL. More advancements in federated aggregation techniques could lead to better model performance across a range of datasets. Developing aggregation strategies that account for different types of data, such as weight- or clustering-based approaches, will help ensure that the system remains efficient and equitable for different user groups.
4. **Improved Communication Efficiency:** The communication overhead between local devices and central servers is a major obstacle to FL. By lowering this overhead, methods like compressed model updates or local training optimizations can enhance the system's performance and scalability without compromising the precision of the calorie estimate.

5. **Utilizing Multi-Modal Data:** Integrating multi-modal data—such as images, sensors, or audio inputs—could greatly enhance calorie estimation systems. By combining visual data with other inputs like food labels or health sensors, future FL systems can generate more comprehensive and accurate calorie estimates, allowing for more informed dietary tracking.
6. **Federated Learning at the Edge:** The next wave of FL systems will leverage edge computing, allowing more processing to be done on local devices, such as smartphones or wearables. This will reduce reliance on central servers and provide faster, real-time calorie estimations. Advances in edge computing will be key to enabling efficient, on-device learning.
7. **Regulatory and Ethical Considerations:** As FL applications involve sensitive personal data, compliance with regulations like GDPR or HIPAA will be critical. In the future, systems will need to integrate privacy-aware learning models and ensure that they follow ethical AI guidelines, maintaining fairness and transparency in model predictions.
8. **Collaboration with Healthcare Systems:** Future FL systems could integrate with healthcare platforms, providing personalized nutritional recommendations and dietary tracking based on individual medical needs. Collaborating with healthcare providers will expand the utility of these systems, especially for individuals managing conditions like diabetes or obesity.
9. **Continuous Learning and Adaptability:** FL systems could evolve into continuous learning frameworks that adapt over time. As user dietary habits shift, the system can dynamically learn and update to maintain high accuracy in calorie estimation. This adaptive learning will improve the model's long-term effectiveness.
10. **Cross-Domain Federated Learning:** The future could also see the integration of FL systems across different domains, such as fitness and health tracking. Sharing insights across these domains will create a more holistic view of users' health, allowing for comprehensive calorie and activity tracking.
11. **Computer vision and deep learning** is used for digital quantitative volume estimation for the optimization of dietary assessment [20].

2.7 Conclusion

FL provides a novel approach to food calorie estimation by resolving privacy issues and utilizing a variety of data sources to enhance model precision. In contrast to conventional centralized methods, FL enables models to be trained locally on users' devices, protecting private sensitive data like dietary preferences and medical records. In applications where privacy is crucial, such as health monitoring, this decentralized method is especially helpful.

The ability of FL to integrate data from a variety of users without transferring it to a central server is one of its main advantages. By using a variety of datasets that reflect various demographics, eating habits, and cultural preferences, this improves the accuracy of the model. As a result, a more accurate model with accurate calorie estimates for a range of populations is produced. Nevertheless, there are still obstacles in the way of fully utilizing FL in food calorie estimation. To maximize model performance, concerns like data heterogeneity, communication overhead, and system scalability must be addressed. Furthermore, it is imperative to conduct additional research to guarantee the robustness, objectivity, and efficiency of models trained through FL across various user groups and devices.

Subsequent investigations ought to concentrate on surmounting these constraints, like refining methods for handling heterogeneous data, maximizing communication protocols, and augmenting aggregation techniques. By overcoming these obstacles, FL has the potential to develop into an effective food calorie estimate tool that provides a safe, practical, and inclusive solution for real-world health monitoring applications.

AI disclosure statement: The author acknowledges the use of AI-based tools for minor language editing or grammatical correction. However, all intellectual contributions, analyses, and conclusions presented in this chapter are entirely the author's own.

References

1. Truong, N., Sun, K., Wang, S., Guitton, F., Guo, YK., Privacy preservation in federated learning: An insightful survey from the GDPR perspective. Submitted to Elsevier for A Conference On computers & security, 102402, 2021.
2. Abadi, M., Chu, A., Goodfellow, I., McMahan, H.B., Mironov, I., Talwar, K., Zhang, L., Deep learning with differential privacy, in: *Proceedings of the*

2016 ACM SIGSAC Conference on Computer and Communications Security, pp. 308–18, 2016.
3. Aono, Y., Hayashi, T., Wang, L., Moriai, S., *et al.*, Privacy-preserving deep learning: Revisited and enhanced, in: *International Conference on Applications and Techniques in Information Security*, Springer, pp. 100–10, 2017.
4. Ateniese, G., Mancini, L.V., Spognardi, A., Villani, A., Vitali, D., Felici, G., Hacking smart machines with smarter ones: how to extract meaningful data from machine learning classifiers. *Int. J. Secur. Netw.*, 10, 3, 137–50, 2015.
5. Konstantinos, D., Stergiou, K., Psanni, E., Federated Learning Approach Decouples Clients from Training a Local Model and With the Communication with the Server. *IEEE Trans. Netw. Serv. Manage.*, 19, 4, DECEMBER 2022.
6. https://timesofindia.indiatimes.com/life-style/food-news/know-the-calorie-content-of-the-most-common-food-items/photostory/81949358.cms.
7. https://www.linkedin.com/pulse/federated-learning-motivation-pankaj-mishra/.
8. Rocha, A., Hauagge, D.C., Wainer, J., Goldenstein, S., Automatic fruit and vegetable classification from images. *Comput. Electron. Agric.*, 70, 96–10, 2010.
9. Latif, G., Alsalem, B., Mubarky, W., Mohammad, N., Alghazo, J., Automatic Fruits Calories Estimation through Convolutional Neural Networks.
10. Pouladzadeh, P., Shirmohammadi, S., Senior Member, IEEE, Al-Maghrabi, R., Measuring Calorie and Nutrition from Food Image. *IEEE Trans. Instrum. Meas.*, 63, 8, August 2014.
11. F.S., Konstantakopoulos, Georga, E., II, Member, IEEE, Fotiadis, D., II, A Review of Image-Based Food Recognition and Volume Estimation Artificial Intelligence Systems. *IEEE Rev. Biomed. Eng.*, 17, 2024.
12. Akhi, A.B., Akter, F., Khatun, T., Uddin, M.S., Recognition and Classification of Fast-Food Images Recognition and Classification of Fast-Food Images.
13. Wen, J., Zhang, Z., Lan, Y., *et al.*, A survey on federated learning: challenges and applications. *Int. J. Mach. Learn. Cyber.*, 14, 513–535, 2023, https://doi.org/10.1007/s13042-022-01647-y.
14. Kasyap, V.B. and Jayapandian, N., Food Calorie Estimation using Convolutional Neural Network, in: *2021 3rd International Conference on Signal Processing and Communication (ICPSC)*, pp. 666–670, 2021, doi: 10.1109/ICSPC51351.2021.9451812.
15. Tripathi, P., *et al.*, Applications of Machine Learning in Agriculture, in: *Smart Village Infrastructure and Sustainable Rural Communities*, M.A. Khan and (Eds.), pp. 99–118, IGI Global, 2023, https://doi.org/10.4018/978-1-6684-6418-2.ch006.
16. Rai, M., *et al.*, *et al.*, Advancement in Agricultural Techniques With the Introduction of Artificial Intelligence and Image Processing, in: *Smart Village Infrastructure and Sustainable Rural Communities*, M.A. Khan (Ed.), pp. 47–68, IGI Global, 2023, https://doi.org/10.4018/978-1-6684-6418-2.ch004.

17. Wasif, S.M., *et al.*, Food calorie estimation using machine learning and image processing. *Int. J. Adv. Res., Ideas Innov. Technol.*, 5, 2, 1627–1630, 2019.
18. Tripathi, P., *et al.*, Applications of Deep Learning in Agriculture, in: *Artificial Intelligence Applications in Agriculture and Food Quality Improvement*, M.A. Khan and (Eds.), pp. 17–28, IGI Global, 2022, https://doi.org/10.4018/978-1-6684-5141-0.ch002.
19. Chen, J., *et al.*, The use of a food logging app in the naturalistic setting fails to provide accurate measurements of nutrients and poses usability challenges. *Nutrition*, 57, 208–2165, 2019.
20. Tay, W., *et al.*, Current developments in digital quantitative volume estimation for the optimisation of dietary assessment. *Nutrients*, 12, 4, 1167, 2020.
21. Latif, G., Mohammad, N., Alghazo, J., DeepFruit: A dataset of fruit images for fruit classification and calories calculation. *Data Brief*, 50, 109524, 2023. ISSN 2352-3409, https://doi.org/10.1016/j.dib.2023.109524.
22. Vishnu, H.S., Sindhushree, B., Punith, A., Aishwarya, K., Praveen, G., Fruit Recognition System for Calorie Management. *Int. J. Eng. Res. Technol. (IJERT)* 8, 2020.
23. Anushadevi, S., Calorie Measurement of Food from Food Image. *Int. J. Appl. Inf. Commun. Eng.*, 2015, https://www.ijert.org/fruit-recognition-system-for-calorie-management.
24. O'Shea, K. and Nash, R., An Introduction to Convolutional Neural Networks. ArXiv e-prints, 2015.
25. Wang, B. and Kindratenko, V., RL-Pruner: Structured Pruning Using Reinforcement Learning for CNN Compression and Acceleration. 10.48550/arXiv.2411.06463, 2024.
26. Yang, M., Qian, H., Wang, X., Zhou, Y., Zhu, H., Client Selection for Federated Learning With Label Noise. *IEEE Trans. Veh. Technol.*, 71, 2, 2193–2197, 2022.
27. Sogari, G., Velez-Argumedo, C., Gómez, M.I., Mora, C., College Students and Eating Habits: A Study Using An Ecological Model for Healthy Behavior. *Nutrients*, 10, 12, 1823, 2018 Nov 23. doi: 10.3390/nu10121823. PMID: 30477101; PMCID: PMC6315356.

3

Federated Learning for Food Safety and Compliance

Ramit Sehgal and Nitendra Kumar*

Amity Business School, Amity University, Noida, Uttar Pradesh, India

Abstract
This chapter explores the application of federated learning (FL), a decentralized technology, to improve food safety, quality standards, and product compliance management worldwide. FL supports the processing of confidential information while ensuring privacy through technologies such as privacy differentiation and advanced integration. This chapter highlights the challenges of integrating FL into businesses with diverse information sources (e.g., food safety, healthcare, and finance) under stringent regulations such as the General Data Protection Regulation. It explores how self-protecting technology can prevent data leaks and improve stakeholder collaboration by leveraging blockchain for traceability. The chapter also highlights the limitations of current FL architectures and suggests future directions for developing blockchain, privacy, and machine learning solutions to enable governance management and sensitive information.

***Keywords*:** Federated learning, food safety, quality standards, general data protection regulation (GDPR)

3.1 Introduction

In a globalized society, food safety/quality grows into one of the most critical requirements for societies and individuals, as well as for the economic aspect of food safety/quality management. With people's numbers growing constantly, the need for food products also grows, which results in supply chains becoming

**Corresponding author*: nkshukla.kumar4@gmail.com

Padmesh Tripathi, Bhanumati Panda, Shanthi Makka, Reeta Mishra, S. Balamurugan and Sheng-Lung Peng (eds.) Federated Learning for Smart Agriculture and Food Quality Enhancement, (57–84) © 2026 Scrivener Publishing LLC

more intricate and often international, including many players. Therefore, food security and quality are not issues that could be tended to locally, but they need a focused and wide-ranging approach in the food industry.

Foodborne diseases bear major risks to public health; according to the World Health Organization (2021), at least 600 million people, or nearly 10% of the global population, get sick from foodborne diseases each year. All these illnesses not only have an impact on the health of individuals and families but also carry a heavy economic impact in terms of costs that governments and businesses bear through the costs of medical expenses, loss of productivity, and loss in business due to damage control resulting in loss of reputation. Such products include foods. For instance, food safety recalls fundamentally lead to millions of worth of losses as well as long-term impacts on consumer confidence [1]. Therefore, the sanitary quality and the safety of foods have become one of the main goals on the local and global agenda among governments, producers, and consumers.

3.1.1 Challenges in Traditional Food Safety Monitoring

In the usual concepts of surveillance of food safety, they are those schemes in which data is gathered, sorted out, and delivered to the main regulating body or authority. This approach is useful, yet it has its inherent difficulties. Another emerging challenge specifically linked to the context of the work presented in this chapter is the problem of data silos or the widespread practice of isolating data in separate organizations. For instance, a farm records how much pesticide it has applied, and a processor records how many cars of product were contaminated, but, due to the lack of data integration, some valuable information can be missing when performing risk evaluations [2].

However, compliance with regulations is generally a very elaborate and time-consuming activity. Currently, countries and regions have their own food safety regulations, which prove complex to meet by international producers. This is further compounded by the requirement to provide real-time data due to the fact that delayed reporting of incidences puts the public health facility at risk. Standardization may also be a challenge given that conventional approaches are slow to adapt to changes in food production systems and dynamics in consumer demand.

To these, one can also tie systemic problems linked to data sharing, in particular issues of privacy. In organizations, many employees are not willing to share important information like production procedures, supplier information, and customer data because this information is sensitive or because it may infringe on the law [3]. This unwillingness to share data hampers efficiency and works against the generation of efficient safety supervision and control mechanisms affecting the quality of the food that consumers get in their kitchens.

3.1.2 The Role of Federated Learning

In view of these challenges, there is a new concept that can sufficiently offset many drawbacks of conventional food safety monitoring systems federated learning. Federated learning (FL) is a form of machine learning in which multiple parties can train a model with features of other models without direct access to each other's data. However, in each of the participants, one develops a local model based on its data, and each transmits only the model updates to a central server that in its turn applies the updates to enhance a global model [4]. There are several huge benefits that this architecture provides for applications in the food industry; here are the most obvious ones:

First, it should be noted that FL improves data protection and security. In that way, the organizations can work together on creating the related predictive models while the data remains local and thus secure. This is especially important in the food industry, where often there are several secrets that should be preserved at any cost. In addition, FL is GDPR compliant and helps organizations adapt to privacy legislation across jurisdictions to achieve effective analysis of data cooperatively [5].

Secondly, FL enables the aggregation of different data sources that may be more diverse and therefore helps to improve model robustness. This means that in the case of food safety, it can be a great advantage because food safety hazards can occur due to numerous factors such as environmental conditions, other companies' operations, production processes, and supply chain management. This way, FL models can be more accurate, making the risk assessment and management more effective and efficient when drawing on a broader spectrum of inputs [7].

Moreover, FL cuts out the data transmission for bandwidth needs, thus saving traffic costs. Conventional filing and data-sharing techniques involve moving vast amounts of confidential data to a central repository that can put a load on the network and entail a steep price. Indeed, FL only sends model updates instead of raw data, leading to a decreased amount of data transferred and, therefore, reducing communication overhead [8].

3.1.3 Objectives and Structure of the Chapter

There is the need to discuss various ways and the extent to which FL can be used to enhance food safety and quality standards in this chapter.

Based on the key benefits of FL, the food industry stakeholders can improve the existing approaches to food quality control, safety threat forecasting, and management of the new risks and threats arising. The chapter will address the following objectives:

i. **Discuss the principles and mechanisms of federated learning:** This includes a critical insight into how FL works and the components involved, as well as the as well as the strengths of FL over the traditional centralized machine learning models.
ii. **Explore specific applications of federated learning in food safety:** This section will discuss how FL can enhance traceability, predictive analysis, and anomaly detection with respect to the food supply chain.
iii. **Examine the enhancement of food quality standards through federated learning:** To this end, the chapter will also reveal how FL can enhance quality control activities and analyze data from sensors and consumers' feedback for better enhancement.
iv. **Identify challenges and limitations:** Despite the advantages that FL holds, there are some important issues that have to be solved to unleash its potential in the fields of the food industry.
v. **Discuss future directions and innovations:** The chapter will end by discussing new opportunities for current future FL development in food safety and quality assurance systems.

Based on the insight derived from this paper and understanding of FL, actors in the food industry can only improve the food safety and quality. The subsequent parts will provide more detailed information about the particular use cases and consequences of FL in this crucial area.

3.2 Principles and Mechanisms of Federated Learning

FL is a brand innovative model for machine learning that allows decentralized data training with full protection to the data. This section uses an information-processing approach to, firstly, explain the principles and mechanics of FL and, secondly, to explain how this ground-breaking approach works and why it is superior to traditional centralized learning approaches. It will be important to consider these fundamentals at a fundamental level in order to better understand where FL can come into play in enhancing food safety and quality.

3.2.1 Conceptual Framework of Federated Learning

In its essence, FL enables several stakeholders—farms, manufacturers, and retailers—to build a centralized machine learning model by training it together without exposing local data [4]. This is quite opposite to the conventional machine learning setup where data from all depths is fed to a central server for further learning. The fundamental principles of FL can be summarized as follows:

i. **Data Locality:** In FL, the data is local, and different participants remain as different owners of the data set. Updates—storing a local model generated from the local data—only flow in the opposite direction to a central server; no global data is transmitted. This approach reduces the privacy issue and data leakage because, with this system, data never travels over the network [8].

ii. **Collaborative Learning:** FL allows separate parties to combine and contribute to the building of a machine learning model. During the training process, each participant calculates model updates with reference to a portion of the data set. These updates are accumulated in the central server to further improve the global model by learning from the combined expertise of all the participants as formulated in [7].

iii. **Asynchronous Updates:** Participants are enabling to train the models at different times and rates through FL. It is advantageous, especially where sources of data have different sizes and capabilities of computations. Low-frequency synchronous updates provide that the global model keeps on being augmented even if the training regime of the participants is not well coordinated [4].

iv. **Model Aggregation:** The most fundamental concept of FL is to reduce the various participants' model updates. To do so, this server uses an aggregation algorithm, which can be associated with weighted averaging, to collate the local updates into a single, better global model. This step is important because it guarantees the share of each learner during the learning phase in relation to the amount shared by other members [8].

3.2.2 Technical Components of Federated Learning

To implement FL effectively, several technical components are necessary:

i. **Local Training:** The general training procedure consists of each participant training its model on a given local dataset. There are standard procedures used in most practice, including stochastic gradient descent procedures for finding the model parameters [4].

ii. **Communication Protocol:** Another scheme ensures that local devices can send updates of models to the central server, and vice versa. This protocol has to be efficient to consume as little bandwidth as possible, particularly in those networks whose accessibility is restricted [7].

iii. **Privacy-Preserving Mechanisms:** FL uses different methods to protect the privacy of information shared among different users. For example, there are techniques like differential privacy that add some noise to the updates of the model so that one cannot get the particular data back, for example, the same as adding salt to fish so that it cannot be detected again [9]. This in turn guarantees that the accumulated update scheme does not endanger the local data of any participant.

iv. **Federated Optimization:** FL uses optimization techniques suitable for decentralized environments. These algorithms address the issues with the non-independent and non-identically distributed (IID) data distribution across the participants; the global model gets converged because of a different data distribution [4, 6, 7].

v. **Security Protocols:** Due to the distributed nature of FL, it must be guaranteed that data remains truthful and secure throughout the update file exchange. For instance, the use of secure multi-party computation (MPC) or homomorphic encryption to seal updates to the model to unauthorized parties [3].

3.2.3 Advantages of Federated Learning

FL offers several advantages over traditional centralized machine learning approaches, particularly in the context of food safety and quality standards:

i. **Enhanced Data Privacy:** By keeping data local, FL significantly reduces the risk of data breaches and enhances participant trust. Organizations can collaborate without fear of exposing sensitive information, making FL an attractive option in industries where privacy is paramount [3].
ii. **Increased Collaboration:** FL fosters collaboration among various stakeholders in the food industry. Farms, manufacturers, and retailers can contribute to a collective knowledge base, improving model performance and enabling more effective responses to safety incidents [7].
iii. **Reduced Bandwidth Consumption:** Traditional centralized models often require the transmission of large datasets to a central server, consuming significant bandwidth. In contrast, FL only transmits model updates, substantially reducing the amount of data exchanged and the associated costs [8].
iv. **Improved Model Robustness:** By leveraging diverse datasets from multiple sources, FL models are more robust and generalizable. This is particularly valuable in the food industry, where safety incidents can arise from various factors across different regions and production practices [4].
v. **Scalability:** FL is inherently scalable, allowing organizations to add new participants and data sources without the need for extensive reconfiguration. This scalability is crucial in the rapidly evolving food industry, where new players frequently enter the market [7].

3.3 Applications of Federated Leaning in Food Safety and Quality Standards

The application of FL in food safety and quality standards has garnered significant attention due to its ability to enhance data collaboration without compromising privacy. This section examines several key applications of FL in the food industry, providing insights into how it can address specific challenges and improve safety and quality across the supply chain.

3.3.1 Enhancing Traceability in the Food Supply Chain

Traceability is a critical component of food safety management, allowing stakeholders to track the journey of food products from farm to table. Effective traceability systems help identify sources of contamination and facilitate swift recalls when necessary [10]. FL can enhance traceability by enabling decentralized data sharing among various participants, including farmers, processors, distributors, and retailers.

Case Study: The Role of Federated Learning in Traceability
A notable case is the implementation of FL in a distributed agricultural network. Imagine a scenario where multiple farms contribute data on their agricultural practices, soil health, and crop yields while retaining ownership of their sensitive information. Through FL, a global model can be trained to identify optimal farming practices based on insights drawn from the collective data.

For example, if 50 farms in a region employ FL, then they can share model updates regarding soil quality and pesticide usage without directly disclosing their proprietary data. As a result, the global model can identify patterns and provide recommendations for improving crop quality while ensuring that no individual farm's data is exposed [4].

Figure 3.1 illustrates that farm (value: 1) represents the starting stage of the supply chain where data is collected. Processing (value: 0.75): The value decreases as it moves to the next stage due to processing activity. Distribution (value: 0.5): Distribution value reflects a mid-level reduction (smaller flow). Retail (value: 1): Final value is restored to 1, reflecting completion of the supply chain.

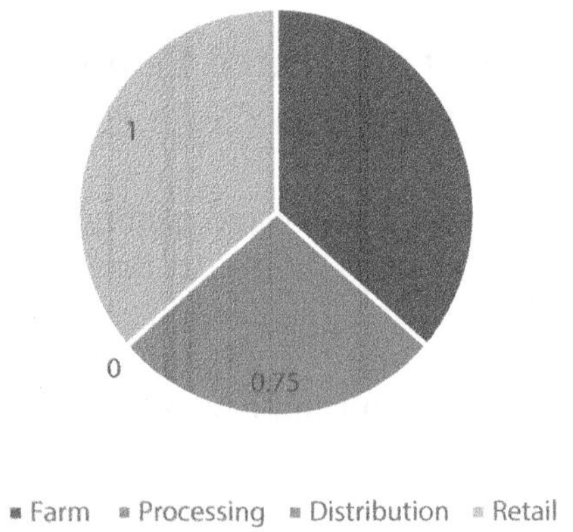

Figure 3.1 Federated learning in enhancing traceability across the food supply chain.

3.3.2 Monitoring Food Quality in Real Time

Real-time monitoring of food quality is essential for ensuring safety and compliance with regulations. FL can enable participants in the food supply chain to collaboratively develop predictive models for quality assessment without exchanging sensitive data.

Data Analysis: Impact of Real-Time Monitoring

To illustrate the impact of FL on real-time monitoring, we can analyze a dataset involving temperature and humidity sensors in a cold storage facility. Let us assume that we have data from three different storage units, each belonging to a different company. Each unit collects temperature and humidity readings every hour as shown below in Table 3.1.

Figure 3.2 illustrates the average temperature and humidity readings from each storage unit over a week.

By utilizing FL, these storage units can collaboratively develop a predictive model that learns from their collective data. For instance, if a significant spike in temperature is detected in one unit, then the model can alert all participants, enabling them to take preventative measures to avoid spoilage.

In this scenario, the FL model would leverage local updates from each storage unit to improve predictions regarding optimal temperature and humidity levels, ensuring that food quality is maintained throughout the supply chain [7].

Table 3.1 Average temperature and humidity readings from storage units a, b, and c.

Day	Storage unit a (°C)	Storage unit b (°C)	Storage unit c (°C)
Monday	4	5.5	4.5
Tuesday	3.8	5.6	4.7
Wednesday	3.9	5.3	4.4
Thursday	4.1	5.4	4.6
Friday	4	5.5	4.5
Saturday	4.2	5.7	4.8
Sunday	4	5.4	4.5

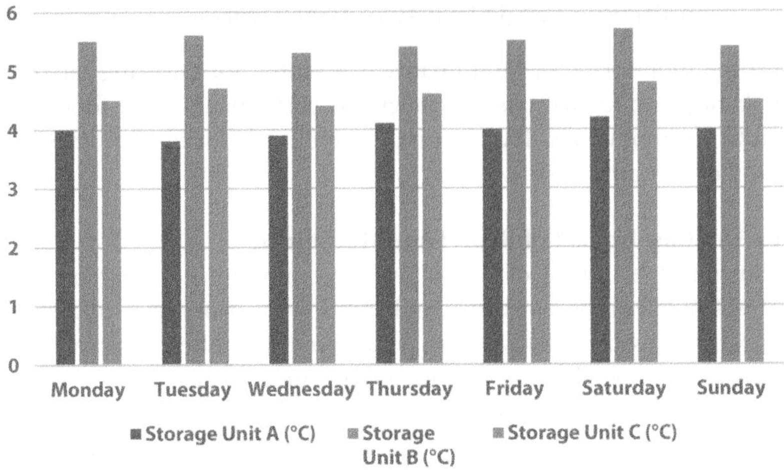

Figure 3.2 Average temperature and humidity readings from storage units a, b, and c.

3.3.3 Predicting Food Safety Risks

FL can also play a crucial role in predicting food safety risks, allowing stakeholders to identify potential hazards before they escalate into serious issues. By analyzing data from various sources, such as production facilities, transportation logs, and retail environments, FL models can identify patterns that indicate emerging risks.

Example: Predicting Contamination Events

Consider a scenario in which multiple food processing plants collaborate using FL to predict contamination events. Each plant collects data on operational parameters, cleaning protocols, and quality control checks. By applying FL, these plants can develop a robust predictive model that alerts them to potential contamination risks.

For instance, if a model trained on data from several plants detects a correlation between a specific cleaning schedule and increased contamination reports, then it can recommend adjustments to the cleaning protocols across all plants without disclosing the underlying data [3].

The process of predictive modeling for food safety risks involves a series of structured steps, as illustrated in Figure 3.3. The process begins with data collection, where relevant information is gathered from various sources. The collected data undergoes data processing, which ensures it is clean and ready for analysis. Following this, feature extraction identifies and selects the most impactful attributes for the model. The model is then trained using

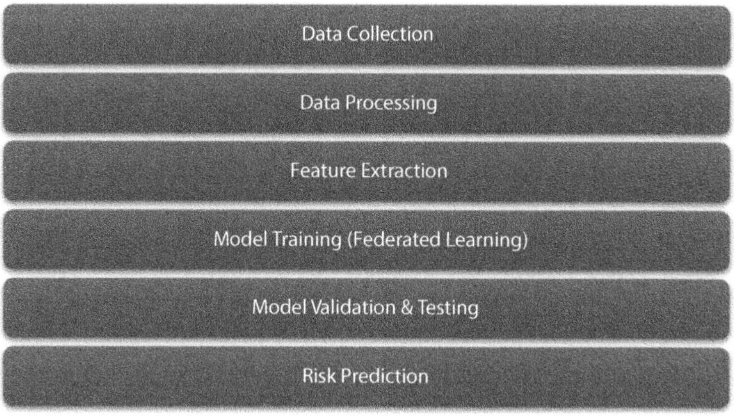

Figure 3.3 Overview of predictive modeling for food safety risks.

FL, ensuring data privacy and distributed learning across multiple systems. Once trained, the model is subjected to validation and testing to assess its accuracy and reliability. Finally, the validated model is used for risk prediction, providing actionable insights to mitigate potential food safety risks.

3.3.4 Optimizing Supply Chain Operations

FL can optimize supply chain operations by enabling participants to share insights into inventory levels, demand forecasts, and transportation logistics. This collaborative approach can lead to more efficient resource allocation and improved response times to market fluctuations.

Data Analysis: Inventory Optimization

To exemplify the impact of FL on supply chain optimization, we have considered a scenario below as shown in Table 3.2 involving three distribution centers that share inventory data while retaining local control. By employing FL, these centers can collaboratively develop a model that forecasts demand for various products based on historical sales data.

The demand forecasting for distribution centers provides critical insights into daily fluctuations across various locations. Figure 3.4 illustrates the forecasted demand trends for distribution centers A, B, and C over a 5-day period. Distribution center B shows the highest demand on day 1, followed by a decline and subsequent recovery on day 5. In contrast, distribution center A experiences a relatively steady demand pattern, with a noticeable dip on day 2 before gradually rising again. Distribution center C displays

Table 3.2 Demand data for a specific product over the last month.

Day	Distribution center A	Distribution center B	Distribution center C
Day 1	150	200	180
Day 2	130	220	170
Day 3	160	210	175
Day 4	140	200	160
Day 5	150	230	185

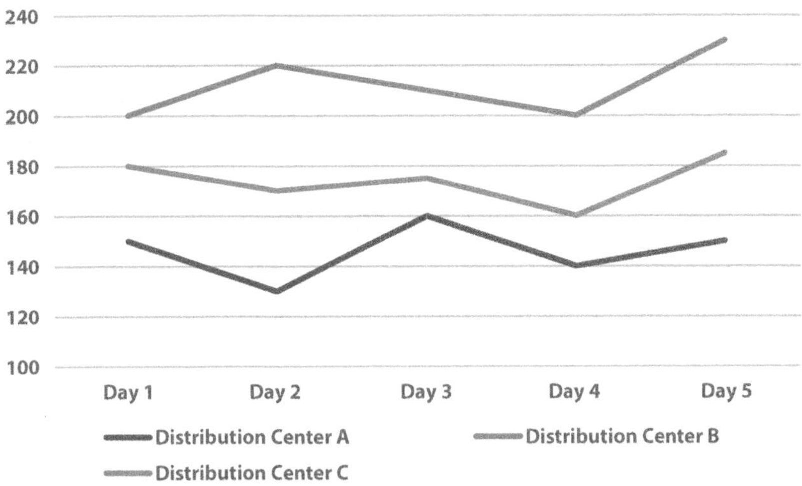

Figure 3.4 Demand forecast for distribution centers a, b, and c.

moderate demand with minimal fluctuations, maintaining a more stable trend throughout the observed period. These trends highlight the varying demand patterns and the importance of tailored resource allocation for each center to optimize supply chain efficiency.

Using FL, the distribution centers can aggregate their demand forecasts and improve accuracy while ensuring sensitive sales data remains private. This optimized forecasting can lead to reduced stockouts, minimized waste, and increased customer satisfaction [7].

3.3.5 Enhancing Regulatory Compliance

FL can assist food businesses in maintaining compliance with food safety regulations by providing insights derived from collaborative data analysis. By sharing insights without revealing sensitive information, businesses can collectively monitor compliance and identify areas for improvement.

For example in a scenario of compliance monitoring with FL, where, multiple food manufacturers utilize FL to monitor compliance with hygiene standards, each manufacturer can contribute data regarding their sanitation practices. The federated model can identify trends that indicate non-compliance, allowing participants to take corrective actions before violations occur.

For instance, if a model indicates that a specific cleaning method correlates with increased microbiological contamination, then manufacturers can revise their practices based on the collective insights generated through FL [4].

3.4 Challenges and Limitations of Federated Learning in Food Safety and Quality Standards

While FL holds great promise for improving food safety and quality standards, its implementation is not without challenges and limitations. This section delves into the various obstacles that organizations may encounter when integrating FL into their food safety management systems, along with potential strategies to mitigate these challenges.

3.4.1 Data Heterogeneity

One of the primary challenges of FL is data heterogeneity, which refers to the differences in data distributions across different participants. In the food industry, data collected by different entities (e.g., farms, processors, and retailers) can vary significantly in terms of quality, format, and relevance. This heterogeneity can hinder the performance of FL models [11].

For instance, consider a scenario where several farms contribute data regarding their pesticide use, but the data collection methods differ widely. One farm may use automated sensors to track pesticide application, while another relies on manual logs. The resulting discrepancies can lead to biased model updates and affect the overall accuracy of the predictions.

Data Analysis: Impact of Data Heterogeneity

To illustrate the impact of data heterogeneity, consider a dataset containing pesticide usage information from five different farms. Table 3.3 shows the differences in data collected by each farm over a 2-week period.

The analysis of pesticide usage across various farms reveals significant differences in both the volume applied and the variability of usage. Figure 3.5 illustrates the pesticide usage (L/Ha) for farms A through E, alongside the corresponding variability percentages. Farm C demonstrates the

Table 3.3 Data of each farm over a 2-week period.

Farm	Data collection method	Average pesticide usage (L/Ha)	Variability (%)
A	Automated Sensors	12.5	5
B	Manual Logs	10	15
C	Automated Sensors	15	10
D	Manual Logs	8	20
E	Automated Sensors	11.5	8

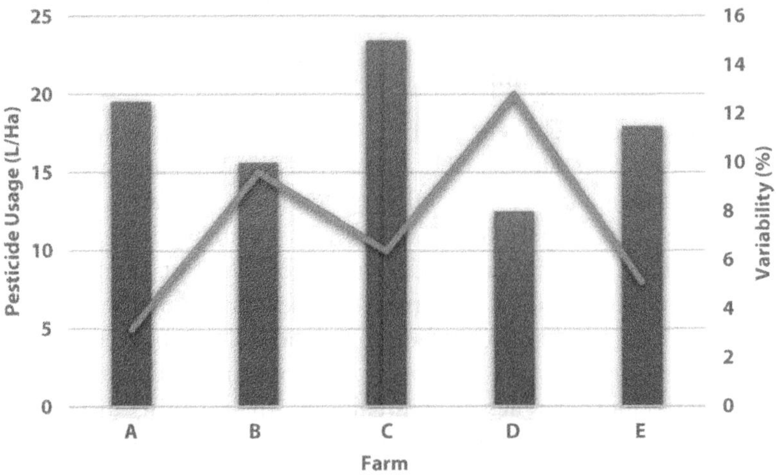

Figure 3.5 Differences in pesticide usage data across farms.

highest pesticide usage, reaching approximately 25 L/Ha, with moderate variability. In contrast, farm D shows the lowest usage at around 10 L/Ha, coupled with the least variability, indicating consistent application practices. Farms A, B, and E exhibit intermediate usage levels, but their variability percentages vary significantly, suggesting inconsistent usage patterns. This data highlights the need for targeted strategies to optimize pesticide usage and reduce variability, particularly for farms with higher application levels. The variability in data collection methods leads to challenges in developing a robust FL model that accurately captures the nuances of pesticide usage across different farms.

3.4.2 Communication Overhead

FL relies on continuous communication between local devices and a central server to update models based on local data. This communication can lead to significant overhead, especially when dealing with large datasets and complex models. The need for frequent synchronization can result in delays and increased bandwidth consumption [7].

3.4.3 Privacy and Security Concerns

Although FL is designed to enhance data privacy, concerns remain regarding the security of model updates shared among participants. Malicious actors may attempt to exploit vulnerabilities in the system to access sensitive information or manipulate model outputs. Ensuring the security of the communication channels and the integrity of the federated model is crucial [3].

Example: Privacy Breaches in Federated Learning
Consider a scenario where several food manufacturers participate in a FL system to improve their quality control processes. Each manufacturer provides model updates based on their internal data. If these updates are not adequately encrypted, then a malicious entity could intercept and analyze the updates to extract sensitive information about each manufacturer's production practices.

3.4.4 Regulatory Compliance Challenges

Implementing FL in the food industry requires adherence to various regulatory standards concerning data privacy, safety, and traceability. Ensuring compliance with regulations such as the Food Safety Modernization Act (FSMA) in the United States and the General Data Protection Regulation (GDPR) in Europe can complicate the deployment of FL systems [7].

Data Analysis: Compliance Mapping

To understand how regulatory requirements affect the implementation of FL, consider a compliance mapping exercise involving three key regulations:

- **FSMA:** This requires food manufacturers to establish hazard analysis and risk-based preventive controls.
- **GDPR:** This mandates that organizations protect personal data and privacy.
- **HACCP:** This requires systematic preventive approaches to food safety.

Table 3.4 summarizes the key compliance requirements and how they intersect with FL implementations.

3.4.5 Technical Complexity and Implementation Costs

Finally, the technical complexity associated with implementing FL can be a significant barrier for many organizations. Developing the necessary infrastructure, algorithms, and frameworks requires substantial investment in both time and resources [11]. Additionally, organizations may need to upskill their workforce to effectively manage FL systems.

Cost Analysis: Federated Learning Implementation

To quantify the costs associated with implementing a FL system in the food industry, consider the following breakdown:

- Infrastructure Development: \$100,000–\$500,000
- Algorithm Development: \$50,000–\$150,000

Table 3.4 Key compliance requirements and implications for federated learning.

Regulation	Key requirements	Implications for Federated learning
FSMA	Hazard analysis, preventive controls	Data must be accurate and timely, affecting model updates.
GDPR	Data protection and privacy	Requires strong encryption and consent management
HACCP	Preventive measures, documentation	Requires robust traceability of data contributions

- Training and Upskilling: $20,000–$100,000.
- Ongoing Maintenance: $10,000–$50,000 per year

Estimated Total Costs: $180,000–$800,000 (initial setup and first year)

3.5 Challenges and Opportunities in Implementing Federated Learning for Food Safety

FL presents a transformative approach to data analysis in the context of food safety and quality standards. By enabling decentralized data processing, FL allows stakeholders in the food supply chain to collaborate without compromising sensitive data. However, implementing FL in food safety entails a myriad of challenges and opportunities. This section explores these challenges, including data heterogeneity, regulatory compliance, technical complexities, and the need for stakeholder collaboration. It also discusses the potential opportunities that FL offers to enhance food safety protocols, improve compliance with quality standards, and foster innovation within the food industry.

3.5.1 Challenges in Implementing Federated Learning

- **Data Heterogeneity:** One of the primary challenges in implementing FL for food safety is the inherent heterogeneity of data. Different stakeholders in the food supply chain, such as farmers, processors, distributors, and retailers, generate diverse data types and formats. For instance, data from farms may include soil health metrics, weather conditions, and pesticide usage, while processing plants may focus on batch processing times, temperatures, and contamination levels. This variance complicates the integration of data across different entities, hindering the development of robust predictive models [12].

  Moreover, the quality of data collected can vary significantly between stakeholders. For instance, smallholder farms may have limited access to sophisticated data collection tools compared to larger agricultural enterprises, resulting in discrepancies in data quality and quantity. As a result, the effectiveness of FL models can be adversely affected, leading to inaccurate insights and decisions.

- **Regulatory Compliance:** The food industry is subject to stringent regulations that govern food safety and quality standards. Regulations such as the FSMA in the United States and the General Food Law in the European Union necessitate compliance from all stakeholders in the food supply chain. FL must navigate these regulatory frameworks to ensure that data sharing and model training processes adhere to legal requirements [13].

  For instance, the need for transparency in data usage and the requirement for obtaining informed consent for data sharing can create complexities in implementing FL systems. Additionally, various jurisdictions may have different regulatory requirements, complicating the development of a standardized FL approach across international supply chains. Companies may face legal repercussions if they fail to meet these compliance standards, which can create a disincentive for adopting FL solutions.
- **Technical Complexities:** Implementing FL involves significant technical complexities. Developing and maintaining FL frameworks requires specialized expertise in distributed computing, secure data sharing, and machine learning algorithms [14–16]. Furthermore, ensuring the security of data during transmission and processing is paramount, particularly when dealing with sensitive information related to food safety.

  Challenges related to communication overhead must also be considered. In FL, models are trained locally, and updates are sent to a central server, which aggregates the updates to improve the global model. However, this process can lead to significant communication overhead, particularly when the number of participating nodes is large. High communication costs can hinder the timely deployment of FL solutions, making it essential to develop efficient communication protocols that minimize bandwidth usage while maintaining model accuracy.
- **Stakeholder Collaboration:** Effective implementation of FL for food safety relies on collaboration among diverse stakeholders, including farmers, processors, retailers, and regulatory bodies. Achieving collaboration can be challenging due to differing priorities and incentives across the food supply chain. For example, farmers may prioritize

maximizing yields, while retailers may focus on minimizing costs. Aligning these divergent goals to foster a collaborative environment for data sharing and model training is essential for the successful adoption of FL solutions [17].

Moreover, concerns about data ownership and the competitive landscape can create reluctance among stakeholders to participate in FL initiatives. Trust-building measures and clear agreements regarding data usage and model ownership are crucial to overcoming these barriers.

3.5.2 Opportunities in Implementing Federated Learning

- **Enhanced Food Safety Protocols:** Despite the challenges, FL offers significant opportunities for enhancing food safety protocols. By facilitating the integration of data from various stakeholders, FL can lead to more comprehensive analyses of food safety risks. For instance, farmers can share data on pesticide usage, whereas processors can provide insights into contamination incidents, allowing for more accurate risk assessments and proactive measures to mitigate food safety hazards [18].

  Furthermore, FL can enable real-time monitoring of food safety parameters across the supply chain. By leveraging data from multiple sources, stakeholders can respond more rapidly to potential food safety issues, thereby reducing the risk of widespread outbreaks and recalls.
- **Improved Compliance with Quality Standards:** FL can also enhance compliance with quality standards. By allowing stakeholders to collaboratively develop and refine quality assurance models, FL can ensure that food products meet safety and quality standards throughout the supply chain. This collaborative approach enables continuous improvement, as insights gained from one stakeholder can benefit others in the network.

  For example, FL can facilitate the development of predictive models that assess the likelihood of food spoilage based on environmental conditions and handling practices across different points in the supply chain. By adhering to quality standards, stakeholders can enhance consumer trust and improve market access.

- **Fostering Innovation in the Food Industry:** The collaborative nature of FL encourages innovation within the food industry. By sharing insights and best practices, stakeholders can identify emerging trends and challenges in food safety and quality. This knowledge-sharing can lead to the development of innovative solutions, such as advanced monitoring systems and predictive analytics, to address food safety concerns [19].

  Additionally, FL can facilitate cross-industry collaborations, enabling stakeholders to leverage expertise from various sectors. For example, partnerships between the food industry and technology firms can drive the development of sophisticated FL algorithms that enhance food safety analytics.

3.6 Future Directions and Innovation in Federated Learning for Food Safety

The future of FL in food safety is characterized by a rapidly evolving landscape where advancements in technology and data science converge to enhance food safety protocols and quality standards. As food supply chains become increasingly complex and interconnected, the need for innovative solutions that leverage distributed data processing grows. This section explores future directions and innovations in FL for food safety, focusing on technological advancements, integration with emerging technologies, potential applications, and strategies for scaling FL initiatives across the food industry.

3.6.1 Technological Advancements in Federated Learning

Technology has an immense prospect to capture the future of food safety in the advanced frameworks of FL. Some of these categories are as follows: communication, models, and privacy, respectively.

- **Improved Communication Protocols:** FL has some inherent issues related to the exchange of model updates between the nodes or substations and the fusion center. Next steps will involve refining the communication processes to lower the amount of bandwidth needed in order to make the learning models more efficient. It will be learned that simple approaches like model compression, quantization, and communication-efficient federated averaging (FedAvg) [11].

For instance, the sizes of model updates will be cut down by quantization in weight; it is also dramatically cutting down the amount of data to be transmitted. But when these methods are incorporated, FL systems should be more implementable even in environments of low resources, such as in small-scale farming or a processing company.

- **Enhanced Model Efficiency:** The future advancements in model making will be an improvement in the algorithms in addressing the food safety data with emphasized proper characterization. This may imply using lightweight architectures where there is only a small level of required computational resources and accuracy is still paramount. Transfer learning and meta-learning approaches may help bring-in solutions to different data distribution across the organizations [17]. Moreover, improvements in federated optimization algorithms like FedAvg and federated proximal (FedProx) will improve the convergence rate of FL models, making the updating and real-time decision in food safety possible.

3.6.2 Integration with Emerging Technologies

FL integration with other advanced innovation will unveil next levels of opportunities in food safety improvement. The main technologies are the Internet of Things (IoT), blockchain, and artificial intelligence (AI).

- **IoT:** Smart devices within the IoT environment allow the perpetual assessment of factors that may affect food safety. Stakeholders are, therefore, able to use FL trained on IoT sensors and monitoring devices used in the food supply chain. This integration opens the possibility for the construction of real-time risk modeling that considers safety risks concerning foods [20]. For instance, devices that constantly examine the temperature and humidity during transportation also supply such information to a FL system to evaluate food quality in real-time. Consequently, stakeholders can foster management strategies for specific food safety problems, which create a minimal chance of contamination and spoilage.

- **Blockchain Technology:** Blockchain applied to FL models can increase the federation's data accountability and origin. Through the creation of proof-of-origin, blockchain makes stakeholders in the food supply chain place their trust in the records contributed through that technology. Smart contracts may also facilitate the automation of sharing data, differing agreements while effectively meeting the regulatory standards not to disclose some information [21]. For instance, a FL social structure based on the blockchain can help farmers relay some of the pesticide usage details to the processor without having to worry about a data breach or compromise of anonymity. This approach advances integration rather than isolation while protecting important information and knowledge.
- **AI:** FL models could be utilized to amalgamate various insights, and then the AI algorithms can be used to analyze the compiled insights. Thus, using big data and evolving machine learning methodologies, organizations and their stakeholders can move over large and diverse patterns, generating practical understandings for enhancing food safety [22]. For instance, AI can be applied to epidemiological data that pinpoints relationships with certain forms of farming or processing. The evidence generated from this study may help develop specific strategies to improve food safety practices within these links.

3.6.3 Potential Applications of Federated Learning in Food Safety

With the future advancements of FL, its use in food safety will increase as well. Key areas of focus include:

- **Risk Assessment for Food Components:** By applying FL, it is possible to design models for predicting risks associated with food safety using historical and current data acquisition. Recognizing that FL models can add value to a vast amount of data collected from different actors, it can be said that these models can reveal such patterns and trends that suggest possible food safety concerns [18]. For instance, a FL system could learn from the usage of pesticides, climatic conditions, and composition of the soil to determine the possibility of contamination in crops produced. The reader is hereby reminded that early indicators are very useful because they can allow stakeholders to avoid problems.

- **Quality Assurance as well as Compliance Monitoring:** With FL, it is possible to carry out constant supervision of food quality combined with compliance with the regulatory requirements. Through the sharing of quality assurance models, the concerned stakeholders can work together to achieve the conformity of food products to safety and quality standards along the supply chain [19]. For example, FL can let processors convey information on production processes while receiving information from farmers on farming practices. This model can improve the quality of food products as they are produced jointly, as well as minimize the occurrence of non-compliance.

3.6.4 Strategies for Scaling Federated Learning Initiatives

For FL to make a difference in food safety, various actors must implement measures that ensure scalability. Key strategies include the following:

- **Building Trust among Stakeholders:** Stakeholder trust is a desirable principle for FL, and its application entails creating trust among the participants. The ownership of data, how data has to be used, and which rules and regulations have to be followed are the essential requirements, which need to be clarified in order to have good collaboration. Data sharing threats should be discussed, and stakeholders should seek to find common ground in the issue.
- **Developing User-Friendly Interface:** Therefore, interfaces for stakeholders in the food supply chain need to be created as part of FL promotion. These interfaces should make simple the process of contributing data and training a model, such that lay people can address FL systems.
- **Pilot Program and Case Studies:** These pilot programs and case studies can actually illustrate how FL may be employed to improve food safety. Promoting FL's best success stories helps participants establish trust in such a strategy and promote a wider implementation of FL within the food organization sector.

3.7 Challenges and Limitations of Federated Learning in Food Safety

Although the FL has the potential to improve the ambition of food protection's idea of getting quality food to the population and helping it become safer, there are several challenges and limitations that need to be addressed. These challenges are outlined as data heterogeneity, privacy, communication difficulties, legal fencing, and the call for heterarchical cooperation. It is, therefore, important to ascertain these limitations to formulate measures of handling them and enhance the use of FL in food safety.

3.7.1 Data Heterogeneity

Data heterogeneity is defined as the dissimilar distribution of the data between the different actors of the FSC. This may be as a result of variations in data collection techniques, units of measurement, and organizational environment. While FL enables privacy-preserving collaborative information processing for participants with different local datasets, such heterogeneity raises serious difficulties for FL algorithms, especially related to generalization capability across the datasets.

Non-IID Data: Check the non-IID nature of data gathered from multiple singular sources. Because non-IID data can result in biases in data distribution or even skewed model forecasts given that models can be trained on specific data distribution. For instance, a model trained on images from large-scale industrial farming may not decipher well images from smallholder organic farming as a result of differences in cropping patterns, crop types, and climatic conditions prevailing on the two farming systems.

This is not easy as all the features of the datasets used in the models may not be identically distributed, hence the need to design FL algorithms suitable for non-IID data sets. This can be mitigated by using some techniques, such as personalized FL, where the model update being sent to each participant is unique to their data. However, such personalized approaches can increase the difficulty of performing FL.

3.7.2 Privacy Concerns

This is really one of the main benefits of FL, where stakeholders are empowered to maintain their private data. Currently, privacy has emerged as a major issue, especially when handling or processing issues to do with food safety and the proprietary nature of some data. Data Leakage Risks: Similar to the original FL model, there is still a chance of leakage of data during the model transmission. Opponents can use the communication path to obtain further information from the model parameters exchanged [23]. For instance, if a model learns to identify certain uses of pesticides, then the adversaries can deduce various risky usages of agricultural pesticides from the updates of the model. In order to avoid such risks, there needs to be a proper integration of privacy-preserving protocols for FL, such as differential privacy and secure multiparty computation. These techniques either introduce noise into the model updates or use encryption to shield data being transmitted [24].

3.7.3 Communication Inefficiencies

The biggest challenge associated with communication in FL is the need to ensure efficient communication when there are many participants in the food supply chain. The fact that between-model communication can be very heavy, particularly in networks where bandwidth is limited, can present a concern. Bandwidth Limitations: Typically, where FSQM is implemented in small to moderately sized food processing facilities or in remote agricultural settings, internet connectivity is not optimal. Overloading of nodes with a large number of communications to the central server may cause delays due to the available bandwidth [14]. This leads to inefficiencies; the models become outdated, undermining the goal of using real-time food safety monitoring. To overcome the communication problem, the FL systems allow model compression, in which the model updates are compressed before transmission, and the use of communication-efficient algorithms, which reduce the number of rounds needed for convergence. For example, much work has been done on methods that need to be optimized to reduce communication overhead, such as FedAvg [4].

3.7.4 Regulatory Compliance

One of the greatest challenges of the food industry is that it is highly regulated, and compliance standards hinder the use of FL. Each industry may even have certain rules for conducting the business, sharing the data, and reporting the occurrences related to food safety. Data Sharing Restrictions could still be limitations placed on the dissemination of certain content to the stakeholders, even when the learning is in a federated manner. For instance, food safety data may involve discretionary information about production methods of certain products and enterprises; such data could be concealed, even if the data are aggregated. Such restrictions may reduce the amount of data as well as the data distribution fed into FL algorithms.

This is then made possible by stakeholders engaging with regulatory authorities to agree with the framework for sharing the data, given that there is legal compliance in the environment. Joint cooperation can result in the creation of norms that make the utilization of FL safe in terms of food quality.

3.7.5 Need for Stakeholder Collaboration

For the FL to be implemented in food safety, the following stakeholders have to be involved: the farmers, the processors, the distributors, as well as the regulators. But reaching for collaboration can often be a problem by having conflicts of interest with experts in different technical knowledge and with differences on the concept of data value. Building Trust and Engagement Stakeholder: Trust is critical in collaborating in a FL setup. Hearing or reporting concerns of data privacy, ownership, and incentives to participate in data sharing initiatives may pose challenges among the stakeholders [19]. For instance, small farmers may consider losing competitive position if they share data with the large processors. In order to avoid these issues, it is recommended that stakeholders get into open and honest discussion as well as put in place mutual understandings on who gets to own data and benefits from it. Basics of trust can also be established through success stories that revolve around the pilot projects that seek to show how coordination breeds improvement in food safety.

3.8 Conclusion

The food supply chain can benefit from FL as implementing change on a large scale becomes possible through decentralized data sharing and collaborative learning. In this chapter, we have looked at a number of options and opportunities for using FL; these include its capability in enhancing predictive models, monitoring in real-time, improving traceability, and finally promoting collaboration among several actors in the industry. Therefore, FL can be a solution to improve the development of models based on food safety data, which presents substantial heterogeneity.

Albeit this has great potential for the food industry, there are some issues related to the application of FL. Huge challenges that are associated with its adoption include data heterogeneity, privacy, data communication, legal and regulatory compliance, and necessary collaboration, which always pose a problem. Bridging these gaps is important for enhancing the utility of FL in enhancing food safety results.

For the advancement, stakeholders require working on conditioning well-organized FL methodologies with the discretion act and controlling the communication complexity. Food chain integration and coordination of farmers and processors with regulators and equipment providers to develop common understanding, beliefs, and structural relationships regarding data ownership and the computation of value. In that respect, through addressing the multiple challenges tied to FL, the food industry shall be able to implement its benefits for more production of better and safer food, which is good for the consumer and overall health of the population.

Therefore, we have established that FL is a useful platform for the food industry and others to gain insights from data without compromising customers and regulatory measures. Based on ongoing development of research and applications in the future, FL can be an important factor affecting the future development of food safety and quality standards [14, 19].

AI disclosure statement: The author acknowledges the use of AI-based tools for minor language editing or grammatical correction. However, all intellectual contributions, analyses, and conclusions presented in this chapter are entirely the author's own.

References

1. Hoffmann, S.A., Batz, M.B., Morris, J.G., Annual cost of foodborne illness in the United States. *Econ. Food Saf.*, 3, 2, 1–13, 2012.
2. Thakur, R. and Sinha, A., Food safety: Issues and challenges. *Food Saf. Manage.*, 1, 2, 1–11, 2021.
3. Shokri, R., Stronati, M., Song, L., Shmatikov, V., Privacy-preserving deep learning. *Proceedings of the 2017 ACM SIGSAC Conference on Computer and Communications Security*, pp. 1310–1328, 2015.
4. McMahan, H.B., Moore, E.J., Ramage, D., Communication-efficient learning of deep networks from decentralized data. *Proceedings of the 20th International Conference on Artificial Intelligence and Statistics (AISTATS)*, pp. 1–11, 2017.
5. European Commission, General Data Protection Regulation. Retrieved from https://ec.europa.eu/info/law/law-topic/data-protection/reform-data-protection-law_en, 2016.
6. Yurdem, B., Kuzlu, M., Gullu, M. K., Çatak, F. Ö., Tabassum, M., Federated learning: Overview, strategies, applications, tools and future directions. *Heliyon*, 10, 19, e38137, 2024. https://doi.org/10.1016/j.heliyon.2024.e38137.
7. Konečny, J., McMahan, H.B., Ramage, D., Federated optimization: Distributed machine learning for on-device intelligence. *Proceedings of the 5th International Conference on Machine Learning and Applications (ICMLA)*, pp. 1–9, 2016.

8. Bonawitz, K., Eichner, H., Gruber, M., Sweeney, L., Towards federated learning at scale: System design. *Proceedings of the 2nd SysML Conference*, 2019.
9. Dwork, C., Roth, A., Steinke, T., Concentrated differential privacy. *Proceedings of the 2016 ACM SIGSAC Conference on Computer and Communications Security*, pp. 1–15, 2014.
10. Mena, C., Adenso-Díaz, B., Yurt, O., The role of traceability in the food supply chain: A review. *Food Control*, 45, 97–108, 2014.
11. Li, T., Sahu, A.K., Talwalkar, A., Federated Learning: Opportunities and Challenges. *IEEE Signal Process Mag.*, 37, 3, 50–60, 2020.
12. Zhang, Y., Wang, J., Zhang, X., Data heterogeneity in federated learning: Challenges and solutions. *IEEE Trans. Neural Netw. Learn. Syst.*, 32, 5, 2033–2047, 2020.
13. Gao, Y., Chen, M., Zhang, Z., Regulatory compliance in federated learning: Challenges and opportunities. *J. Food Saf.*, 41, 2, e12836, 2021.
14. Kairouz, P., McMahan, H.B., Avent, J., Advances and Open Problems in Federated Learning. *Found. Trends® Mach. Learn.*, 14, 1-2, 1–210, 2021.
15. Tripathi, P., Kumar, N., Rai, M., Khan, A., Tripathi, P., Kumar, N., Rai, M., Khan, A., Applications of deep learning in agriculture. Https://Services.Igi-Global.Com/Resolvedoi/Resolve.Aspx?Doi=10.4018/978-1-6684-5141-0.Ch002. https://doi.org/10.4018/978-1-6684-5141-0.ch002, 2022.
16. Tripathi, P., Kumar, N., Rai, M., Shukla, P.K., Verma, K.N., Tripathi, P., Kumar, N., Rai, M., Shukla, P.K., Verma, K.N., Applications of machine learning in agriculture [Chapter]. Https://Services.Igi-Global.Com/Resolvedoi/Resolve.Aspx?Doi=10.4018/978-1-6684-6418-2.Ch006. https://doi.org/10.4018/978-1-6684-6418-2.ch006, 2024.
17. Wang, J., Wu, X., Li, Q., Stakeholder collaboration in federated learning for food safety: A conceptual framework. *Int. J. Food Sci. Technol.*, 55, 1, 1–11, 2020.
18. Yin, Z., Liu, Y., Chen, X., Enhancing food safety through federated learning: A case study. *Food Control*, 125, 107912, 2021.
19. Liu, S., Li, H., Zhang, L., Opportunities for innovation in food safety through federated learning. *Food Qual. Preference*, 87, 104045, 2021.
20. Zhang, Y., Wang, J., Zhang, X., IoT-enabled federated learning for food safety: A comprehensive review. *Sensors*, 21, 9, 3117, 2021.
21. Kumar, A., Singh, R., Sharma, A., Blockchain technology in the food supply chain: A review of applications and challenges. *J. Food Sci. Technol.*, 58, 4, 1183–1196, 2021.
22. Sun, Q., Zhao, X., Zhang, J., Application of machine learning in food safety: A review. *Food Control*, 110, 106964, 2020.
23. Hitaj, B., Ateniese, G., Perez-Cruz, F., Deep Models Under the GAN: Information Leakage from Graduated Datasets. *Proceedings of the 2017 ACM SIGSAC Conference on Computer and Communications Security*, pp. 603–618, 2017.
24. Abadi, M., Andersen, D.G., Basart, S., *et al.*, Deep Learning with Differential Privacy. *Proceedings of the 2016 ACM SIGSAC Conference on Computer and Communications Security*, pp. 308–318, 2016.
25. World Health Organization, Food safety. Retrieved from https://www.who.int/health-topics/food-safety, 2021.

4

Federated Learning and Its Applications in Smart Agricultural Processes

Mahesh Kumar Singh[1]*, Pushpa Choudhary[2], Akhilesh Kumar Singh[3], Arun Kumar Singh[4] and Om Prakash Rishi[5]

[1]Department of Computer Science and Engineering, SRM Institute of Science and Technology, Delhi NCR Campus, Modi Nagar, Ghaziabad, Uttar Pradesh, India
[2]Department of Computer Science and Engineering, Galgotias College of Engineering and Technology, Greater Noida, Uttar Pradesh, India
[3]School of Computing Science and Engineering, Galgotias University, Greater Noida, Uttar Pradesh, India
[4]Department of Computer Science and Engineering, Greater Noida Institute of Technology, Gr Noida, Uttar Pradesh, India
[5]Department of Computer Science and Informatics, University of Kota, Kota, Rajasthan, India

Abstract

Federated learning (FL) technology uses technology data and computing resources to teach machine learning models without importing original local data. FL is important for supporting agricultural management and promoting smart agriculture in the agricultural context. This course covers FL, its advantages over deep learning and machine learning, FL categories, and methods used in the farming industry. It also examines and compares scientific issues and problems related to state learning models and algorithms used in agriculture. Briefly, it discusses the use of state education in agricultural technology. Additionally, it explores the challenges and factors that need to be considered when implementing FL in agriculture, such as model fusion, communication limitations, and disparate data. Despite these issues, FL could change the agricultural industry by allowing farmers to obtain informed consent while protecting their personal information.

**Corresponding author*: maheshkrsg@gmail.com

Padmesh Tripathi, Bhanumati Panda, Shanthi Makka, Reeta Mishra, S. Balamurugan and Sheng-Lung Peng (eds.) Federated Learning for Smart Agriculture and Food Quality Enhancement, (85–110) © 2026 Scrivener Publishing LLC

Keywords: Federated learning, edge computing, sensors, agriculture processes automation, machine learning, deep learning, convolutional neural networks

4.1 Introduction

Agriculture is a key player in the economic growth of every nation. It offers considerable resources produced by multiple industries. Farmers and agricultural collaborators have access to numerous crucial sources of information, including pest sightings, meteorological conditions, crop productivity, and soil health. Nevertheless, data sharing for collaborative analysis and design is hindered by concerns related to ownership and privacy. State education in agriculture offers a solution that blends specialized knowledge with a shared education and information-sharing approach. Federated learning (FL) utilizes knowledge-sharing groups by allowing learning models to save data locally on farmers' farm equipment or servers while also safeguarding the confidentiality of crucial agricultural data. FL allows for collaborative training of learning models using different data sources, focusing on data privacy and ownership while promoting the sharing of knowledge and understanding experiences among stakeholders. This collaboration is spurring creativity and speeding up the implementation of data-driven solutions to the intricate problems in agriculture. Government education can bring about a significant change in smart agriculture by focusing on common requirements for modeling and data sharing, with a strong emphasis on data privacy. As agriculture embraces this technology, FL allows farmers and other agricultural stakeholders to utilize the vast knowledge in their data while ensuring blockchain privacy and ownership *via* decentralized data technology.

4.1.1 Machine Learning (ML)

The use of ML in agriculture could help create knowledge-based farming systems, leading to more precise and efficient farming practices. ML algorithms can analyze extensive agricultural databases containing data on soil, weather [14], and crop yields to uncover patterns and enhance understanding of the information. Knowledge-driven farming systems utilize this data to empower farmers to make better-informed choices. The agricultural sector, essential to the economy, is poised for a revolution due to the widespread adoption of ML technology. Data in agriculture can enhance the precision of predictions. Collecting data from sensors on crops, soil, weather, and other factors in agriculture at different times can improve the effectiveness of ML, improve agricultural practices, increase productivity, and promote sustainability, efficiency, and utilization of ML in farming.

- ML is used to examine sensor data in real time for monitoring crop health, identifying diseases, and optimizing irrigation and fertilization timing in crop monitoring and management. These technologies are making educated choices and enhancing crop production in formers.
- Using historical data, predictive analytics can forecast crop yields with ML models, considering weather, soil conditions, and planting methods for yield optimization. This allows for improved planning and distribution of resources.
- Precision agriculture process utilizes ML tools to improve field operations by accurately administering inputs (like water, fertilizers, and pesticides) according to specific localized requirements. This minimizes trash and increases efficiency.
- ML can monitor and manage livestock by tracking their health, behavior, and feeding habits. It aids in identifying diseases, making breeding choices, and managing the entire herd.
- ML enhances logistics, inventory management, and distribution to optimize the supply chain, guaranteeing the timely delivery of agricultural products from the farm to the table.

4.1.2 Deep Learning (DL)

DL is the subfield of ML by adding more layers and complexity to models [15–17]; it enhances conventional ML and makes learning at different levels of abstraction easier. Hierarchical learning is supported by the many processing layers that make up DL. DL's architecture allows for nonlinear learning, as each layer gradually incorporates more data. DL has significantly changed the agriculture industry, tackled important difficulties, and increased output. The following are some ways that the industry is being impacted by DL:

- The early detection of diseases in crops is done using DL models like convolutional neural networks (CNNs) to analyze images. For example, they can recognize diseases in plant leaves, allowing for prompt action and reducing crop damage.

- Crop health and yield forecasts are monitored and optimized using DL algorithms that analyze data from sensors, drones, and satellites to allocate resources efficiently. This assists farmers in making knowledgeable choices regarding irrigation, fertilization, and pest management. Precision Agriculture: DL aids precision farming by accurately administering inputs (such as water and fertilizers) according to specific localized requirements. This helps decrease waste and enhance overall productivity.
- Identifying and Handling Weeds: DL algorithms can distinguish between crops and weeds, enabling precise weed management without causing harm to the crops.
- Assessment of Soil Health: DL models examine soil information to evaluate soil condition, nutrient quantities, and moisture levels. This data directs how soil is managed. DL methods are used to classify various crop types and generate precise field maps to assist with land management and planning.

4.1.3 Edge Computing

Edge computing is a computing framework focused on analyzing data from clients close to their source, at the edge of the network. Instead of sending data to a central data center for processing and analysis, the data is managed where it is generated. Edge computing aims to move computing capabilities and resources away from centralized locations and closer to data generation, improving privacy and security. Processing information on local devices effectively reduces costs by eliminating the need to transfer data to central servers or require large storage space. DL can be easily integrated into edge computing systems, enabling effective monitoring and maintenance of cutting-edge computing, and, thus, the adoption of edge computing has increased significantly due to its suitability for DL applications.

Edge software is transforming the agricultural sector by enhancing efficiency, stability, and decision-making capabilities. Edge aggregation entails processing data near its source, typically at the network's edge. This approach does not depend solely on centralized data centers; instead, it allows for local data analysis. For instance, data gathered by Internet-of-Things (IoT) devices on farms can be categorized and partially processed within small on-site data centers. Reduced latency: Edge computing shortens the time it takes for data to move from its source to processing by decreasing the distance between them. Edge computing allows for instant decision-making in precision farming.

- **Irrigation Management:** Farmers can improve water usage by examining environmental data specific to their area.
- **Monitoring Livestock:** Immediate insights assist in the management of animal health and behavior.

4.2 Federated Learning (FL)

FL solves a ML problem by having multiple clients work together with a central server. FL is a decentralized ML technique that avoids central data aggregation by training models locally on several devices or locations utilizing distributed datasets [12–14]. This innovative strategy tackles a significant issue facing the agriculture industry, striking a balance between data sharing and privacy protection. Each partner in FL receives the model—typically a pre-trained base model—from a cloud data center. After that, each party modifies its dataset.

These updates are being distributed for consolidation. The updates are gathered by the server and sent back to the relevant parties. The parties continue to modify their interpretation of the model and go through the same steps again.

In agriculture, field research is costly and frequently concentrates on narrow areas or specific purposes. Agricultural costs can also be tracked *via* remote sensing [5–8]. These tools rely on an internet connection and produce actual data. It is critical to lower the danger of security and privacy violations. Farmers believe that protecting data is crucial for big businesses [9]. First, at the server end, a global random sample of a neural network—like a CNN—is created [1–4]. The first phase in the FL training process is a central server that simulates a real-world scenario. A portion of the clients will be chosen for the FL round.

Client-side tools have access to global instance options. Each client computes a local update of the model, typically *via* Gradient Descent, after training its model solely with its data. To compute a combined update for the model, the master server gathers updates from neighboring devices. To improve communication efficiency, loss reduction can be incorporated into the integration process [10, 11]. The formula structure is updated by the central server, which also updates the global model. After that, the client receives all model parameters from the server for the subsequent learning model. The global instance will be sent to the designated client-side tools.

4.3 Types of Federated Learning

Based on architecture and procedures of working of FL, it is characterized into five categories.

4.3.1 Centralized Federated Learning

In this method, the training procedure is controlled by a central server: the model is trained using data provided by the clients and the updated model parameters are sent to the central server. As shown in Figure 4.1, the server collects variables and updates the model before distributing it to clients. Typically, client-side devices share a global instance. To improve the overall server model, each client trains the model using its data and passes the model parameters to the backend server for aggregation. After collecting the model parameters, the backend server provides the clients with the latest overall parameters. The global parameters are used to initialize the local model and then recycled. This process continues until either the iteration limit is reached or the server instance is flipped in a non-trivial way, i.e., the master server controls the learning process.

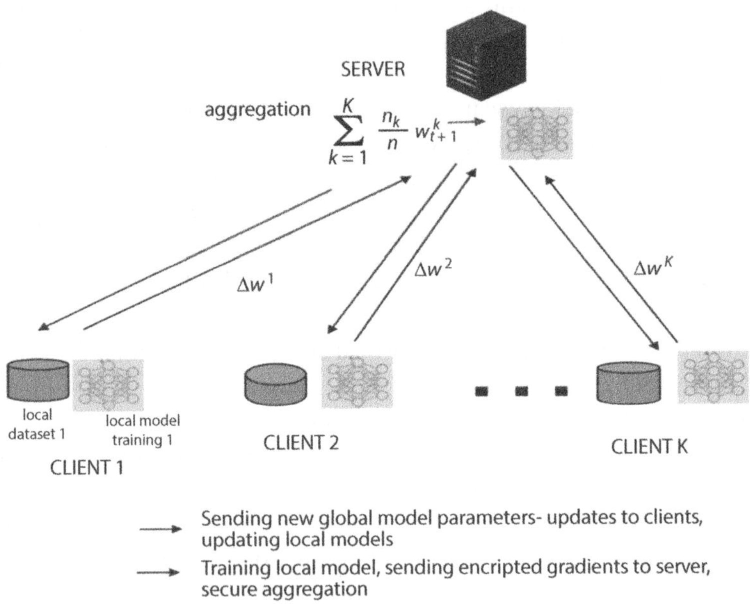

Figure 4.1 Architecture for a centralized FL system.

- **Consolidation:** Model updates are combined in a centralized location.
- **Control:** The central server has control over the training.

Example Situation: Hospitals collaborate in a healthcare partnership to provide localized training on patient data to improve disease prediction models using a centralized server for data aggregation.

4.3.2 Decentralized Federated Learning

A central server is not used in this method. Conversely, local tools are compatible with servers. Using local data, each participant creates his own model and shares updates with other participants. A client is selected at the start of each session to handle all server functions in a distributed framework. This agent oversees allocating the parameters to other agents and updating the global model. As seen in Figure 4.2, decentralized FL systems can be constructed using a variety of technologies, including peer-to-peer (P2P) networks, graphs, and blockchain systems. Depending on the circumstances, the computing resources used to guide the partners' interactions and orientation within the communication network will adjust. Common traits and closeness can help customers connect.

- No central server eliminates the need for a central governing entity.
- Distributed training requires decentralized nodes to update models iteratively.
- Data remains confidential by keeping it on a local basis.

4.3.3 Horizontal Federated Learning

Data samples are distributed across multiple devices or servers, with the model going through collaborative training, as shown in Figure 4.3. Attendees share similar traits while possessing different data points. Sample-based FL, also called horizontal FL, is used with datasets that share common features but have different samples. For example, two community banks might serve different groups of customers in their areas, resulting in a little crossover in their customer base. However, their business is very similar, so the feature spaces are the same.

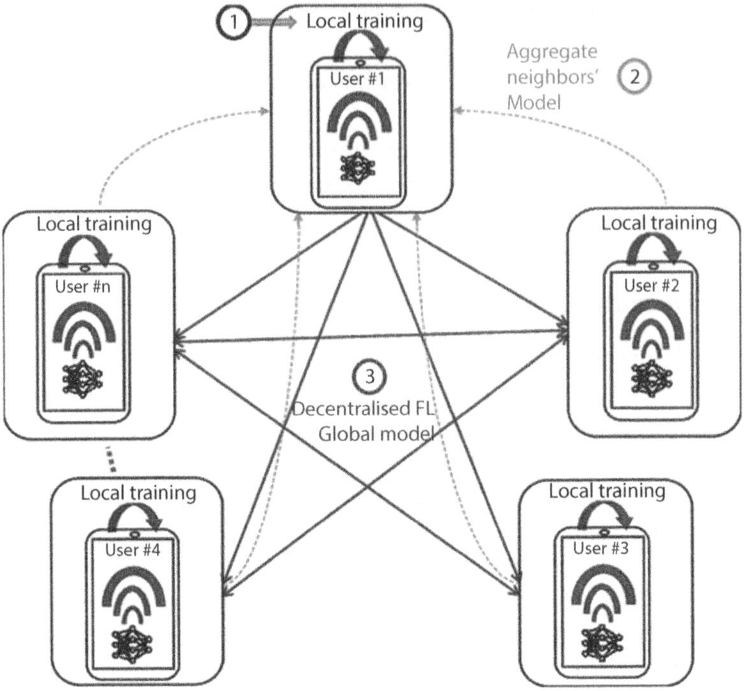

Figure 4.2 Architecture for a decentralized FL system.

Step 1: Participants compute training gradients on their devices, use encryption, variances, or shared secrets to hide some gradients, and send these hidden attributes to the server.

Step 2: The server combines the information without learning about any of the participants.

Step 3: The server sends the merged results to the participants.

Step 4: In the next step, participants correct their models by combining the extracted gradients. Data privacy is maintained by keeping information in a small area.

4.3.4 Vertical Federated Learning

Objects are trained in tasks that complement one another, and units are given distinct assignments. Although each person is unique, they all have comparable traits. Figure 4.4 illustrates vertically combined learning, also known as cross-sectional combined learning, which uses two datasets with varied spacing but the same ID space. Consider, for instance, two distinct

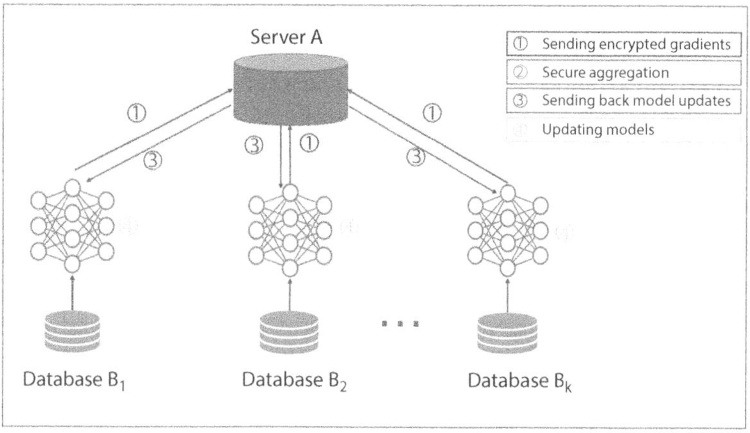

Figure 4.3 Horizontal federated learning.

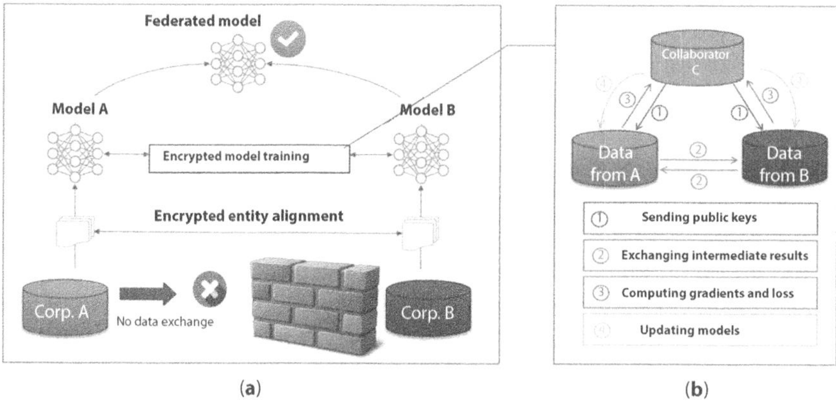

Figure 4.4 Vertical federated learning.

businesses—one an e-commerce company and the other a bank—that are both situated in the same city. There is a lot of overlap in the user base because most of the local population is probably in their user group. However, the combination of an e-commerce platform that measures a user's online activity and purchase history and a bank that keeps track of a user's income, spending patterns, and credit score is separate data. Consider a situation where both parties need to have predictive models for product sales based on user and product data. Data is stored locally for privacy.

Step 1: Partner C creates encryption pairs and shares the public key between A and B.
Step 2: Participates in encryption and shares temporary results between A and B to calculate gradients and losses.
Step 3: In the third step, A and B calculate the signaled gradients using additional masks, and B calculates the signal loss. A and B send the encrypted data to C.
Step 4: Decrypt, C sends the decryption sends the decrypted data and passes the loss to A and B. A and B remove the slopes and make the necessary adjustments to the parametric model.

4.3.5 Hybrid Data Partitioning and Transfer Federated Learning

In this category, pre-existing models are adjusted using data from various sources for purposes. It decreases the requirement for large amounts of local data by utilizing information from current models.

Federated cross-learning is effective when two datasets differ not only in sample size but also in feature space as shown in Figure 4.5. Consider two entities: one of which is a bank from China, and one is a private company for online business from America. Due to geographic limitations, the user bases of the two organizations have little overlap. However, due to the wide scope of the current industry, there is little convergence of the characteristics of the two sides. Transfer learning methods can be used to provide solutions for all students and work areas in a group in this situation (Figure 4.5). In fact, a common indicator is derived using common patterns of two sets of features and is used to predict patterns with specific features on the other hand. Table 4.1 shows a comparative overview of federated learning applications in agriculture. It highlights methods, objectives, datasets, models, and challenges across various studies. Key focus areas include disease detection, intrusion prevention, and data privacy in Agri-IoT.

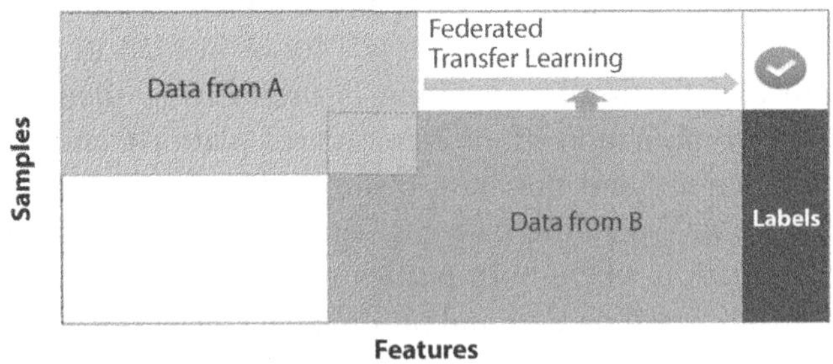

Figure 4.5 Hybrid data partitioning and transfer federated learning.

Table 4.1 Description of the survey papers related to agriculture using federated learning.

Area of agriculture [References]	The federated learning method and architecture applied	Objectives of the work	Research issues and challenges	Algorithms used	Dataset used	Used trained model
Detection of various diseases and pests [18] detection of intrusions	Horizontal FL, centralized architecture	To avoid costly data transfer and storage, unbalanced and insufficient orchard data, various diseases and pests, and challenging detection using traditional cloud–based deep learning	Avoid the communication costs associated with a large data upload in order to address the issue of inconsistent and inadequate data	FedAvg	Images, 445 photos of apple orchards, only 152 of which show five diseases	Reinforced regional convolutional neural network (R-CNN) with faster performance
Agricultural-IoT infrastructure security using an intrusion detection system [19]	Hybrid data partitioning, centralized architecture	Securing agricultural-IoT infrastructures	Securing agricultural IoT infrastructures, protects data privacy	FedAvg	Real-world traffic datasets - CSE-CIC-IDS2018, MQTTset, and InSDN	Classifiers: Deep neural network (DNN), convolutional neural network (CNN), and recurrent neural network (RNN)

(Continued)

Table 4.1 Description of the survey papers related to agriculture using federated learning. (*Continued*)

Area of agriculture [References]	The federated learning method and architecture applied	Objectives of the work	Research issues and challenges	Algorithms used	Dataset used	Used trained model
Cross-silo FL's function in promoting data exchange [20]	Horizontal, central server-based FL	Encouraging data exchange throughout the agri-food industry's supply chain	Data privacy	FedBN (FLon non-id features *via* local batch normalization), extends FedAvg	Datasets for predicting agricultural yields derived from both tabular (weather and soil data) and image (remote sensing) sources	CNN and RNN
Agricultural production [21]	Hybrid data partitioning, centralized architecture	Guiding agricultural production	High modeling accuracy under resource restrictions, minimal communication costs, and quick convergence rate	Proposed a greedy approach to identify the best answer together with a coupled FL Framework for Edge-Internet of Agricultural Things (edge-IoAT)	The images of real-world soybean iron deficiency chlorosis (IDC) dataset	Greedy algorithm,

(*Continued*)

Table 4.1 Description of the survey papers related to agriculture using federated learning. (*Continued*)

Area of agriculture [References]	The federated learning method and architecture applied	Objectives of the work	Research issues and challenges	Algorithms used	Dataset used	Used trained model
Automated detection of animal behavior using distributed data in a heterogeneous data environment [22]	Horizontal FL, centralized architecture	Automated animal activity recognition (AAR)	Data privacy	FedAAR with gradient-refinement–based aggregation	Six horses with neck-attached IMUs provided 87,621 two-second samples for the publicly available, centralized dataset.	CMI-Net
Crop classification, smart farming [23]	Horizontal FL, centralized architecture	Data privacy in smart farming	The application of FL to smart farming	Federated averaging model	The dataset includes temperature, humidity, pH, and precipitation as independent variables.	CNN

(*Continued*)

Table 4.1 Description of the survey papers related to agriculture using federated learning. (*Continued*)

Area of agriculture [References]	**The federated learning method and architecture applied**	**Objectives of the work**	**Research issues and challenges**	**Algorithms used**	**Dataset used**	**Used trained model**
Crop yield estimation [24]	Horizontal FL, centralized architecture	Crop yield prediction	Data ownership, privacy preservation	FedAvg	Soybean yield dataset: weather, soil components, and crop data.	CropResNet-based, regression models, such as ResNet-16 and ResNet-28
FL for smart sensors: A multi-function control approach for enhancing agricultural production [25]	Horizontal	Improving productivity	FL from sensing data	Amendable multi-function sensor control method (AMFSC)	Crop and soil data	AMFSC

(*Continued*)

Table 4.1 Description of the survey papers related to agriculture using federated learning. (*Continued*)

Area of agriculture [References]	**The federated learning method and architecture applied**	**Objectives of the work**	**Research issues and challenges**	**Algorithms used**	**Dataset used**	**Used trained model**
EfficientNet deep model, classifying nine types of pests [26]	Horizontal FL, centralized architecture	Diagnoses of plant diseases	Minimal processing power for classifying pests in an agricultural setting	FedAvg	The pest images and IoT platforms, in conjunction with unmanned aerial vehicles (UAVs)	Pre-trained EfficientNetB3 is combined with thick layers to create a dense convolutional neural network (CNN) model.
Diagnosis of diseases in food crops [27]	Horizontal FL, centralized architecture	Leaf disease prediction	Data privacy	FedAvg	Plant village	Five CNNs: AlexNet, SqueezeNet, ResNet-18, VGG-11, and ShuffleNet

(*Continued*)

Table 4.1 Description of the survey papers related to agriculture using federated learning. (*Continued*)

Area of agriculture [References]	**The federated learning method and architecture applied**	**Objectives of the work**	**Research issues and challenges**	**Algorithms used**	**Dataset used**	**Used trained model**
Deep privacy–encoding–based FL framework for smart agriculture [28]	FL that is horizontal and concentrated; FL server and peripherals like a gateway/ router that connects to numerous IoT devices	Intrusion detection	Minimizing the risk of security and data privacy violation	FedGRU	ToN-IoT dataset	GRU

4.4 Applications of Federated Learning

A novel modeling technique called FL allows data from several sources to be used to train a single model while maintaining the confidentiality and privacy of the data. Many industries, including sales and banking, are eagerly awaiting this functionality because issues with privacy, data security, and intellectual property rights hinder direct data aggregation for ML model training. To solve these problems, we need FL and transfer learning. The power of FL allows us to build three-way ML models without sharing enterprise data. Both parties benefit because it ensures complete data security and privacy protection while providing customers with personalized and targeted service. By handling data variations and training a model on data from many sources, cross-training is chosen as a new modeling technique in this context to overcome the shortcomings of traditional artificial intelligence methods. FL thus offers vital technical assistance as we establish a common big data and AI ecosystem that cuts across organizational, data, and geographical borders. Multiple groups can query databases using blended learning frameworks without disclosing private information. Suppose, for example, that we are trying to find several accounts in the financial scenario. This risk has historically been substantial for the financial sector. This occurs when certain users wish to take out a bank loan to repay a bank loan [29–31]. Borrowing from various sources poses a risk to sustainability. If any of these illegal activities happen, then the financial system will collapse. We can use a mixed-learning architecture to identify these users without sharing the list of users between banks A and B. We can use the list of users in each session through learning encryption methods. After that, find the net of the encrypted list in the merge. The decryption process shows the list of credit numbers involved in the multi-party transaction without revealing the identity of other "good" users.

FL facilitates the training of ML models without data fusion, maintaining privacy and security, and FL finds many applications in field type. These applications demonstrate the flexibility and power of blended learning by creating ML models that are safer and more efficient in many areas [32]. There are several key applications.

- **Healthcare:** Without disclosing private information, FL is used to train models using patient data from several institutions. This technology is used in the creation of more effective diagnostic instruments and customized treatment programs.
- **Finance:** By training models on decentralized transaction data, banks, and other financial organizations utilize FL to identify fraud and enhance risk management.

- **Mobile Devices:** By training on user data locally, FL improves language models and predictive text features for cellphones, improving user data locally, and improving user experience without compromising privacy.
- **Autonomous Vehicles:** By combining data from several vehicles, FL enhances safety and efficiency in the training of self-driving car models.
- **IoT and Smart Devices:** By training models on data from numerous sensors and devices, FL is used in smart homes and cities to optimize energy use, traffic management, and other services.
- **Telecommunications:** By training models using decentralized data from several regions, telecom operators employ FL to enhance customer service and network performance.
- **Pharmaceuticals:** FL enables pharmaceutical companies to work together on research projects without exchanging confidential data, which helps with medication discovery and development.

4.4.1 Crop Yield Prediction and Management

The objective of FL for predicting and managing crop yield is to develop models that can make accurate predictions and enhance agricultural management methods [33]. FL is a novel method for predicting and managing crop yields that tackles issues such as data privacy, availability, and transfer costs. FL is effective in predicting crop yields in the specified ways.

Decentralized data training involves training models across decentralized devices or servers instead of gathering data in a central repository. This is especially beneficial in agriculture, as data from different sources like weather stations, soil sensors, and farm management systems may be fragmented and isolated. Figure 4.6 shown below illustrates the architecture diagram of crop yield prediction. This diagram is adapted from Crop Yield Prediction and Management, highlighting the key processes involved in forecasting agricultural yields.

- The Federated Averaging Algorithm combines locally trained models to create a global model while maintaining data privacy and security, without the need to transfer raw data.
- Model optimization in federated settings has demonstrated optimal outcomes using techniques such as the ResNet-16 regression model with Adam optimizer.

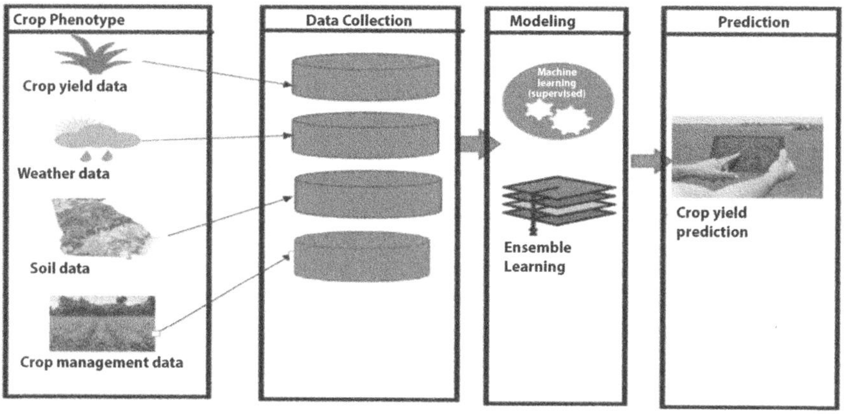

Figure 4.6 Architecture diagram of crop yield prediction [34].

- **Improved Data Privacy:** By keeping sensitive data on local devices, the likelihood of data breaches is minimized.
- **Enhanced Model Precision:** Utilizing a variety of datasets from various locations, FL can enhance the precision of crop yield forecasts.
- Decreases the necessity for extensive data transfer and storage infrastructure, thus increasing cost efficiency.

4.4.2 Pest and Disease Identification

Recognizing and managing pest infestations and plant diseases is essential for crops to remain healthy and productive (see Figure 4.7). Numerous investigations have investigated the application of FL for the early diagnosis and detection of pests and plant diseases. FL is a cutting-edge method for agricultural pest and disease diagnosis. The help of region-specific

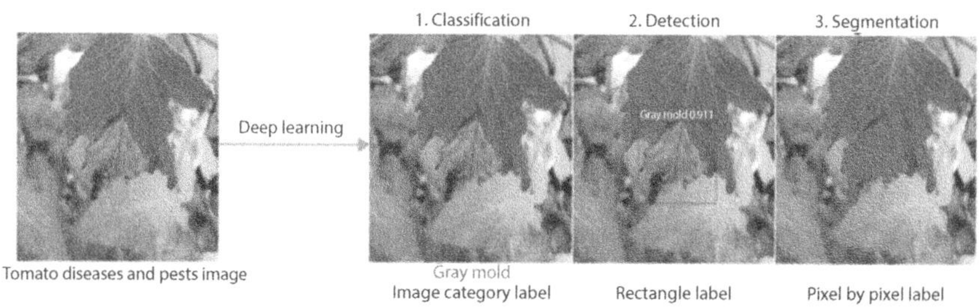

Figure 4.7 Plant diseases and pests identifications [35].

datasets enables the training of several local models, exchanging expertise without transferring raw data. This technique lowers computational costs while addressing data privacy concerns.

4.4.3 Resource Optimization in Agriculture Using Federated Learning

FL has also been researched to maximize the use of fertilizer, water, energy, and other agricultural resources. A few studies investigated the application of FL for adaptive irrigation management. This theory proposes using ML models trained on decentralized data sources to modify irrigation plans in real time in response to weather forecasts and soil moisture levels [36]. FL can significantly enhance resource optimization in agriculture by enabling cooperative model training across multiple farms or regions without needing sensitive data exchange. Here are a few significant uses for it.

- **Effective Data Usage:** By utilizing a variety of datasets from many sources, FL enhances the precision and resilience of predictive models. By anticipating the precise requirements of crops depending on local conditions, this can aid in the optimization of the use of resources like water, fertilizers, and pesticides.
- **Cost Reduction:** FL lessens the requirement for centralized data processing, which can be expensive and resource-intensive, by utilizing local data and computer resources. Additionally, the decentralized method reduces the cost of data transfer.
- **Better Decision-Making:** Farmers can allocate resources more wisely when they have access to real-time information about crop health, soil conditions, and weather patterns from FL models. Better agricultural yields and more environmentally friendly farming methods may result from this.
- **Scalability:** The framework for FL can be readily.

4.4.4 Rice-Leaf Disease Classification

Asia grows rice, a grain that has a major impact on agricultural productivity worldwide. More than half of the world's population, mostly in Asia, Africa, and America, mostly eats rice. It supplies roughly 20% of daily calories and 13% of the protein that an individual consumes daily. Many methods have been developed over the years in the field of using image analysis to identify

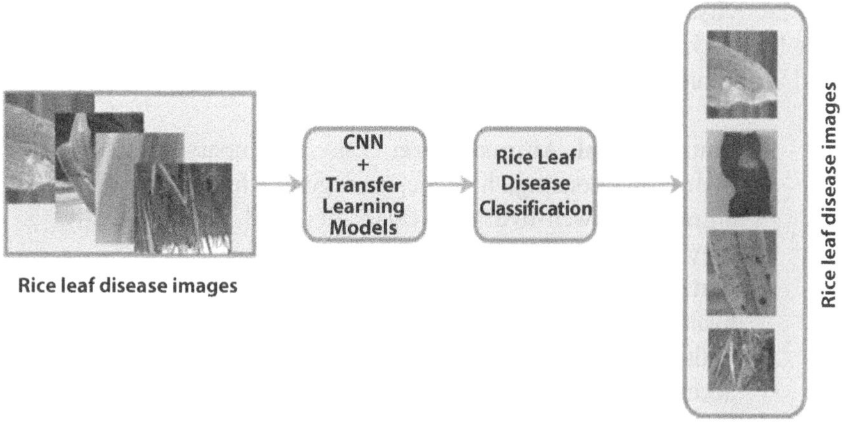

Figure 4.8 Rice-leaf disease classification using CNN and transfer learning [35].

rice-leaf diseases. Because rice leaf diseases can seriously affect crop quality and productivity, early detection is crucial. Diseases that harm the leaves of rice plants are among the many ailments that might strike them. Early detection of these illnesses is essential for efficient disease management and reducing the detrimental effects on rice yield (Figure 4.8).

Farmers can reduce expenses and increase yields by receiving tailored guidance on crop management, pest control, and efficient resource allocation. Agriculture experts may make data-driven judgments and give farmers tailored advice by utilizing the common knowledge built into the federated approach. The availability of useful data on supply chain dynamics and market trends enables stakeholders in the agriculture sector to plan and manage risks more effectively. In general, federated transfer learning technology can boost agricultural productivity, sustainability, and profitability while preserving privacy and data security [37].

4.4.5 Livestock Management

FL could enhance agricultural animal management techniques while concentrating on crop-related applications. Using wearable sensors and IoT devices, some researchers have created a system for tracking and forecasting animal health [38]. Their methodology made it possible to train decentralized models using data from individual animals, facilitating the prompt identification of health problems and prompt action by farmers or vets. In agriculture, livestock management includes organizing and overseeing livestock activities to ensure optimal animal health and productivity. In addition to ensuring the well-being of the animals, effective livestock management increases the farming industry's sustainability and profitability.

- **Foundational Animal Care:** This encompasses offering hydration, housing, hygiene, and gentle handling. When animals are well-hydrated, clean, dry, and handled gently, they are in better health and more productive.
- **Nutrition:** Maintaining a balanced diet is essential for the health and efficiency of livestock. This involves giving animals nutritious feed and making sure they constantly have access to fresh water.
- Healthcare is crucial for the upkeep of the herd's well-being, requiring regular health check-ups, vaccinations, and treatments to prevent illnesses.
- **Breeding:** Successful breeding techniques enhance the genetic quality of animals, resulting in increased productivity and overall health.
- **Sustainability in Livestock Farming:** It is crucial to adopt sustainable practices like effective manure handling and minimizing greenhouse gas emissions to protect the environment and ensure the continued success of livestock farming.
- **Technology Usage:** A lot of livestock supervisors utilize agricultural programs to monitor feeding, breeding, treatments, and other tasks. This aids in keeping comprehensive records and enhancing overall management.

4.5 Conclusion

FL is a strong method for training ML models on edge devices without sharing raw local data or utilizing data and computational resources. FL has become a focal point in agriculture because it enhances agricultural practices and facilitates intelligent farming. This paper explored FL by providing a concise overview of its various types and how they are used. It also provided a thorough analysis of different FL models applied across various agricultural sectors. FL shows the potential to improve agriculture by utilizing decentralized data while maintaining confidentiality. It is a fascinating field of study with real-world applications for eco-friendly agriculture and effective organization. This paper also discussed the applications of FL in different aspects, especially in agriculture operations.

AI disclosure statement: The author acknowledges the use of AI-based tools for minor language editing or grammatical correction. However, all intellectual contributions, analyses, and conclusions presented in this chapter are entirely the author's own.

References

1. Wani, J.A., Sharma, S., Muzamil, M., Ahmed, S., Sharma, S., Singh, S., Machine Learning and Deep Learning Based Computational Techniques in Automatic Agricultural Diseases Detection: Methodologies, Applications, and Challenges. *Arch. Comput. Methods Eng.*, 29, 641–677, 2022.
2. Benos, L., Tagarakis, A.C., Dolias, G., Berruto, R., Kateris, D., Bochtis, D., Machine Learning in Agriculture: A Comprehensive Updated Review. *Sensors*, 21, 3758, 2021.
3. LeCun, Y., Bengio, Y., Hinton, G., Deep Learning. *Nature*, 521, 436–444, 2015.
4. Schmidhuber, J., Annotated History of Modern AI and Deep Learning. arXiv 2022, arXiv:2212.11279.
5. Schmidhuber, J., Deep learning in neural networks: An overview. *Neural Netw.*, 61, 85–117, 2015.
6. Ghosh, M. and Grolinger, K., Edge-Cloud Computing for Internet of Things Data Analytics: Embedding Intelligence in the Edge with Deep Learning. *IEEE Trans. Ind. Inform.*, 17, 2191–2200, 2021.
7. Wang, X., Han, Y., Leung, V.C.M., Niyato, D., Yan, X., Chen, X., Convergence of Edge Computing and Deep Learning: A Comprehensive Survey. *IEEE Commun. Surv. Tutor.*, 22, 869–904, 2020.
8. Wójtowicz, M., Wójtowicz, A., Piekarczyk, J., Application of remote sensing methods in agriculture. *Commun. Biometry Crop Sci.*, 11, 31–50, 2016.
9. Kaur, J., Hazrati Fard, S.M., Amiri-Zarandi, M., Dara, R., Protecting farmers' data privacy and confidentiality: Recommendations and considerations. *Front. Sustain. Food Syst.*, 6, 903230, 2022.
10. Kairouz, P., McMahan, H.B., Avent, B., Bellet, A., Bennis, M., Bhagoji, A.N., Bonawit, K., Charles, Z., Cormode, G., Cummings, R., et al Advances and Open Problems in Federated Learning. *Found. Trends® Mach. Learn.*, 14, 1–210, 2021.
11. Mothukuri, V., Parizi, R.M., Pouriyeh, S., Huang, Y., Dehghantanha, A., Srivastava, G., A survey on security and privacy of federated learning. *Future Gener. Comput. Syst.*, 115, 619–640, 2021.

12. Beltrán, E.T., Pérez, M.Q., Sánchez, P.M., Bernal, S.L., Bovet, G., Pérez, M.G., Pérez, G.M., Celdrán, A.H., Decentralized FederatedLearning: Fundamentals, State-of-the-art, Frameworks, Trends, and Challenges. arXiv 2022, arXiv:2211.08413.
13. Yuan, L., Sun, L., Yu, P.S., Wang, Z., Decentralized Federated Learning: A Survey and Perspective. arXiv 2023, arXiv:2306.01603.
14. Yang, Q., Liu, Y., Chen, T., Tong, Y., Federated Machine Learning: Concept and Applications. *ACM Trans. Intell. Syst. Technol. (TIST)*, 10, 2, 1–19, 2019.
15. Phong, L.T., Aono, Y., Hayashi, T., Wang, L., Moriai, S., Privacy-Preserving Deep Learning *via* Additively Homomorphic Encryption. *IEEE Trans. Inf. Forensics Secur.*, 13, 5, 1333–1345, 20182018.
16. Shokri, R. and Shmatikov, V., Privacy-Preserving Deep Learning, in: *Proceedings of the 22Nd ACM SIGSAC Conference on Computer and Communications Security (CCS '15)*, ACM, New York, NY, USA, pp. 1310–1321, 2015, https: //doi.org/10.1145/2810103.2813687.
17. Bonawitz, K., Ivanov, V., Kreuter, B., Marcedone, A., McMahan, H.B., Patel, S., Ramage, D., Segal, A., Seth, K., Practical Secure Aggregation for Privacy-Preserving Machine Learning, in: *Proceedings of the 2017 ACM SIGSAC Conference on Computer and Communications Security (CCS '17)*, ACM, New York, NY, USA, pp. 1175–1191, 2017, https://doi.org/10.1145/3133956.3133982.
18. Deng, F., Mao, W., Zeng, Z., Zeng, H., Wei, B., Multiple Diseases and Pests Detection Based on Federated Learning and Improved Faster R-CNN. *IEEE Trans. Instrum. Meas.*, 71, 3523811, 2022.
19. Friha, O., Ferrag, M.A., Shu, L., Maglaras, L., Choo KK, R., Nafaa, M., FELIDS: Federated learning-based intrusion detection system for agricultural IIn Proceedings of the internet of Things. *J. Parallel Distrib. Comput.*, 165, 17–31, 2022.
20. Durrant, A., Markovic, M., Matthews, D., May, D., Enright, J., Leontidis, G., The role of cross-silo federated learning in facilitating data sharing in the agri-food sector. *Comput. Electron. Agric.*, 193, 106648, 2022.
21. Yu, C., Shen, S., Zhang, K., Zhao, H., Shi, Y., Energy-aware device scheduling for joint federated learning in edge-assisted internet of agriculture things, in: *Proceedings of the 2022 IEEE Wireless Communications and Networking Conference (WCNC)*, pp. 1140–1145, 10–13 April 2022.
22. Mao, A., Huang, E., Gan, H., Liu, K., FedAAR: A Novel Federated Learning Framework for Animal Activity Recognition with Wearable Sensors. *Animals*, 12, 2142, 2022.
23. Idoje, G., Dagiuklas, T., Iqbal, M., Federated Learning: Crop classification in a smart farm decentralized network. *Smart Agric. Technol.*, 5, 100277, 2023.
24. Manoj, T., Makkithaya, K., Narendra, V.G.A., Federated Learning-Based Crop Yield Prediction for Agricultural Production Risk Management, in: *Proceedings of the 2022 IEEE Delhi Section Conference (DELCON)*, pp. 1–7, 11–13 February 2022.

25. Abu-Khadrah, A., Ali, A.M., Jarrah, M., An Amendable Multi-Function Control Method Using Federated Learning for Smart Sensors in Agricultural Production Improvements. *ACM Trans. Sen. Netw.*, 2023.
26. Khan, F.S., Khan, S., Mohd MN, H., Waseem, A., Khan MN, A., Ali, S., Ahmed, R., Federated learning-based UAVs for diagnosing Plant Diseases, in: *Proceedings of the 2022 International Conference on Engineering and Emerging Technologies (ICEET)*, pp. 1–6, 27–28 October 2022.
27. Antico, T.M., Moreira LF, R., Moreira, R., Evaluating the Potential of Federated Learning for Maize Leaf Disease Prediction, in: *Anais do XIX Encontro Nacional de Inteligência Artificial e Computacional; Sociedade Brasileira de Computação*, pp. 282–293, 2022.
28. Kumar, P., Gupta, G.P., Tripathi, R., PEFL: Deep privacy-encoding-based federated learning framework for smart agriculture. *IEEE Micro*, 42, 33–40, 2021.
29. Tripathi, P., Kumar, N., Rai, M., Shukla, P.K., Verma, K.N., Applications of machine learning in agriculture, in: *Smart Village Infrastructure and Sustainable Rural Communities*, pp. 99–118, IGI Global, 2023.
30. Singh, M.K. and Rishi, O.P., Dynamic Knowledge-based Recommendation System for E-commerce using Deep Neural Network, 1023–1036, 2021. ISSN: 2041-3165Mathematics in Engineering, Science and Aerospace (MESA). http://nonlinearstudies.com/index.php/mesa/article/view/2754.
31. Iniyan, S., Mutual Information Feature Selection (MIFS) Based Crop Yield Prediction on Corn and Soybean Crops Using Multilayer Stacked Ensemble Regression (MSER). *Wirel. Pers. Commun.*, 126, 1–30, 2021. 10.1007/s11277-021-08712-9.
32. Liu, J. and Wang, X., Plant Dis. pests' detection based on deep learning a review. . *Plant Methods*, 17, 1, 22, 2021.
33. Rai, M., Tyagi, N., Tripathi, P., Kumar, N., Kumari, P., Advancement in agricultural techniques with the introduction of Artificial Intelligence and Image Processing, *in: Smart Village Infrastructure and Sustainable Rural Communities*, pp. 47–68, IGI Global, 2023.
34. Aggarwal, M., Khullar, V., Goyal, N., Gautam, R., Alblehai, F., Elghatwary, M., Singh, A., Federated Transfer Learning for Rice-Leaf Disease Classification across Multiclient Cross-Silo Datasets. *Agronomy*, 13, 2483, 2023. https://doi.org/10.3390/agronomy13102483.
35. Idoje, G., Dagiuklas, T., Iqbal, M., Federated Learning: Crop classification in a smart farm decentralized network. *Smart Agric. Technol.*, 5, 100277, 2023. ISSN 2772-3755, https://doi.org/10.1016/j.atech.2023.100277.
36. Ilić, M. and Ivanović, M., Federated Learning-Opportunities and Application Challenges. *International Conference on Computational Collective Intelligence*, Springer Nature Switzerland, Cham, 2023.
37. Tripathi, P., Kumar, N., Rai, M., Khan, A., Applications of Deep Learning in Agriculture, in: *Artificial Intelligence Applications in Agriculture and Food*

Quality Improvement, pp. 17–28, IGI Global, USA, 2022, doi: 10.4018/978-1-6684-5141-0.ch002.

38. Singh, M.K., Singh, A.K., Singh, P., Kalpana, Rishi, O.P., Artificial Intelligence Enabled IoT System for Football Identification in a Football Match, in: *Advanced Computing. IACC 2022. Communications in Computer and Information Science*, vol. 1782, D. Garg, V.A. Narayana, P.N. Suganthan, J. Anguera, V.K. Koppula, S.K. Gupta (Eds.), Springer, Cham, 2023, https://doi.org/10.1007/978-3-031-35644-5_37.

5

Federated Learning in Food Inspection and Grading

Reeta Mishra[1], Padmesh Tripathi[2*], Reddy Saisindhutheja[3], Gagandeep Arora[3] and Bhanumati Panda[4]

[1]School of Computer Science and Engineering, IILM University, Greater Noida, India
[2]Department of Computer Science and Engineering, Delhi Technical Campus, Greater Noida, India
[3]Department of Aritifical Intelligence and Machine Learning, Vardhaman College of Engineering, Shamshabad, Hyderabad, Telangana, India
[4]Department of Applied Sciences, ABES Engineering College, Ghaziabad, UP, India

Abstract

Federated learning (FL) is a game-changing technique in the food industry, especially when it comes to improving inspection and grading procedures. By allowing various stakeholders, such as farmers, processors, and retailers, to work together on machine learning models while protecting data privacy, FL creates a decentralized framework that greatly enhances food quality and safety. FL enables organizations to train models using localized data, protecting sensitive information while utilizing a variety of datasets. FL offers numerous benefits, such as increased model accuracy by combining distinct data sources, decreased communication overhead by sending only model updates rather than entire datasets, and scalability that makes it easy for new participants to join the collaborative framework. Despite all of these benefits, FL has a wide range of potential uses in food inspection. It can help enhance quality control procedures in processing facilities, optimize supply chain management, collect customer insights while protecting privacy, and make it easier to monitor crop health on the field. FL can be further enhanced by emerging technologies like edge computing and the Internet of Things, which will make inspection and grading procedures more precise, effective, and flexible. The food industry's strategy for guaranteeing food quality and safety could ultimately be completely transformed by FL,

**Corresponding author*: padmesh01@rediffmail.com

Padmesh Tripathi, Bhanumati Panda, Shanthi Makka, Reeta Mishra, S. Balamurugan and Sheng-Lung Peng (eds.) Federated Learning for Smart Agriculture and Food Quality Enhancement, (111–134) © 2026 Scrivener Publishing LLC

opening the door for more creative and efficient ways to satisfy contemporary needs. This article provides a thorough analysis of the application of FL in food inspection and grading. It examines FL's key concepts, current uses, challenges in integrating FL into food systems, and potential for enhancing food safety and quality in the future. The use of FL to maximize the accuracy, consistency, and scalability of the grading process is investigated through the use of real-world case studies.

***Keywords*:** Federated learning, food inspection, food grading, machine learning, decentralized systems

5.1 Introduction

Several tools and techniques like machine learning [1], deep learning [2], and artificial intelligence [3] have been used in agriculture. The global food industry, a multi-billion-dollar sector, faces continuous scrutiny to maintain high standards of quality and safety. Food inspection and grading play critical roles in ensuring that products meet these standards from the farm to the consumer's table. In [4], a novel hybridization between the MCDM process and horizontal federated learning (FL) is investigated. This strengthens the MCDM approaches' independence in assessing cold chain hazards under the organized hierarchy. Together with common fruit characteristics like color, size, shape, and texture, authors in [5] provided an overview of the fruit classification and grading process using techniques like Speeded Up Robust Features, Histogram of Orientated Gradient, and Local Binary Pattern. Convolutional neural networks (CNNs), K-nearest neighbor, support vector machines, and artificial neural networks (ANNs) are examples of machine learning techniques. The tropical fruits of Maharashtra were the main focus in [6] where image processing techniques were also reviewed. Traditionally, inspection systems have been centralized, where data is collected at specific points and processed using machine learning algorithms to assess food quality and grading.

The role of emotional intelligence in enhancing human–robot collaboration within the domain of human–machine collaboration and emotional intelligence in food inspection and grading was suggested in [7]. However, these conventional approaches come with various challenges, such as concerns over data privacy, exposure to security vulnerabilities, inefficiencies in processing, and limited scalability. Addressing these issues is crucial for enhancing the overall integrity and efficiency of food supply chains.

Traditional food inspection systems are designed to monitor the quality, safety, and regulatory compliance of food products throughout the supply chain, from production to consumption. Figure 5.1 represents the block

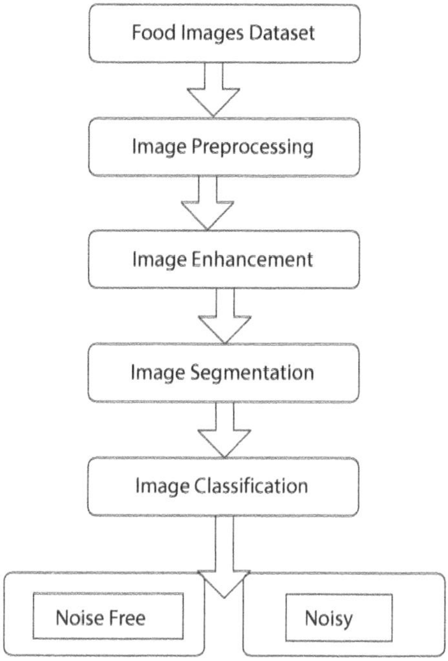

Figure 5.1 Block diagram of food inspection and grading.

diagram of food inspection and grading. Emerging trends in machine learning and intelligent optimization for business analytics were suggested in [8]. They typically involve physical inspections, laboratory testing, and human assessments, which are guided by national and international food safety regulations.

5.1.1 Manual Inspection Methods

The most common traditional method of food inspection involves manual checks conducted by trained inspectors at various points in the supply chain, such as farms, processing facilities, and retail outlets. These inspections often include visual evaluations of food for signs of contamination, spoilage, or defects, such as discoloration, off-odors, or physical damage. Inspectors also assess the hygiene of processing environments and ensure that facilities comply with health and safety regulations.

5.1.2 Sampling and Laboratory Testing

In addition to physical inspections, food samples are often taken for laboratory analysis. These tests may evaluate various attributes, including chemical composition, microbial contamination, pesticide residues, and allergen levels. Laboratory testing is critical in determining whether food meets the necessary safety standards, particularly for identifying invisible contaminants, such as bacteria (e.g., Salmonella and Listeria), toxins, or harmful chemicals. Various machine vision–based postharvest citrus fruit grading and illness prediction techniques published between 2010 and 2021 were examined in [9].

5.1.3 Checkpoints in the Supply Chain

Traditional food inspection systems involve multiple checkpoints along the food supply chain, including:

- **On-farm inspections:** Ensuring that crops and livestock are produced under safe and hygienic conditions.
- **Slaughterhouses and processing plants:** Verifying that meat and poultry are processed in a sanitary manner and ensuring the humane treatment of animals.
- **Packaging and distribution:** Ensuring that food products are stored and transported under conditions that prevent spoilage and contamination.
- **Retail and food service outlets:** Inspecting the final products at grocery stores, restaurants, and other food establishments to ensure they are safe for consumption.

5.1.4 Role of Regulatory Agencies

In most countries, food inspection is regulated by government agencies that set food safety standards and ensure compliance. For example:

- **United States:** The Food and Drug Administration and the United States Department of Agriculture oversee the safety of food products.
- **European Union:** The European Food Safety Authority plays a similar role in coordinating food safety across member states.
- **International:** The Codex Alimentarius, established by the Food and Agriculture Organization and the World Health Organization, provides international standards and guidelines for food safety.

5.2 Traditional Food Inspection Methods: Key Approaches

5.2.1 Visual Inspection

Visual inspections are perhaps the most rudimentary method of food quality assessment. Inspectors visually examine food products for any signs of spoilage, contamination, or damage. This includes checking for mold, discoloration, and foreign objects in packaged foods. While visual inspections are straightforward and cost-effective, they are limited by the inspector's experience and the visibility of certain hazards (e.g., bacterial contamination).

5.2.2 Olfactory and Taste Testing

In certain cases, food inspectors may rely on their sense of smell and taste to identify off-flavors or odors that indicate spoilage or contamination. This method is more common in artisanal food production or industries like wine and olive oil, where sensory characteristics are critical to quality. However, such testing is subjective and may vary between inspectors.

5.2.3 Temperature Monitoring

Ensuring that food is stored and transported at the correct temperature is vital for preventing microbial growth and spoilage. Traditional food inspection systems involve manual temperature checks at various stages of the supply chain. Inspectors use thermometers to ensure that perishable items like meat, dairy, and seafood are maintained at safe temperatures throughout processing, distribution, and retail storage.

5.2.4 Chemical and Microbial Testing in Laboratories

Laboratory testing is an essential component of traditional food inspection systems. It involves the analysis of food samples for chemical contaminants, pathogens, and other harmful substances that are not visible to the naked eye. This method can include the following:

- **Microbiological analysis:** Detecting bacteria, viruses, and parasites in food products.
- **Chemical residue testing:** Measuring levels of pesticides, heavy metals, and food additives.

- **Nutritional analysis:** Verifying that food labeling accurately reflects the nutritional content.

Although laboratory testing is highly accurate, it is resource-intensive, requiring specialized equipment, trained personnel, and time for sample analysis. This can result in delays in product release or recalls in cases of contamination.

5.2.5 Traceability Systems

Traditional food inspection systems also rely on traceability, which involves keeping detailed records of food production, processing, and distribution to ensure accountability at each stage. By tracking batches of food products through the supply chain, inspectors can identify the source of contamination in the event of a foodborne illness outbreak. However, traceability systems are often manual and paper-based, making them prone to errors and inefficiencies.

5.3 Challenges and Limitations of Traditional Food Inspection Systems

While traditional food inspection methods have proven effective, they face several challenges that limit their efficiency and adaptability in the modern food industry.

5.3.1 Resource-Intensive Processes

One of the primary drawbacks of traditional food inspection systems is their reliance on manual labor and laboratory testing. Conducting thorough inspections at every stage of the supply chain requires a significant investment in manpower, time, and resources. For example, laboratory testing can take several days to complete, leading to delays in product release or potential recalls.

5.3.2 Inconsistent Results

Because traditional inspection systems rely on human expertise, they are subject to variability. Different inspectors may have different interpretations of food safety standards, leading to inconsistent results. Sensory tests, such as taste and smell, are inherently subjective and prone to variation between individuals.

5.3.3 Reactive Approach

Traditional food inspection systems are often reactive rather than proactive. Inspectors typically identify contamination or spoilage after it has already occurred, leading to recalls, loss of product, or, in severe cases, foodborne illness outbreaks. This reactive approach makes it challenging to prevent issues before they impact consumers.

5.3.4 Data Collection and Traceability Issues

Many traditional food inspection systems rely on paper-based record-keeping, which is inefficient and error-prone. Traceability is crucial for identifying the source of contamination, but manual systems can lead to gaps in data, making it difficult to track products through the supply chain accurately. This lack of real-time data can slow down response times in the event of contamination.

5.3.5 Globalization and Complex Supply Chains

The global nature of modern food production presents challenges for traditional inspection systems. Food products are often sourced from multiple countries, with ingredients passing through several hands before reaching consumers. This complexity makes it difficult for inspectors to monitor every stage of the supply chain, increasing the risk of contamination or fraud.

In recent years, FL has emerged as a promising technology in many sectors, including healthcare, finance, and automotive industries. In the food industry, FL has gained attention as a decentralized approach where data remains on the edge devices (e.g., farms, food processing units), and only model updates are shared, thus safeguarding privacy and improving scalability. FL enables multiple stakeholders in the food supply chain to collaboratively train machine learning models without sharing sensitive data, which is particularly important in addressing issues such as intellectual property and regulatory compliance. A thorough analysis of the suitability of the strategy at various seriousness levels was presented in [10]. Client MP_5 performed better, achieving 96% accuracy. Client mp_4 (77%), Client mp_6 (88%), Client mp_3 (82%), and Client mp_1 (to be named) came next. However, certain well-known multiplayer games may suffer from these additional expenses in terms of quality or gameplay. The study concludes that CNN and FL can be effectively used to classify soybean leaf diseases. This paper aims to provide an in-depth review of the application of FL in food inspection and grading, from initial data collection to model optimization and decision-making. It will explore existing frameworks, discuss their implementation challenges, and propose future directions to improve the accuracy, reliability, and speed of food quality inspection and grading systems. A method based on edge-enabled FL that allows visual inspectors to identify invisible flaws through secure model exchange and on-device model fine-tuning was suggested in [11]. According to [12], the CNN model is dependable and adaptable in assessing the degree of Cucurbit leaf disease. It also demonstrates how FL can handle a variety of datasets. Its findings have important implications for precision agriculture since they offer a scalable and privacy-preserving approach to disease management and control in a variety of agricultural scenarios.

5.4 Federated Learning: Overview and Applications in Food Systems

5.4.1 What is Federated Learning

FL is a machine learning paradigm where multiple clients (such as devices, institutions, or organizations) collaboratively train a model while keeping the training data decentralized. Instead of sending raw data to a central server, FL involves training models locally and sharing only the model parameters or updates. The central server aggregates these updates to improve a global model. In the context of food inspection and grading, FL enables the collaboration of different food producers, distributors, and inspectors without sharing sensitive or proprietary data. This approach enhances data privacy, reduces latency, and scales effectively to larger datasets and numerous collaborators. Authors in [13] provided a thorough strategy for accurately determining the severity of papaya leaf diseases utilizing CNN and FL. According to the data and tests, the model is reliable and effective, which will assist develop more effective farm disease control solutions that are not reliant on a single facility. In [14], it has been claimed that distributed computing capacity could improve disease classification efficiency and accuracy for almond farming.

5.4.2 Features of Federated Learning

FL is an advanced machine learning approach that enables decentralized data processing while maintaining data privacy. This method is increasingly relevant in various fields, including healthcare, finance, and the food industry, due to its unique features. Starting with an overview of the latest developments in FL and Internet of Things (IoT), followed by a discussion of their integration, authors in [15] offered a thorough analysis of the developing uses of FL in IoT networks. We specifically examine and evaluate FL's potential to provide a variety of IoT services, such as mobile crowd sensing, attack detection, localization, data sharing, data offloading and caching, and IoT privacy and security. Key features of FL include the following.

5.4.3 Applications of Federated Learning in Food Inspection and Grading

The application of FL in food inspection and grading spans several use cases:

- **Quality Control at Farms:** FL allows different farms to use local environmental, soil, and crop data to improve machine learning models that predict crop quality without sharing data that could compromise competitive advantage. Figure 5.2 below shows the features of federated learning.

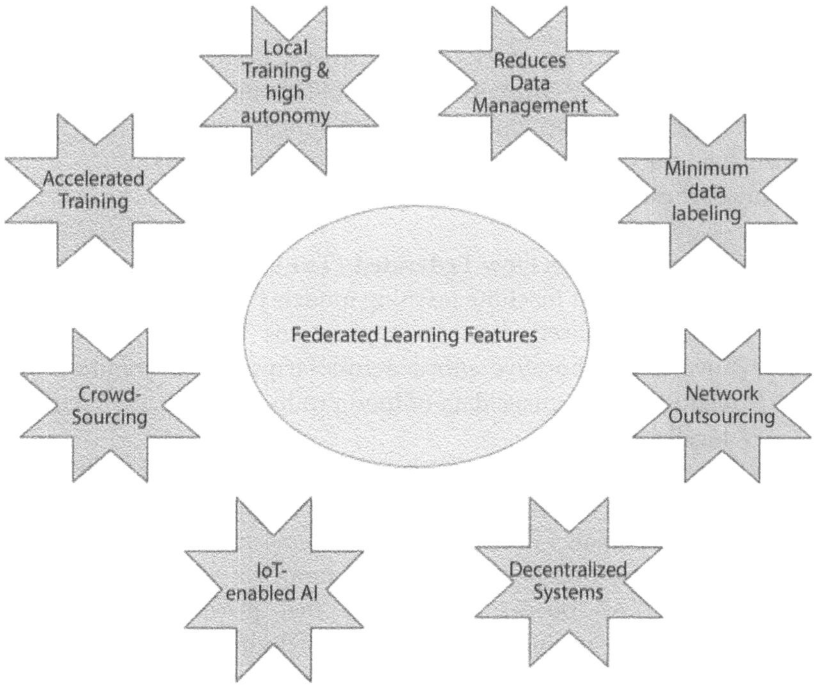

Figure 5.2 Federated learning features.

- **Food Processing Units:** FL can be used in food processing facilities where real-time inspection of food items is required. Multiple facilities can collaborate in refining models that assess food quality based on parameters like appearance, texture, and freshness.
- **Supply Chain Monitoring:** FL can optimize grading standards across the supply chain, ensuring that food quality is maintained from production to consumer delivery without exposing proprietary data about handling procedures.
- **Retail and Distribution:** Retailers can use FL to better assess the quality of incoming food shipments based on historical inspection data from different suppliers. This ensures consistency in grading standards while preserving the confidentiality of the suppliers.

5.5 Existing Frameworks and Implementations in Food Inspection

5.5.1 Federated Learning Frameworks

There are several FL frameworks currently being used or adapted for food inspection and grading. The current state of FL from five perspectives: systems heterogeneity, communication architecture, machine learning model, privacy method, and data partitioning was provided in [16].

Some of these include the following:

- **Google's TensorFlow Federated (TFF):** TFF allows developers to build machine learning models that can be trained on decentralized data. Its applications in food grading involve collaborative efforts across farms and distribution centers where image data of food products can be analyzed locally.
- **Flower:** This FL framework is lightweight and adaptable, making it useful for food inspection applications where multiple smaller devices, such as cameras in food processing units, need to collaboratively train models.
- **PySyft:** An open-source library that provides privacy-preserving machine learning tools, including FL. It has the potential to be used in food safety, where strict regulatory compliance requires strong data protection.

5.5.2 Use Cases of FL in Food Grading

Several industries have begun to implement FL in food inspection. Authors in [17] examined the methods that put out in research articles from 2010 to 2019 with an emphasis on the advancement of state of the art. The research is contrasted with several methods for fruit identification, classification, and grading. For example, dairy farms have adopted FL to standardize milk grading without sharing sensitive production data. Similarly, meat and fish processing plants are exploring FL to enhance inspection and grading without violating privacy laws or industrial confidentiality agreements.

5.6 Comparison Between Existing and Future Federate Learning System for Food Inspection and Grading

FL has emerged as a key enabler for modern food inspection and grading systems, offering decentralized machine learning capabilities that address data privacy, security, and scalability issues. While existing FL systems have made significant strides in improving food inspection and grading processes, future advancements will further enhance their capabilities by incorporating real-time processing, better collaboration, and more efficient aggregation methods. According to [18], the federated CNN model produced very high accuracy, with recall ranging from 92.83% to 97.55% and consumer accuracy ranging from 92.76% to 97.56%. The F1 scores, which regulate memory and accuracy, were consistently high and ranged from 92.78% to 97.55%. The important data representation ranged from 391 to 721.75 instances, as indicated by the support values, which showed the number of actual instances of each class. For every customer, consistency demonstrated outstanding model accuracy, never falling below 96%. According to [19], the majority of the literature has been on the illness known as wheat stripe rust (60.81%), and the most popular prediction method is ANNs (13.32%). The most important performance indicator, according to their findings, is accuracy (67%).

5.7 Case Studies in Federated Learning for Food Inspection and Grading

Several industries are already experimenting with FL for food inspection and grading, yielding promising results since 2019. Table 5.1 below

Table 5.1 Comparison between existing and future FL systems.

Aspect	Existing FL systems	Future FL systems
Data Privacy	Data remains local, improving privacy and reducing risks of data breaches.	Enhanced privacy with advanced techniques like differential privacy and secure multiparty computation
Model Personalization	Localized models for different stakeholders but limited to basic customization	More advanced personalization with improved aggregation, ensuring better performance across diverse datasets
Scalability	Scalable but with limitations due to communication overhead and model update frequency	More efficient scaling with reduced communication overhead and better aggregation techniques
Real-time Processing	Limited real-time applications; models often operate in batch mode	Integration with edge computing for real-time grading decisions and feedback.
Data Heterogeneity	Difficulty handling diverse data types across different stakeholders	Advanced FL models capable of managing various data types (images, chemical data, etc.) effectively
Aggregation Techniques	Basic aggregation techniques, leading to bias in handling heterogeneous data	Improved aggregation methods that reduce bias and enhance model accuracy for underrepresented data
Global Collaboration	Limited collaboration across international supply chains due to data privacy concerns	Decentralized global collaboration with cross-border compliance and model sharing without data transfer

(*Continued*)

Table 5.1 Comparison between existing and future FL systems. (*Continued*)

Aspect	Existing FL systems	Future FL systems
Network Latency and Bandwidth	High communication overhead due to frequent model updates between clients and the central server	Optimized communication protocols, reducing latency and bandwidth usage, enhancing scalability
IoT and Blockchain Integration	Minimal integration with IoT devices; blockchain use is still in exploratory stages	Full integration with IoT for real-time data collection and blockchain for secure, transparent traceability
Cost of Implementation	High, especially for small stakeholders due to specialized hardware and software needs	Potentially reduced costs with technological advancements and more affordable edge devices
Global Standardization	Models are often localized and limited to regional practices	Global standardization of food grading practices with seamless cross-border collaboration
Adaptability	Models are slow to adapt to new food products or environmental changes	Adaptive learning models that evolve dynamically with changing conditions and new food types
Security	Secure, but challenges remain in fully preventing model poisoning and other attacks	Enhanced security with blockchain integration and more robust FL protocols to prevent malicious attacks
Ethical and Legal Compliance	Compliance with local privacy laws, but global compliance remains a challenge	Improved global compliance with legal frameworks and ethical guidelines across international borders

expresses the comparison between existing and future FL systems. This section will explore a few case studies in-depth, providing statistical insights into the impact of FL on food quality control. Authors in [20] suggested the four distinct categories into which the different levels of severity of chili leaf infections fall: low (1%–25%), mild (26%–50%), moderate (51%–75%), and severe (76%–100%). Six clients' data are incorporated into the study, allowing for a thorough analysis across numerous locales. The model suggested in [21] uses deep learning skills to recognize and detect disease symptoms in potato leaves, efficiently differentiating between normal and abnormal conditions. Following a rigorous testing and analysis process, our CNN model achieves an impressive 98.99% accuracy rate in classifying potato leaf pictures into three groups: healthy, early, and late blight.

5.7.1 Case Study 1: Federated Learning for Dairy Quality Control in the EU (2020–2021)

In a 2020 initiative across several dairy farms in the European Union (EU), a consortium of dairy producers implemented FL to enhance milk quality grading. Each farm used its local data (e.g., fat content, bacteria counts, and sensory analysis) to train a decentralized model, contributing to a global model used for real-time milk quality evaluation across different farms.

- **Participants:** 50 dairy farms across the EU.
- **Data type:** Chemical analysis, bacterial contamination levels, and sensor data for environmental monitoring.
- **Outcome:** By the end of 2021, FL-driven grading reduced milk quality assessment errors by 25% compared to traditional centralized systems. The federated model achieved a 92.5% accuracy in grading, compared to 84.7% for previous centralized models.
- **Benefits:** Reduced data transfer costs by 15% and improved regulatory compliance by enabling local data retention, which was crucial for sensitive datasets.

5.7.2 Case Study 2: Federated Learning for Fruit and Vegetable Quality in China (2019–2022)

In China, several fruit and vegetable farms collaborated on an FL project to develop a decentralized system for grading produce based on factors such as color, texture, and freshness. This initiative involved real-time image analysis of fruits and vegetables as they were harvested, packaged, and distributed to markets. Research in [22] contains four disease levels and one healthy level in a FL CNN model for identifying and categorizing tomato leaf illnesses across five severity levels. The model was analyzed using data from four clients, reviewing each client's accuracy, precision, recall, and F1-score performance metrics. With an accuracy range of 96% to 98% across all severity levels, the collected results demonstrate that the model consistently produced high-quality results. The potential use of a CNN model based on FL for the classification of tomato leaf diseases was investigated in [23]. The research includes four customers from each of the five classes, four of which represent the four tomato leaf illnesses and one of which represents the healthy class. The model's performance is evaluated using precision, recall, F1-score, and accuracy. The data shows consistently strong performance across all customers and courses, with overall accuracies ranging between 98% and 99%. The majority of precision, recall, and F1-score values are over 94%, indicating the model's ability to correctly identify healthy and diseased tomato leaf images with few false positives. According to the approach suggested in [24], the model becomes more dependable and generalized as it accumulates experience from various data sources. This strategy yields measurable results, as demonstrated by improved model performance among clients. The analysis produced a number of noteworthy findings. In NA_6 and NA_1, the macro average accuracy of disease diagnosis through clients is 97.55% and 89.71%, respectively.

- **Participants:** 120 farms and 30 distribution centers in Hebei and Shandong provinces.
- **Data type:** High-resolution images, environmental sensor data (temperature and humidity), and historical market data for ripeness and freshness prediction.
- **Outcome:** Between 2019 and 2022, the FL model improved the consistency of fruit and vegetable grading across farms by 18%. The model was able to detect defects in produce with 96% accuracy, compared to 85% accuracy in systems relying on traditional, centralized methods.
- **Economic Impact:** The initiative led to a 12% reduction in wastage during the supply chain due to better real-time defect detection and grading. Sales prices for high-quality produce increased by 7% as a result of more accurate grading.

5.7.3 Case Study 3: Seafood Quality Inspection Using Federated Learning in Japan (2020–2021)

A large-scale FL system was implemented in Japan to monitor the quality of seafood, particularly tuna and salmon, from processing plants to retail markets. The initiative was aimed at improving the freshness and texture of the fish while reducing spoilage during transport.

- **Participants:** 15 major seafood processing plants and 50 retail chains in Japan.
- **Data type:** Image data, temperature monitoring during transport, pH levels of seafood, and historical sales data.
- **Outcome:** The FL system, deployed from 2020 to 2021, showed a 14% improvement in detecting substandard fish (e.g., those with degraded texture or color) compared to centralized machine learning models. The accuracy of seafood grading reached 94.3%, up from 81% using the older systems.
- **Economic Impact:** The initiative saved $2.5 million annually in spoilage-related losses across the retail supply chain due to improved grading accuracy and real-time monitoring.

5.7.4 Case Study 4: Federated Learning in Coffee Bean Grading in Brazil (2019–2022)

According to [25], CNN cooperative learning can be used to assess the severity of wheat diseases, boost agricultural productivity, and halt disease transmission. Brazil, being the largest coffee exporter globally, undertook a FL initiative to enhance the quality grading of coffee beans. The project involved multiple coffee farms and cooperatives sharing model updates to grade coffee based on factors like bean size, color, moisture content, and defect count.

- **Participants:** 200 coffee farms across Brazil's Minas Gerais and São Paulo regions.
- **Data type:** Sensor data for moisture content, high-resolution images for bean size and defect identification, and local environmental data.
- **Outcome:** From 2019 to 2022, FL-enhanced grading systems increased the accuracy of coffee bean classification by 20%, achieving an accuracy rate of 93.2% compared to 77.1% in prior systems.
- **Economic Impact:** Coffee farms using FL experienced a 9% increase in their average sales price due to improved quality assurance. Additionally, coffee defects during export were reduced by 15%, leading to fewer rejections and financial losses in international markets.

5.7.5 Case Study 5: Federated Learning for Meat Quality Control in the United States (2021–2022)

In 2021, several meat processing companies in the United States initiated a FL program to inspect and grade meat products, focusing on beef and pork. The model trained on factors such as texture, marbling, and color of the meat to ensure consistent quality standards.

- **Participants:** 25 meat processing plants and distribution centers across the United States.
- **Data type:** Visual inspection data, chemical composition, and environmental factors such as storage temperature and humidity during transport.
- **Outcome:** By the end of 2022, the federated model achieved 90% accuracy in detecting defects in meat products, up from 82% in centralized systems. Additionally, spoilage was reduced by 10%, and consumer complaints related to meat quality decreased by 5%.
- **Benefits:** The FL system improved supply chain transparency and facilitated compliance with federal food safety regulations. Processing plants reported a 7% reduction in operational costs due to the increased efficiency in grading and inspection processes.

5.7.6 Case Study 6: Olive Oil Quality Grading in Spain (2019–2021)

In Spain, a collaboration between olive oil producers and research institutions explored the use of FL to improve the grading process of extra virgin olive oil. The project aimed to assess the oil's quality based on sensory evaluation, chemical analysis, and production conditions.

- **Participants:** 35 olive oil producers across Andalucía.
- **Data type:** Sensory data (flavor, aroma, and bitterness), chemical composition (fatty acids and phenolic compounds), and environmental factors (temperature, humidity, and harvesting techniques).
- **Outcome:** From 2019 to 2021, the FL system improved the consistency of olive oil grading by 17% compared to traditional centralized systems. The accuracy of quality grading based on sensory data reached 90%, up from 75%.

- **Economic Impact:** The project led to a 6% increase in export sales as the quality of the extra virgin olive oil improved. Additionally, the cost of quality inspection was reduced by 12% due to more efficient, decentralized model training.

5.7.7 Case Study 7: Wine Quality Assessment in Italy (2020–2022)

FL was implemented in Italy's wine industry to enhance quality assessment for high-end wine producers. Using local vineyard data, winemakers collaborated on a model to assess wine quality based on grape characteristics, environmental factors, and fermentation conditions without sharing proprietary data.

- **Participants:** 50 vineyards across Tuscany, Piedmont, and Veneto.
- **Data type:** Grape sugar content, fermentation temperature, soil quality, and historical harvest data.
- **Outcome:** Over 2 years, from 2020 to 2022, the federated model improved the accuracy of wine quality prediction by 22% compared to centralized systems. The model achieved an 88% accuracy rate in predicting wine flavor profiles, with a significant reduction in grading discrepancies among different vineyards.
- **Economic Impact:** Participating vineyards experienced a 10% increase in premium wine sales due to more accurate grading and improved consistency. This also led to a 5% reduction in overall quality control costs.

5.7.8 Case Study 8: Federated Learning for Poultry Inspection in India (2021–2023)

In 2021, poultry producers in India implemented a FL system to improve the inspection of chicken and egg quality. The project aimed to assess product freshness and detect potential issues such as contamination and spoilage without violating data privacy across different production sites.

- **Participants:** 60 poultry farms and 10 processing units across Tamil Nadu and Kerala.
- **Data type:** Real-time environmental sensor data (temperature and humidity), image data of poultry products, and historical records of contamination levels.

- **Outcome:** From 2021 to 2023, the FL model achieved a 15% improvement in detecting early signs of contamination and spoilage, reaching 93% accuracy in product grading. This system also reduced the occurrence of product recalls by 20%.
- **Economic Impact:** The initiative led to an estimated savings of $3 million per year due to the reduction in spoilage-related losses and improved compliance with national food safety standards.

5.7.9 Case Study 9: Tea Leaf Quality Grading in Sri Lanka (2019–2022)

Sri Lanka's tea industry adopted FL to enhance the grading of tea leaves at various stages of production, from plucking to final packaging. By using FL, multiple tea plantations could collaboratively refine quality models without sharing sensitive data about cultivation methods or tea processing techniques.

- **Participants:** 70 tea plantations and 15 processing units in the central and southern regions of Sri Lanka.
- **Data type:** Leaf texture and color, environmental factors (rainfall and temperature), and historical data on tea flavor profiles.
- **Outcome:** The FL model deployed from 2019 to 2022 improved grading accuracy by **18%**. The model accurately predicted the quality of tea leaves with 91% accuracy, compared to 76% under the previous centralized system.
- **Economic Impact:** Tea plantations saw a 5% increase in international market demand for high-grade tea, and operational costs related to quality assessment dropped by 8% due to the decentralized grading approach.

5.7.10 Case Study 10: Federated Learning for Cocoa Bean Grading in West Africa (2020–2022)

West African cocoa farms collaborated on a FL project to improve the quality control of cocoa beans, which are a critical component in chocolate production. This initiative involved grading beans based on size, color, and moisture content, aiming to reduce the rejection rate of shipments to international buyers.

- **Participants:** 100 cocoa farms across Ghana and Côte d'Ivoire.
- **Data type:** Cocoa bean size, color, moisture levels, and fermentation conditions.
- **Outcome:** Between 2020 and 2022, the FL model increased grading accuracy by 19%, with 89% accuracy in detecting substandard beans. The number of rejected shipments to international buyers was reduced by 12%, leading to significant improvements in quality control.
- **Economic Impact:** The initiative resulted in a $5 million increase in annual revenue for participating farms due to the reduction in rejections and higher prices for premium-quality beans.

5.8 Future Directions and Potential of Federated Learning in Food Systems

The future of FL in food inspection and grading is promising, with potential for widespread adoption as technology matures. Some of the future directions include the following:

5.8.1 Enhanced Privacy and Security Protocols

As FL continues to evolve, more advanced privacy-preserving techniques will be integrated into FL frameworks to protect against data leakage and model tampering. Techniques such as homomorphic encryption and secure multi-party computation are expected to play a key role in making FL more secure for food systems.

5.8.2 Integration with IoT and Edge Computing

The integration of FL with IoT devices is expected to revolutionize food inspection. By combining IoT sensors, edge computing, and FL, food producers can collect real-time data and train models locally, improving the accuracy and speed of food grading.

5.8.3 Real-Time Quality Monitoring

FL can enable real-time monitoring and grading of food quality throughout the supply chain. This would allow stakeholders to identify quality degradation at any stage and take corrective action immediately, reducing waste and ensuring higher quality standards for consumers.

5.8.4 Regulatory Compliance and Standardization

In the future, FL will likely be integrated into regulatory frameworks for food safety. Governments and industry organizations may adopt FL as a standard method for ensuring compliance with food safety regulations, particularly in industries where data privacy is a key concern.

5.9 Conclusion

FL is revolutionizing the food industry's inspection and grading processes by enabling collaboration among diverse stakeholders without compromising data privacy, thereby fostering a decentralized approach that enhances food quality and safety. This innovative paradigm allows organizations to train machine learning models collaboratively using decentralized data, addressing critical needs for accuracy and efficiency in food inspection. While challenges such as data heterogeneity, communication latency, and privacy concerns exist, FL presents significant advantages, including improved model accuracy from diverse data sources, reduced communication overhead by only transmitting model updates, and enhanced scalability as new participants can easily join the model training. Potential applications include farm-level monitoring for crop health, quality control in processing plants, optimization of the food supply chain, and gathering consumer insights—all of which can be further empowered by advancements in technologies like the IoT and edge computing. As the food industry embraces these technological innovations, FL is poised to play a central role in making inspection and grading processes more accurate, efficient, and adaptable to the evolving landscape of food safety and quality assurance.

AI disclosure statement: The author acknowledges the use of AI-based tools for minor language editing or grammatical correction. However, all intellectual contributions, analyses, and conclusions presented in this chapter are entirely the author's own.

References

1. Tripathi, P., Kumar, N., Rai, M., Shukla, P.K., Verma, K.N., Applications of Machine Learning in Agriculture, in: *Smart Village Infrastructure and Sustainable Rural Communities*, M. Khan, B. Gupta, A. Verma, P. Praveen, C. Peoples (Eds.), pp. 99–118, IGI Global, 2023, https://doi.org/10.4018/978-1-6684-6418-2.ch006.
2. Tripathi, P., Kumar, N., Rai, M., Khan, A., Applications of Deep Learning in Agriculture, in: *Artificial Intelligence Applications in Agriculture and Food Quality Improvement*, pp. 17–28, IGI Global, USA, 2022, doi: 10.4018/978-1-6684-5141-0.ch002.
3. Rai, M., Tyagi, N., Tripathi, P., Kumar, N., Kumari, P., Advancement in Agricultural Techniques With the Introduction of Artificial Intelligence and Image Processing, in: *Smart Village Infrastructure and Sustainable Rural Communities*, M. Khan, B. Gupta, A. Verma, P. Praveen, C. Peoples (Eds.), pp. 47–68, IGI Global, 2023, https://doi.org/10.4018/978-1-6684-6418-2.ch004.
4. Lau, H., Tsang, Y.P., Nakandala, D., Lee, C.K., Risk quantification in cold chain management: a federated learning-enabled multi-criteria decision-making methodology. *Ind. Manage. Data Syst.*, *121*, 7, 1684–1703, 2021.
5. Chaudhari, D. and Waghmare, S., Machine vision based fruit classification and grading—a review, in: *ICCCE 2021: Proceedings of the 4th International Conference on Communications and Cyber Physical Engineering*, Springer Nature Singapore, Singapore, pp. 775–781, 2022, May.
6. Khoje, S.A. and Bodhe, S.K., A Comprehensive survey of fruit grading systems for tropical fruits of maharashtra. *Crit. Rev. Food Sci. Nutr.*, *55*, 12, 1658–1671, 2015.
7. Mishra, R., Tripathi, P., Kumar, N., Application of Emotional Intelligence in improvement of Human-Robot Collaboration, in: *Human-Machine Collaboration and Emotional Intelligence in Industry 5.0*, Kumar and (Eds.), pp. 251–267, IGI Global, USA, 2024.
8. Mishra, R., Tripathi, P., Kumar, N., Future Directions in the Applications of Machine Learning and Intelligent Optimization in Business Analytics, in: *Intelligent Optimization Techniques for Business Analytics*, S. Bansal, N. Kumar, P. Agarwal (Eds.), pp. 49–76, IGI, Global, USA, 2024.
9. Palei, S., Behera, S.K., Sethy, P.K., A systematic review of citrus disease perceptions and fruit grading using machine vision. *Procedia Comput. Sci.*, *218*, 2504–2519, 2023.
10. Thapliyal, N., Thapliyal, S., Kukreja, V., Mehta, S., Disruptive Tech in Agriculture: Federated Learning CNNs for Soybean Leaf Disease Classification, in: *2024 3rd International Conference for Innovation in Technology (INOCON)*, IEEE, pp. 1–6, 2024, March.

11. Bharti, S., McGibney, A., O'Gorman, T., Edge-enabled federated learning for vision based product quality inspection, in: *2022 33rd Irish Signals and Systems Conference (ISSC)*, IEEE, pp. 1–6, 2022, June.
12. Kaur, K., Chattopadhyay, S., Verma, A., Kukreja, V., Mehta, S., A Synergistic Approach to Cucurbit Leaf Disease Classification with Federated Learning CNNs, in: *2024 IEEE International Conference on Interdisciplinary Approaches in Technology and Management for Social Innovation (IATMSI)*, vol. 2, IEEE, pp. 1–6, 2024, March.
13. Malhotra, S., Garg, A., Kukreja, V., Mehta, S., Agricultural AI: Integrating Federated Learning and CNN in Papaya Leaf Disease Detection, in: *2024 3rd International Conference for Innovation in Technology (INOCON)*, IEEE, pp. 1–6, 2024, March.
14. Vats, S., Khanna, M.S., Kukreja, V., Mehta, S., AI in Agriculture: A Federated Learning CNN Approach to Detecting Almond Leaf Disease, in: *2024 IEEE International Conference on Interdisciplinary Approaches in Technology and Management for Social Innovation (IATMSI)*, vol. 2, IEEE, pp. 1–6, 2024, March.
15. Nguyen, D.C., Ding, M., Pathirana, P.N., Seneviratne, A., Li, J., Poor, H.V., Federated learning for internet of things: A comprehensive survey. *IEEE Commun. Surv. Tutor.*, *23*, 3, 1622–1658, 2021.
16. Zhang, C., Xie, Y., Bai, H., Yu, B., Li, W., Gao, Y., A survey on federated learning. *Knowl.-Based Syst.*, *216*, 106775, 2021.
17. Behera, S.K., Rath, A.K., Mahapatra, A., Sethy, P.K., Identification, classification & grading of fruits using machine learning & computer intelligence: a review. *J. Ambient Intell. Hum. Comput.*, 15, 117, 1–11, 2020.
18. Chattopadhyay, S., Verma, A., Kukreja, V., Mehta, S., Agriculture Innovations: Broccoli Leaf Diseases Classification through Federated Learning CNN's, in: *2024 11th International Conference on Reliability, Infocom Technologies and Optimization (Trends and Future Directions) (ICRITO)*, IEEE, pp. 1–6, 2024, March.
19. Kumar, D. and Kukreja, V., Deep learning in wheat diseases classification: A systematic review. *Multimed. Tools Appl.*, *81*, 7, 10143–10187, 2022.
20. Jindal, V., Kukreja, V., Rana, D.S., Mehta, S., Thapliyal, S., An Ensemble Approach to Classify Severity Levels in Chili Leaf Diseases Through Federated Learning CNN, in: *2023 IEEE International Conference on Cloud Computing in Emerging Markets (CCEM)*, IEEE, pp. 202–207, 2023, November.
21. Shandilya, G., Anand, V., Chauhan, R., Pokhariya, H.S., Gupta, S., Deep Learning in Agriculture: Convolutional Neural Network-Based Classification of Potato Diseases for Enhanced Crop Management, in: *2024 2nd World Conference on Communication & Computing (WCONF)*, IEEE, pp. 1–6, 2024, July.
22. Mehta, S., Kukreja, V., Gupta, A., Transforming Agriculture: Federated Learning CNNs for Wheat Disease Severity Assessment, in: *2023 8th International*

Conference on Communication and Electronics Systems (ICCES), IEEE, pp. 792–797, 2023, June.

23. Jindal, V., Kukreja, V., Mehta, S., Manwal, M., Joshi, K., Emerging paradigms: A federated learning CNN approach for tomato leaf disease detection, in: *2023 3rd Asian Conference on Innovation in Technology (ASIANCON)*, IEEE, pp. 1–6, 2023, August.
24. Gaur, P., Aggarwal, P., Joshi, K., Mehta, S., Kukreja, V., Brinjal Leaf Pathology: An Improved Disease Classification in a Federated Learning-CNN Approach, in: *2024 IEEE International Conference on Interdisciplinary Approaches in Technology and Management for Social Innovation (IATMSI)*, vol. 2, IEEE, pp. 1–6, 2024, March.
25. Mehta, S., Kukreja, V., Yadav, R., A Federated Learning CNN Approach for Tomato Leaf Disease with Severity Analysis, in: *2023 Second International Conference on Augmented Intelligence and Sustainable Systems (ICAISS)*, IEEE, pp. 309–314, 2023, August.

6

Federated Learning–Based Approach for Crop Recommendation and Market Stability in Agriculture

Saurabh Kumar[1]*, Tejasva Maurya[1], Mritunjay Rai[2] and Abhishek Saxena[2]

[1]*Department of CSE, Shri Ramswaroop Memorial University, Barabanki, U.P., India*
[2]*Department of EEE, Shri Ramswaroop Memorial University, Barabanki, U.P., India*

Abstract

The agricultural sector, particularly in India, faces significant challenges such as crop overproduction, price volatility, and post-harvest losses, which lead to substantial financial losses for farmers and instability in food markets. To address these challenges, this research introduces a federated learning–based system designed to optimize crop production and enhance market stability by offering personalized, data-driven recommendations while maintaining farmer data privacy. Unlike traditional machine learning models that centralize data, federated learning allows for decentralized data processing where models are trained locally on farmer devices, and only model parameters are shared, preserving the confidentiality of sensitive agricultural data. This dual-layer approach incorporates a variety of data sources including Internet-of-Things sensors, market trends, weather conditions, and farmer inputs, enabling the system to deliver tailored crop recommendations that consider regional conditions, resource availability, and market demand. The proposed system enhances decision-making for both small and large-scale farmers by offering dynamic and real-time crop suggestions that help prevent overproduction and align with current market needs. It integrates feedback from farmers after each harvest, continually refining the model to improve accuracy and relevance. The system also takes into account infrastructure capabilities such as storage and transportation, offering crop recommendations

**Corresponding author*: slsaurabh1234567890@gmail.com

Padmesh Tripathi, Bhanumati Panda, Shanthi Makka, Reeta Mishra, S. Balamurugan and Sheng-Lung Peng (eds.) Federated Learning for Smart Agriculture and Food Quality Enhancement, (135–164) © 2026 Scrivener Publishing LLC

that reduce spoilage and optimize profitability. By analyzing real-time market trends and incorporating price prediction algorithms, the system helps farmers navigate fluctuating prices and avoid market saturation. This research highlights the potential of federated learning in promoting equitable market participation, reducing food wastage, and ensuring a more stable and sustainable agricultural sector. This study's federated learning approach not only preserves privacy and minimizes communication costs but also ensures the scalability of the system across diverse farming regions. Future developments of this system may include the integration of blockchain for enhanced data security and transparency, as well as the incorporation of government policies and support systems to further drive sustainable agricultural practices. The system offers a promising solution to the persistent challenges in agriculture, enabling farmers to make informed, data-driven decisions while maintaining control over their sensitive data.

***Keywords*:** Federated learning, crop recommendation, crop planning, agricultural technology, market stability, post-harvest loss reduction

6.1 Introduction

Agriculture remains the backbone of the Indian economy, contributing approximately 17%–18% to the nation's GDP and employing nearly 50% of its workforce [1]. Despite its critical role, the sector faces significant challenges that hinder its potential for growth and sustainability. Among these challenges, crop overproduction, price volatility, and food wastage stand out as persistent issues that threaten both the livelihoods of farmers and the stability of the agricultural market.

One of the major problems plaguing Indian agriculture is the alarming rate of post-harvest losses [2]. Lack of adequate storage facilities in rural areas has been a significant limiting factor, leading to substantial wastage of produce. Approximately 16% of fruits and vegetables, 10% of oilseeds, 9% of pulses, and 6% of cereals are wasted annually due to insufficient storage infrastructure [1]. These losses not only diminish the income of farmers but also exacerbate food insecurity in a country where millions still struggle with hunger and malnutrition.

The crop overproduction is another issue that contributes to market saturation, leading to plummeting prices and financial losses for farmers. For instance, in 2021, India experienced a significant surplus in onion production, resulting in a sharp decline in prices. Many farmers were forced to sell their produce at prices lower than the cost of production, leading to severe financial distress [3].

Price volatility in agriculture is often influenced by various external factors such as erratic weather patterns, global trade policies, and market speculation. This unpredictability makes it difficult for farmers to plan their production effectively, often leading to either an oversupply or undersupply of crops. For example, the price of tomatoes in India witnessed a 700% increase within a few months in 2022 due to unexpected weather changes that affected crop yields. Such fluctuations not only affect farmers' income but also lead to instability in the food market, impacting consumers as well [4].

Market misalignment is another critical issue where there is a disconnect between what farmers produce and what the market demands. Farmers often lack real-time data on consumer preferences, leading to a mismatch between supply and demand. This misalignment is particularly evident in the case of staple crops like rice and wheat, which are often produced in excess due to government incentives, while other high-demand crops like pulses and oilseeds are underproduced.

The inequity between large and small farmers further exacerbates these challenges. Larger farmers, with better access to resources, are often able to dominate the market, leaving small-scale farmers struggling to compete. This disparity leads to income inequality and limits market access for smaller farmers, who constitute the majority in countries like India.

6.1.1 Major Challenges Faced by the Small Farmers Over Large Farmers

6.1.1.1 Resource Access and Financial Capital

Larger farmers typically have access to more substantial financial capital, which allows them to invest in advanced agricultural technologies, modern machinery, high-quality seeds, and fertilizers. This access to resources enables them to achieve higher productivity and efficiency in their farming operations. For instance, large farms may use mechanized equipment for planting and harvesting, reducing labour costs and increasing the speed and scale of production. On the other hand, small-scale farmers often operate with limited financial resources, restricting their ability to invest in such technologies. They may rely on traditional farming methods, which are often labour-intensive and less efficient. Without access to credit or subsidies, small farmers struggle to adopt modern practices that could enhance their productivity. This lack of investment capability not only limits their yield but also makes them more vulnerable to economic shocks, such as price drops or crop failures.

6.1.1.2 Market Access and Bargaining Power

Large farmers, due to their scale of production, often have better access to markets, both domestic and international. They can afford to produce in bulk, which allows them to meet the demands of large buyers such as supermarkets, exporters, and processing companies [5]. This scale also gives them greater bargaining power, enabling them to negotiate better prices for their produce. Additionally, large farmers often have the means to store their crops and sell them when market prices are more favorable, thereby maximizing their profits. In contrast, small-scale farmers often produce limited quantities that may not meet the demands of large buyers, forcing them to sell their produce in local markets where competition is high and prices are low. They typically lack the bargaining power to negotiate better prices, making them more susceptible to exploitation by middlemen who may offer lower prices due to the small scale of their transactions. Furthermore, small farmers often do not have adequate storage facilities, which compels them to sell their produce immediately after harvest, even when market prices are low, resulting in lower income.

6.1.1.3 Income Inequality and Market Disparity

The disparity between large and small farmers contributes significantly to income inequality in the agricultural sector. Large farmers, with their ability to produce more efficiently and sell at better prices, tend to generate higher incomes. They can reinvest their profits into further improving their operations, creating a cycle of growth and increasing wealth. Over time, this leads to the consolidation of land and resources in the hands of fewer, larger farmers, further exacerbating the divide. Small farmers, however, often remain trapped in a cycle of low productivity and low income. The inability to compete effectively with larger farmers means that their earnings remain limited, hindering their ability to improve their operations or escape poverty. This income disparity not only affects the individual farmers but also has broader socio-economic implications, contributing to rural poverty and limiting economic growth in agricultural regions.

6.1.1.4 Impact on Agricultural Diversity and Food Security

This inequity also affects agricultural diversity and food security. Large farmers may focus on high-value cash crops or monoculture farming practices that are profitable but can lead to soil degradation and reduced biodiversity. Small-scale farmers, who often cultivate a variety of crops to meet their own food needs and sell the surplus, play a crucial role in maintaining agricultural diversity. However, their struggle to compete in the market threatens their ability to continue these practices, potentially leading to a loss of crop diversity, which is vital for ecological balance and long-term food security.

6.1.2 Objective

The objective of this research is to develop a federated learning (FL)-based approach that can address these critical challenges in agriculture. The proposed system aims to:

- Provide personalized crop recommendations to farmers based on regional factors, real-time market data, and individual capabilities [6–8]
- Align crop production with market demand, thereby reducing overproduction and stabilizing prices [9]
- Promote equitable market participation for both large and small farmers, ensuring a more balanced and resilient agricultural sector [10]

Through this innovative approach, the research seeks to enhance the profitability of farmers, reduce food wastage, and contribute to a more stable and sustainable agricultural market. The integration of FL in this system ensures that sensitive farmer data remains private while enabling the model to adapt continuously to changing environmental conditions and market trends [11, 12].

6.1.3 Federated Learning in Agriculture

FL is a decentralized approach to machine learning that allows multiple devices or servers (clients) to collaboratively train a model without sharing their raw data as shown in Figure 6.1. Instead of sending data to a central location, the individual devices locally train models using their specific datasets and only share the model updates, such as weights or gradients, with a central server. The server then aggregates these updates to form a global model, which is sent back to the clients for further improvement. This process ensures that sensitive data remains private and secure, as it never leaves the local environment. FL is particularly valuable in scenarios where data privacy and security are critical, such as healthcare, finance, and agriculture.

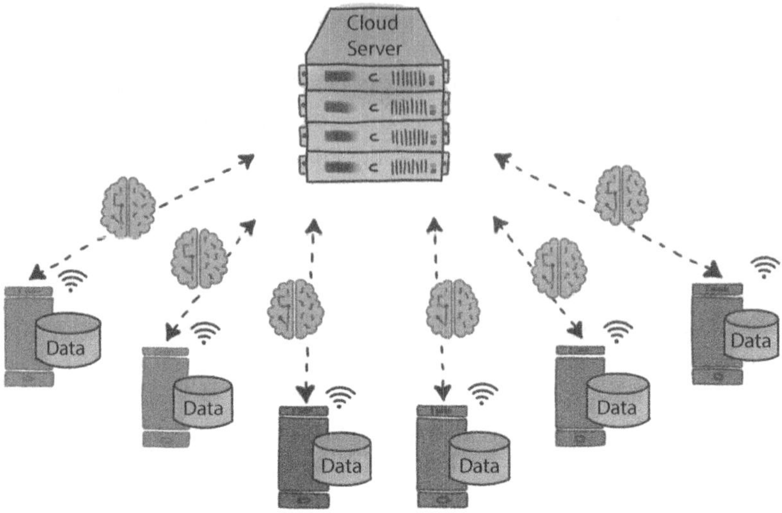

Figure 6.1 Schematic illustration of a federated learning architecture.

In agriculture, FL is revolutionizing the way farmers tackle traditional challenges like crop overproduction, price volatility, and resource inefficiency. By integrating FL, agricultural systems can utilize diverse data sources—such as Internet-of-Things (IoT) sensors, market trends, weather data, and farmer inputs—without centralizing the sensitive data of individual farmers. This approach helps create personalized crop recommendations that consider local environmental conditions and real-time market demands, enabling more precise decision-making. FL also allows for continuous model refinement through iterative updates, ensuring that the system adapts to changing conditions and farmer feedback. This decentralized method improves privacy, scalability, and accuracy, fulfilling modern agricultural needs like sustainable farming, market stability, and reduced food wastage, while providing tailored solutions to both small and large-scale farmers.

6.2 Literature Review

A literature survey was conducted to review the existing works in the field. The following summary presents the key findings from papers related to the proposed research.

Dhillon and Moncur (2023) highlight the critical role that smallholder farmers play in global food security, especially in developing countries. It outlines various challenges facing small-scale farming, such as limited access to technology, infrastructure, and financial resources, exacerbated by climate change, water scarcity, and soil degradation [13]. Although technological advancements like precision agriculture [e.g., unmanned aerial vehicles (UAVs), Internet of Things (IoT), and artificial intelligence] have transformed large-scale farming, smallholders often find these technologies economically inaccessible. The paper also reviews opportunities for technological adoption among smallholders, identifying areas such as irrigation automation, robotics, and smartphone-based applications as promising tools that could help improve efficiency and productivity. The authors conclude that overcoming barriers like market access, education, and infrastructure is crucial for small-scale farming to thrive and contribute to sustainable food systems.

Patil *et al.* (2023), in their work toward Sustainable Development Goal 2, propose a machine learning-based approach to forecasting the prices of key agricultural commodities, including wheat, millet, sorghum, maize, and rice. They introduce a hybrid model, Hybrid SARIMA-LSTM (HySALS), which combines Seasonal autoregressive integrated moving average (SARIMA) and long short-term memory (LSTM) algorithms to analyze price dynamics using data from 2005 to 2022 and forecast prices until 2030 [14]. Focusing on developing nations, the study emphasizes that accurate price predictions can mitigate market volatility and support better decision-making for stakeholders, contributing to global food security and Sustainable Development Goal 2 (Zero Hunger). The authors stress the importance of forecasting in balancing supply and demand and mitigating the effects of natural disasters, political unrest, and economic crises on agricultural markets.

Kumar (2022) provides a comprehensive analysis of the shifts in agricultural growth and cropping patterns in India from 1967–1968 to 2020–2021. The study highlights a significant decline in the area under food grains, particularly coarse cereals, and a shift toward high-value crops (HVCs) such as fruits, vegetables, oilseeds, and spices [15]. While the contribution of cereals to the total value of output has decreased, the share of HVCs has grown, signaling an increasing importance of these crops in India's agriculture. The analysis also points out that, despite technological advancements like improved irrigation and fertilizer use, no crop registered a long-term growth rate above 4% per annum. The paper emphasizes the need for crop diversification, particularly toward pulses, oilseeds, and horticulture, to enhance farmer incomes and address challenges like irrigation, credit, and market access. The study concludes by advocating for improved productivity among small and marginal farmers through technological innovations and integrated farming systems, stressing the role of government policies like MSP in facilitating diversification and sustainable growth.

Shams *et al.* (2024) present an innovative approach to enhancing crop recommendation systems through the integration of explainable artificial intelligence (XAI). Their study introduces the XAI-CROP algorithm, which improves the transparency and interpretability of traditional machine learning models used in agricultural decision-making. By using models like Gradient Boosting, Decision Trees, and Random Forests, the study evaluates key performance metrics such as mean squared error (MSE), mean absolute error (MAE), and R-squared (R^2). XAI-CROP outperforms conventional methods, achieving an MSE of 0.9412 and an R^2 value of 0.94152, signifying a significant improvement in accuracy and interpretability [16]. The authors emphasize that the transparency provided by XAI helps build trust among farmers by explaining the rationale behind crop recommendations, which is crucial for optimizing crop yields under changing environmental conditions. Additionally, this system contributes to sustainable agricultural practices by providing data-driven insights into crop selection and resource management.

Kumar *et al.* (2022) discuss the significance of digital literacy in Indian agriculture, emphasizing both the opportunities and challenges it presents. Their study highlights the critical role that digital platforms play in enhancing agricultural productivity, processing, and marketing, particularly during the COVID-19 pandemic when traditional modes of communication and support were restricted. Despite India's agricultural prominence, digital literacy among its farming population remains low due to factors such as technophobia, inadequate infrastructure, and economic barriers. However, the increased use of smartphones and other digital technologies is gradually bridging the gap [17]. The authors underscore that initiatives like Digital India, along with mobile and internet penetration, can potentially transform Indian farming by providing timely access to information and enhancing efficiency. Nevertheless, challenges such as the digital divide, lack of awareness, and the limited reach of digital literacy programs still hamper widespread adoption. The paper concludes with lessons learned during the pandemic, advocating for increased digital literacy to make Indian agriculture more resilient and competitive.

Žalik and Žalik (2023) provide a comprehensive review of FL in agriculture, exploring its potential to enhance smart farming while maintaining data privacy by training machine learning models directly on edge devices [18]. The authors compare different FL methods such as horizontal, vertical, and hybrid FL, focusing on their data partitioning, architectures, and aggregation algorithms like FedAvg. They highlight the benefits of FL in improving predictions in agricultural tasks like crop yield estimation, disease detection, and resource management. Despite the advantages, the paper identifies communication bottlenecks and the need for efficient aggregation algorithms as key challenges for scaling FL systems in agriculture. The study emphasizes the importance of addressing these issues through methods like sparsification and quantization to reduce communication costs, which are crucial for real-world agricultural applications. The review serves as a valuable reference for understanding the growing role of FL in the agricultural sector and provides insights into future research directions.

Kamal and Bablu (2023) explore the transformative role of mobile apps in enhancing agricultural productivity and livelihoods among smallholder farmers. The research highlights how mobile applications provide real-time access to crucial information such as weather updates, crop prices, and best farming practices, leading to informed decision-making and improved crop yields [19]. Also, the study examines how mobile apps revolutionize market linkages by connecting farmers directly with buyers, reducing reliance on intermediaries, and promoting price transparency, which boosts farmers' income. Financial inclusion is another key impact, with mobile banking and digital payment services enabling efficient access to credit and formal financial systems. Moreover, mobile apps support improved resource management through tools for tracking inputs like water and fertilizer, resulting in more sustainable practices. The study concludes that these technologies not only improve agricultural productivity but also foster rural development and sustainable livelihoods by expanding access to essential services and resources for smallholder farmers in developing countries.

6.3 Proposed Federated Learning–Based Crop Recommendation System Conceptual Approach

6.3.1 Conceptual Overview

The proposed FL-based system aims to empower farmers by providing personalized, data-driven crop recommendations while respecting their data privacy [20]. The system integrates various data sources, including IoT devices, smartphones, and farmers' input, to create a dynamic and real-time model that adapts to individual farm conditions and market trends. The FL approach ensures that the data stays local, only sharing model updates rather than raw data [21–23]. This not only preserves privacy but also allows the system to scale across diverse regions.

6.3.2 Data Collection and Integration

6.3.2.1 Farmer Inputs

The system collects historical data directly from farmers, including

- Crop yield history
- Soil quality and type
- Weather conditions specific to their region
- Available technologies (machinery, irrigation systems, etc.)
- Farmer capabilities (capital, workforce, land area, etc.)

6.3.2.2 External Data Sources

- **Weather Data:** Collected *via* IoT sensors, local weather stations, and satellite imagery to provide real-time updates on weather conditions, which are crucial for crop selection.
- **Market Demand Analysis:** Data on current food trends, consumer preferences, and market demand is gathered from smartphones, e-commerce platforms, and social media to predict which crops are likely to be profitable.
- **Food Trends:** By analyzing data from social media and food delivery apps, the system identifies shifts in consumer preferences (e.g., increasing demand for health supplements, or certain cuisines).
- **IoT Devices:** Sensors installed on farms (soil moisture sensors, temperature sensors, etc.) provide real-time data about environmental conditions.

6.3.3 Federated Learning Model Training

6.3.3.1 Local Model Training

- The system begins by training a local model on each farmer's device or a local server using their specific data.
- This local model learns the conditions specific to that farm, such as soil quality, weather patterns, and historical crop yields.
- This process respects data privacy, as the raw data never leaves the local environment.

6.3.3.2 Model Aggregation

- Once the local models are trained, only the model parameters (not the data itself) are shared with a central server.
- The central server aggregates these parameters to create a global model, which benefits from the collective knowledge of all participating farms [24].

6.3.3.3 Iterative Improvement

- The global model is sent back to the local devices, where it continues to learn and improve with each iteration.
- This iterative process ensures that the model becomes increasingly accurate and tailored to the specific needs of each farm.

6.3.4 Personalized Crop Recommendations

6.3.4.1 Regional and Local Recommendations

- The system provides personalized crop recommendations by analyzing the farmer's specific data and the aggregated data from neighbouring farms.
- It recommends crops that are well-suited to the region's climate, soil, and market demand.
- For instance, if a trend shows that a particular crop is becoming popular in the market, but there is a risk of overproduction, then the system will suggest alternative crops to avoid market saturation.

6.3.4.2 Consideration of Farmer Capabilities

- The system takes into account each farmer's resources, such as land size, access to storage and transportation, and financial capacity [25].
- For example: Small farmers, with limited storage and transportation facilities: the system might recommend growing crops like leafy vegetables that can be sold quickly in local markets, reducing spoilage. Large farmers, with better access to transportation and storage: the system may suggest crops that are in demand in other regions or states, which can be stored and transported efficiently.

6.3.5 Market and Consumer Trend Integration

6.3.5.1 Real-Time Market Analysis

- The system constantly monitors market trends and consumer demands to adjust recommendations in real time.
- If a new health trend emerges, like an increase in demand for superfoods, then the system will notify farmers who can grow the required crops, providing them with a potential market advantage.

6.3.5.2 Price Prediction and Risk Management

- The system predicts potential price fluctuations by analyzing historical price data, market saturation risks, and demand-supply dynamics [26].
- It provides recommendations that minimize risks, such as diversifying crops or shifting to less volatile markets.

6.3.6 Avoiding Market Saturation and Overproduction

6.3.6.1 Collaborative Crop Planning

- The system encourages farmers within the same region to diversify their crop choices to avoid the risk of overproduction, which can lead to price drops.
- It tracks what other farmers in the region are planting and suggests alternative crops that have a stable or growing demand.

6.3.6.2 Regional Crop Diversification

- By recommending a variety of crops within a region, the system ensures a balanced production that meets market demand without oversupply.
- This helps maintain stable prices and reduces waste due to overproduction.

6.3.7 Storage, Transportation, and Market Targeting

6.3.7.1 Tailored Recommendations Based on Infrastructure

- The system analyzes the farmer's available infrastructure, such as storage facilities and transportation options.
- It then recommends crops that align with these capabilities like farmers with limited infrastructure; recommends crops with a short shelf life that can be quickly sold in local markets, reducing the need for long-term storage and extensive transportation and farmers with advanced infrastructure; and recommends crops with longer shelf lives and higher value, suitable for storage and transportation to larger markets or even export.

6.3.8 Feedback Loop and Continuous Improvement

6.3.8.1 Farmer Feedback Collection

- After each harvest, the system collects feedback from farmers on the effectiveness of the recommendations.
- This feedback is crucial for refining the model and improving future recommendations.

6.3.8.2 Continuous Learning

- The system uses the feedback and new data to update the model continuously, ensuring that it remains relevant and effective over time.
- This adaptive learning approach allows the system to evolve with changing market conditions, technological advancements, and environmental factors.

6.4 Workflow for the Proposed System

The workflow for FL-based crop recommendation system (seven-layered approach) is shown in Figure 6.2.

Our conceptual model of the crop recommendation system using FL can be implemented by using the seven-layered approach (shown in Figure 6.2), where each layer has its own predefined function. The first layer, data collection and input layer, gathers data from various sources like IoT devices,

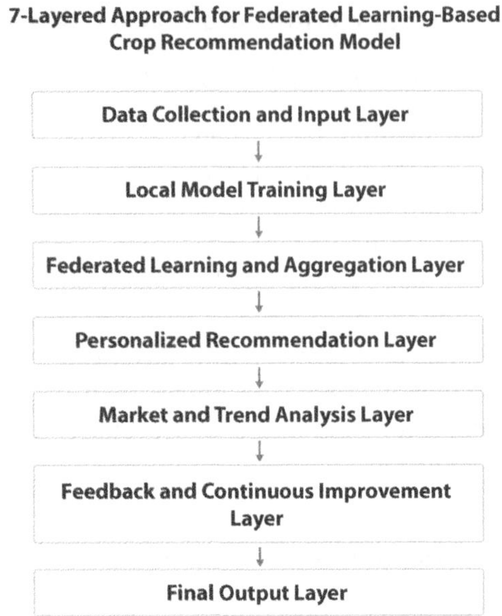

Figure 6.2 Seven-layered approach toward federated learning–based crop recommendation model.

weather information, and soil conditions, ensuring diverse and localized inputs. Next, the local model training layer uses this data to train individual models locally at each farming site, preserving data privacy. The federated learning and aggregation layer then collects model updates from various farms and aggregates them to create a global model without sharing sensitive data. Following this, the personalized recommendation layer provides tailored crop recommendations to farmers based on both local data and global insights from the FL process. The market and trend analysis layer incorporates real-time market conditions and pricing trends, ensuring that the recommendations are aligned with market demands. The feedback and continuous improvement layer collects feedback after each farming cycle to continually refine the model and enhance its accuracy. Finally, the final output layer presents farmers with optimized, actionable crop recommendations, thereby improving agricultural outcomes and market efficiency.

6.4.1 Data Collection and Input Layer

This layer has two types of function as shown in Figure 6.3 and is explained below.

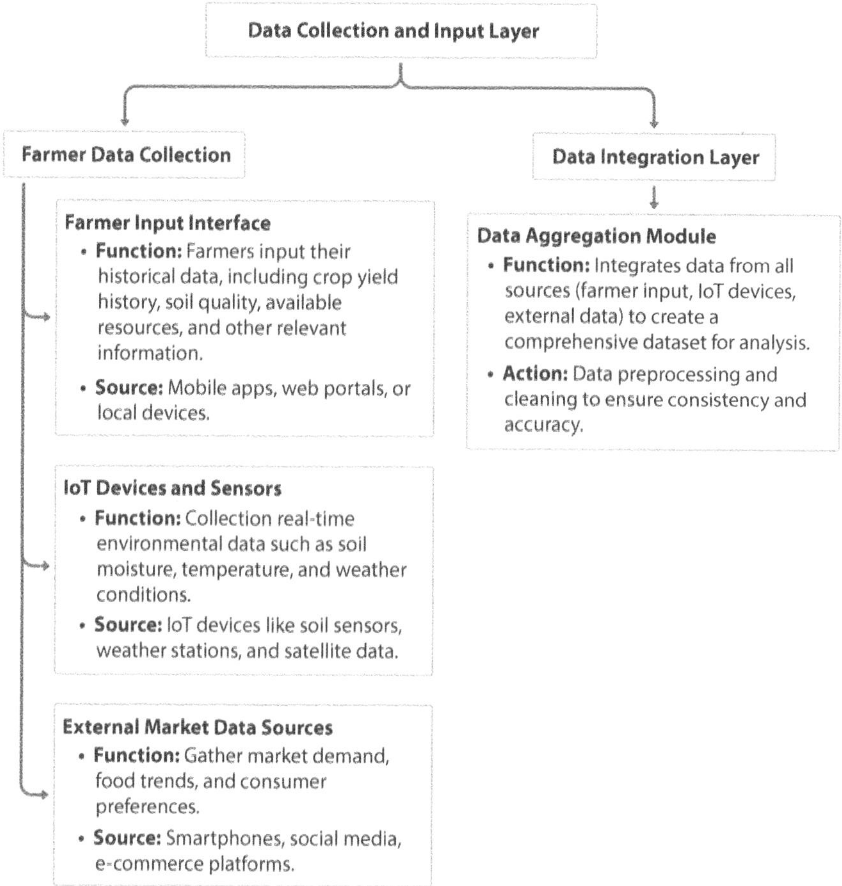

Figure 6.3 Function chart of layer 1, data collection and input layer.

6.4.1.1 *Farmer Data Collection*

This part collects data from various sources which are given below.

6.4.1.1.1 Farmer Input Interface

- **Purpose:** This interface serves as the primary means for farmers to input their data. It could be a mobile app, a web portal, or a local device where farmers can log information about their agricultural practices.
- **Data Collected:** Historical crop yields, soil quality, types of fertilizers used, irrigation practices, and any other relevant data that can influence crop production.

- **Importance:** This data provides a personalized baseline for the recommendations, ensuring that each farmer receives advice tailored to their unique situation.

6.4.1.1.2 IoT Devices and Sensors

- **Purpose:** IoT devices are used to gather real-time environmental data directly from the farm [27].
- **Data Collected:** Soil moisture levels, temperature, humidity, and other critical environmental parameters.
- **Importance:** These real-time metrics help the model understand current growing conditions, allowing for timely and context-aware recommendations.

6.4.1.1.3 External Market Data Sources

- **Purpose:** To stay relevant, the model needs to incorporate data on market trends and consumer behavior.
- **Data Collected:** Popular food trends, market demand for specific crops, and consumer preferences (e.g., increased demand for organic or health-related products).
- **Importance:** This information ensures that the recommendations are aligned with current and future market demands, helping farmers make profitable decisions.

6.4.1.2 Data Integration Layer

This part integrates the collected data to create a dataset.

6.4.1.2.1 Data Aggregation Module

- **Purpose:** This module integrates all the collected data (farmer input, IoT data, and market data) into a unified dataset.
- **Process:** The data goes through preprocessing steps such as cleaning, normalization, and validation to ensure consistency and accuracy.
- **Importance:** Effective data integration is crucial for creating a reliable and holistic model that can generate accurate recommendations.

6.4.2 Local Model Training Layer

This layer working is divided into two parts as shown in Figure 6.4 and is explained below.

6.4.2.1 Local Model Development

This part develops and trains the model on the local data.

6.4.2.1.1 Local Model Training

- **Purpose:** Each farmer's data is used to train a local machine learning model on their own device. This model learns patterns specific to that farmer's conditions.
- **Types of Models:** The model could be a regression, classification, or any other type depending on the data and the specific use case.
- **Privacy:** Because the data never leaves the farmer's device, their privacy is preserved.
- **Importance:** This approach ensures that the model is personalized and can provide highly relevant recommendations based on individual circumstances.

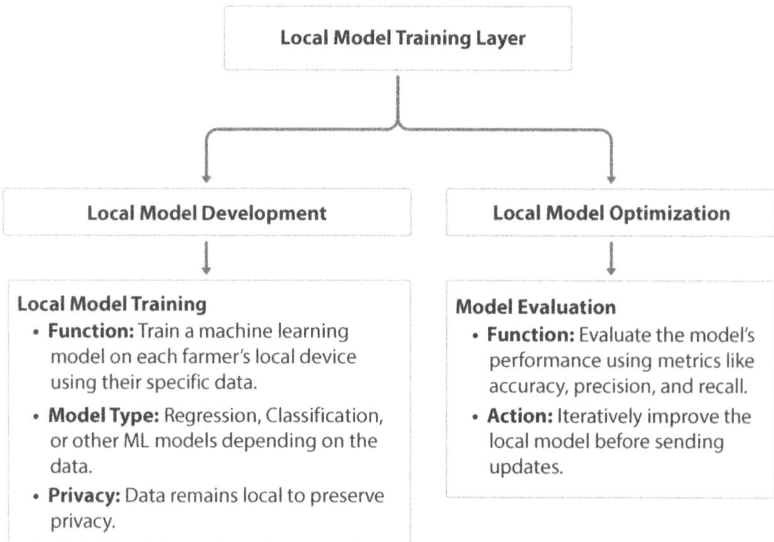

Figure 6.4 Function chart of layer 2, local model training layer.

6.4.2.2 Local Model Optimization

This part evaluates and optimizes the model performance.

6.4.2.2.1 Model Evaluation

- **Purpose:** After training, the model is evaluated to check its performance. Metrics such as accuracy, precision, and recall, are used to measure how well the model predicts outcomes.
- **Process:** The model is iteratively improved until it meets predefined performance standards.
- **Importance:** A well-optimized model is essential for making accurate and actionable recommendations.

6.4.3 Federated Learning and Aggregation Layer

This layer is divided into three parts based on their functions as shown in Figure 6.5 and is explained below.

6.4.3.1 Model Parameter Sharing

This part focuses on sending the local model parameters to the central server.

6.4.3.1.1 Federated Model Update

- **Purpose:** Instead of sharing raw data, the local models share their learned parameters (e.g., weights) with a central server.
- **Process:** The central server receives these updates from multiple local models across different farms.
- **Privacy:** Because only model parameters are shared, not the data itself, farmers' privacy is protected.
- **Importance:** This allows the system to learn from a broad dataset across many farms without compromising individual privacy.

6.4.3.2 Global Model Aggregation

It focuses on aggregating local model parameters from all the devices to create a global model.

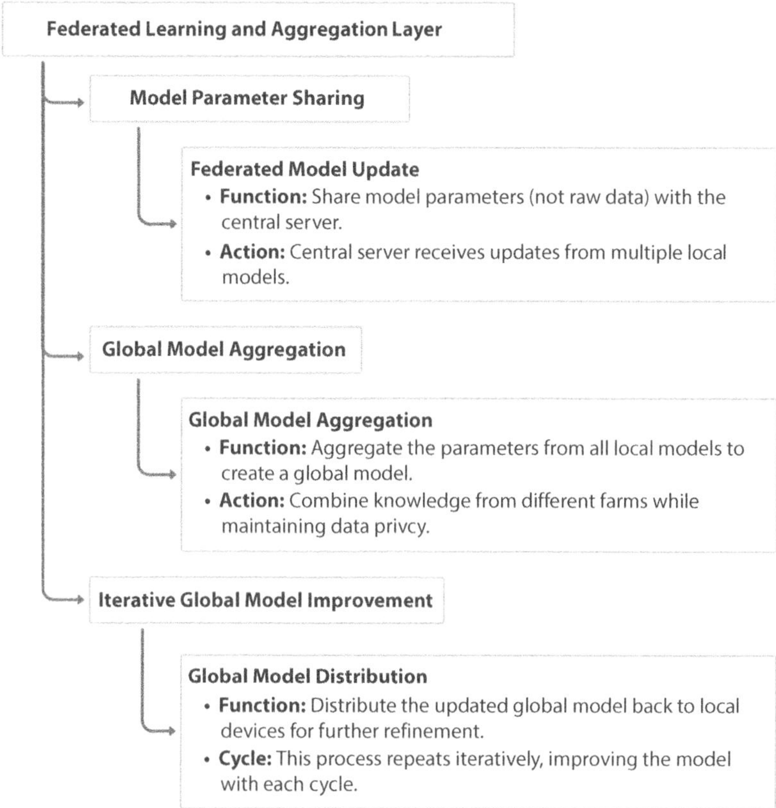

Figure 6.5 Function chart of layer 3, federated learning and aggregation layer.

- **Purpose:** The central server aggregates the parameters from all the local models to create a more generalized global model.
- **Process:** The aggregation process combines the knowledge from different farms to create a robust model that benefits from diverse data sources.
- **Importance:** The global model captures broader patterns and trends, enhancing the accuracy and applicability of the recommendations.

6.4.3.3 *Iterative Global Model Improvement*

It focuses on global model improvement by updating it and distributing it to all the local devices.

6.4.3.3.1 Global Model Distribution

- **Purpose:** The updated global model is then sent back to the local devices, where it is further refined using the farmer's local data.
- **Cycle:** This process is iterative, with each cycle improving the model's accuracy and relevance.
- **Importance:** Continuous improvement ensures that the model remains up-to-date and effective in changing conditions.

6.4.4 Personalized Recommendation Layer

This layer is divided into three parts based on their type of recommendation given to the farmers as shown in Figure 6.6 and is explained below.

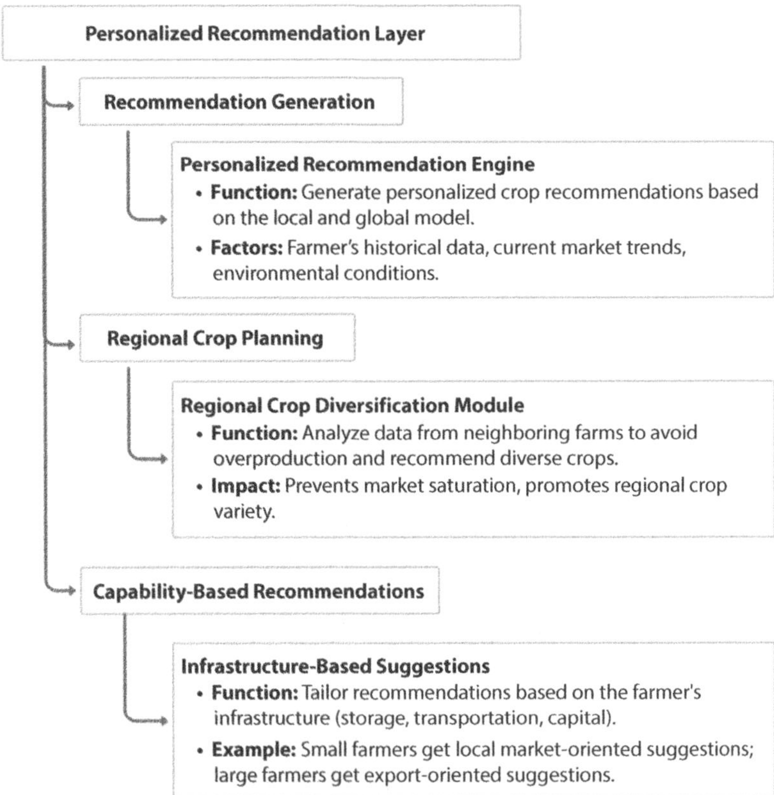

Figure 6.6 Function chart of layer 4, personalized recommendation layer.

6.4.4.1 Recommendation Generation

This part focuses on generating crop recommendation based on general factors, which can affects the crop production.

6.4.4.1.1 Personalized Recommendation Engine

- **Purpose:** This engine uses the trained model to generate crop recommendations tailored to each farmer's specific conditions and needs.
- **Factors Considered:** The engine takes into account factors such as the farmer's historical data, current market trends, and real-time environmental data.
- **Importance:** Personalized recommendations help farmers make better decisions that are aligned with both their capabilities and market opportunities.

6.4.4.2 Regional Crop Planning

This part generates the crop recommendation based on their region.

6.4.4.2.1 Regional Crop Diversification Module

- **Purpose:** This module analyzes data from multiple farms within the same region to avoid overproduction of a single crop.
- **Process:** If many farmers in a region are inclined to grow the same crop, then the system suggests alternative crops to ensure market balance.
- **Importance:** Prevents market saturation, reduces price drops, and encourages biodiversity.

6.4.4.3 Capability-Based Recommendations

This part generates the personalized crop recommendation based on an individual farmers' capability.

6.4.4.3.1 Infrastructure-Based Suggestions

- **Purpose:** The system recommends crops based on the farmer's infrastructure capabilities, such as storage, transportation, and available capital [28].

- **Examples:** Small farmers with limited resources might be advised to grow crops with a short shelf-life and target local markets and large farmers with better infrastructure might be encouraged to grow export-oriented crops.
- **Importance:** Tailored recommendations based on infrastructure help in reducing food waste, optimizing transportation, and ensuring that both small and large farmers can succeed.

6.4.5 Market and Trend Analysis Layer

This layer further divided into two parts based on the function they focus on as shown in Figure 6.7 and is explained below.

6.4.5.1 Market Dynamics Integration

This part works on gathering real-time data based on current market and updates recommendations according to it.

6.4.5.1.1 Market Trend Analyzer

- **Purpose:** This module continuously monitors market trends, consumer preferences, and price fluctuations in real time.

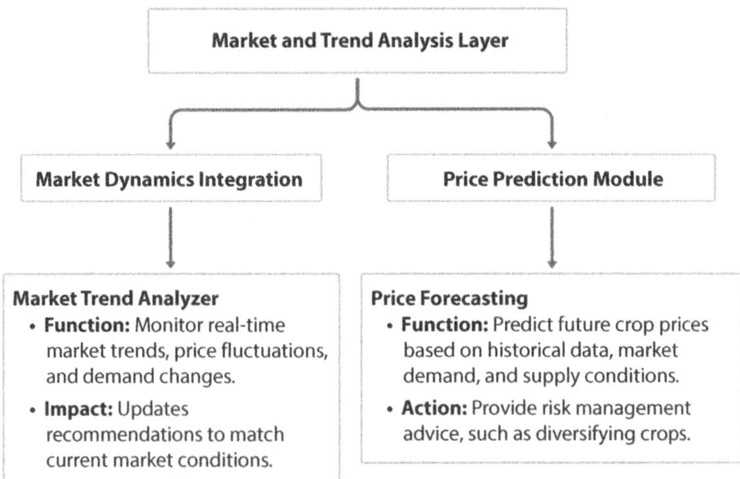

Figure 6.7 Function chart of layer 5, market and trend analysis layer.

- **Process:** It updates the recommendation engine with the latest market data to ensure that the recommendations stay relevant.
- **Importance:** Ensures that farmers receive advice that is in line with current and future market demands.

6.4.5.2 Price Prediction Module

This part focuses on prediction of crop price.

6.4.5.2.1 Price Forecasting

- **Purpose:** Predicts future prices for various crops based on historical data, current supply and demand, and market trends.
- **Action:** Provides farmers with advice on managing risks, such as when to diversify crops or enter into forward contracts.
- **Importance:** Helps farmers plan their production strategy to maximize profits and minimize risks.

6.4.6 Feedback and Continuous Improvement Layer

This layer is divided into two parts based on their functions as shown in Figure 6.8 and is explained below.

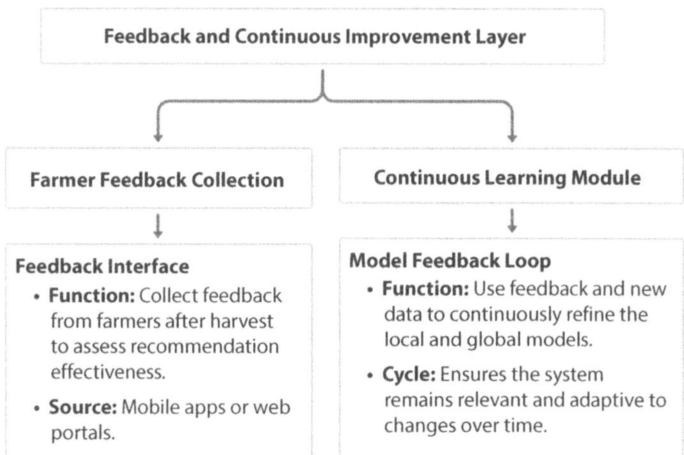

Figure 6.8 Function chart of layer 6, feedback and continuous improvement layer.

6.4.6.1 *Farmer Feedback Collection*

This part focuses on collection of farmers' feedback on models' performance to further improve its performance.

6.4.6.1.1 Feedback Interface

- **Purpose:** After each growing season or harvest, farmers provide feedback on the recommendations they followed.
- **Source:** Feedback can be collected *via* mobile apps, web portals, or other interfaces.
- **Importance:** Collecting feedback is critical for assessing the effectiveness of the recommendations and making necessary adjustments.

6.4.6.2 *Continuous Learning Module*

This part uses feedback and the new data to further refine model.

6.4.6.2.1 Model Feedback Loop

- **Purpose:** This module uses the feedback and new data to continuously refine and improve both the local and global models.
- **Cycle:** This ensures that the system evolves with changing conditions and remains effective over time.
- **Importance:** Continuous learning is essential for maintaining the model's relevance and accuracy.

6.4.7 Final Output Layer

This layer is further divided into two parts based on their function as shown in Figure 6.9 and is explained below.

6.4.7.1 *Final Recommendations*

This part focuses on delivering the crop recommendation to the farmers.

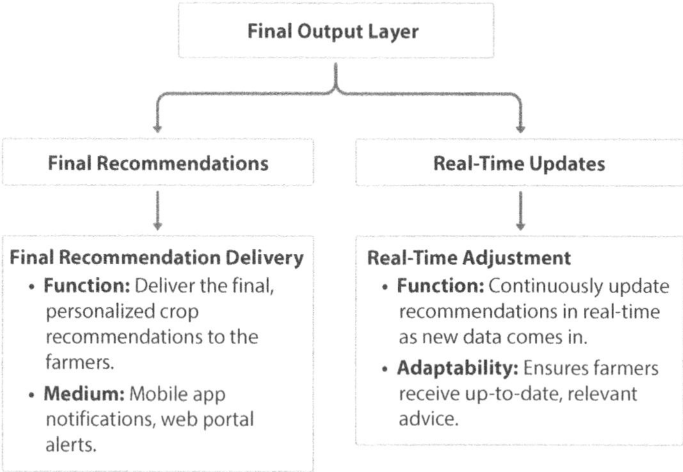

Figure 6.9 Function chart of layer 7, final output layer.

6.4.7.1.1 Final Recommendation Delivery

- **Purpose:** After processing all the data and running the models, the final crop recommendations are delivered to the farmers.
- **Medium:** Notifications can be sent *via* mobile apps, SMS, or other digital platforms.
- **Importance:** Ensures that farmers receive actionable advice they can implement immediately.

6.4.7.2 Real-Time Updates

This part focuses on updating the model's recommendations based on real-time data.

6.4.7.2.1 Real-Time Adjustment

- **Purpose:** The system continuously updates recommendations based on new data and market conditions.
- **Adaptability:** This ensures that farmers receive the most current and relevant advice, even as conditions change.
- **Importance:** Real-time updates are crucial for adapting to sudden changes in weather, market demand, or other factors.

6.5 Conclusion and Future Scopes

The research began by addressing key challenges in Indian agriculture, including crop overproduction, price volatility, and market misalignment, which lead to significant losses for farmers and instability in the agricultural market. The study identified a critical need for personalized solutions that can assist farmers in making data-driven decisions while maintaining data privacy. Recognizing the growing potential of FL, the research proposed a decentralized machine learning approach to offer personalized crop recommendations to farmers, aligning their production with real-time market demand and reducing food wastage. The conceptual approach of the proposed FL system revolves around the integration of diverse data sources, such as IoT devices, farmer inputs, and external market data, to create a dynamic and real-time model. The system begins with local data training on individual farms, ensuring that data privacy is maintained, while only model updates are shared with a central server for aggregation. The iterative process refines the global model, which is redistributed back to the local devices, creating a feedback loop that allows for continuous improvement. The workflow emphasizes personalized crop recommendations based on factors like regional climate, soil quality, and individual farmer capabilities, ensuring that both small and large farmers benefit from the system's tailored advice. Additionally, the model incorporates market trends and consumer preferences, helping farmers avoid overproduction and maximize profits.

Looking ahead, the future scope of this research includes enhancing the FL model by integrating blockchain technology for improved data security and transparency. The system can be expanded to cover more regions and integrate more advanced IoT data, improving precision in environmental monitoring. Predictive analytics for emerging market trends and climate-related risks can provide even greater value to farmers. Additionally, incorporating government policies, subsidies, and support systems within the model will further drive inclusivity and sustainability, enabling the system to adapt to the evolving needs of the agricultural sector globally.

AI disclosure statement: The author acknowledges the use of AI-based tools for minor language editing or grammatical correction. However, all intellectual contributions, analyses, and conclusions presented in this chapter are entirely the author's own.

References

1. Millennium Post, India Wastes Maximum of 16% Fruits, Veggies: Study. Millennium Post, August 12, 2015, 2015. Retrieved from https://www.millenniumpost.in/india-wastes-maximum-of-16-fruits-veggies-study-101329. Accessed September 5, 2024.
2. Tiwari, A., Harischander, H., Rane, M.V., Cold Storage in India for Small Farmers—Current Status and Challenges, *International Refrigeration and Air Conditioning Conference,* Purdue, 2022, https://docs.lib.purdue.edu/iracc/2472.
3. Poor Post-Harvest Storage, Transportation Facilities to Cost Farmers Dearly. Down to Earth, July 5, 2018, 2018. Retrieved from https://www.downtoearth.org.in/agriculture/poor-post-harvest-storage-transportation-facilities-to-cost-farmers-dearly-61047. Accessed September 5, 2024.
4. Press Information Bureau, Post Harvest Loss of Fruits and Vegetables. Press Information Bureau, March 24, 2023, 2023. Retrieved from https://pib.gov.in/PressReleasePage.aspx?PRID=1910360. Accessed September 5, 2024.
5. Nedumaran, S., *et al.,* Digital Integration to Enhance Market Efficiency and Inclusion of Smallholder Farmers: A Proposed Model for Fresh Fruit and Vegetable Supply Chain, *Int. Food Agribus. Manag. Rev.,* 23, 3, 319–337, September 2020, https://doi.org/10.22434/IFAMR2019.0165.
6. Barvin, A.P. and Sampradeepraj, T., Crop recommendation systems based on soil and environmental factors using graph convolution neural network: A systematic literature review. *Eng. Proc.*, 58, 1, 97, 2023. https://doi.org/10.3390/ecsa-10-16010.
7. M.T., Makkithaya, K., V.G.,.N., A Federated Learning-Based Crop Yield Prediction for Agricultural Production Risk Management. *2022 IEEE Delhi Section Conference (DELCON)*, pp. 1–7, 2022.
8. Dwarampudi, A. and Yogi, M.K., Application of Federated Learning for Smart Agriculture System. *Int. J. Inf. Technol. Comput. Eng.*, 4, 03, 36–48, 2024.
9. Nguyen, H., Randall, M., Lewis, A., Factors affecting crop prices in the context of climate change—A review. *Agriculture*, 14, 1, 135, 2024.
10. World Economic Forum, Agri-tech innovation can improve value capture, transform ecosystem for India's small farmers. World Economic Forum, 2021, June 17. https://www.weforum.org/stories/2021/06/agri-tech-innovation-can-improve-value-capture-and-transform-ecosystem-for-india-s-small-farmers/. Accessed November 16, 2024.
11. Idoje, G., Dagiuklas, T., Iqbal, M., Federated Learning: Crop classification in a smart farm decentralised network. *Agric. Technol.*, 2, 100277, 2023.
12. Kondaveeti, H.K., Sai, G.B., Athar, S.A., Vatsavayi, V.K., Mitra, A., Ananthachari, P., Federated Learning for Smart Agriculture: Challenges and Opportunities. *2024 Third International Conference on Distributed Computing and Electrical Circuits and Electronics (ICDCECE)*, pp. 1–7, 2024.

13. Dhillon, R. and Moncur, Q., Small-scale farming: A review of challenges and potential opportunities offered by technological advancements. *Sustainability*, 15, 21, 15478, 2023. https://doi.org/10.3390/su152115478.
14. Patil, A., Shah, D., Shah, A., Kotecha, R., Forecasting prices of agricultural commodities using machine learning for global food security: Towards sustainable development goal 2. *Int. J. Eng. Trends Technol.*, 71, 12, 277–291, 2023.
15. Kumar, V., Five decades of India's agricultural growth across crops: Emerging trends and patterns. National Bank for Agriculture and Rural Development, 2022. Retrieved from https://www.nabard.org/auth/writereaddata/tender/2609222521five-decades-of-indias-agricultural-growth-across-crops-emerging-trends-and-patterns.pdf. Accessed September 5, 2024.
16. Shams, M.Y., Gamel, S.A., Talaat, F.M., Enhancing crop recommendation systems with explainable artificial intelligence: a study on agricultural decision-making. *Neural Comput. Applic.*, 36, 5695–5714, 2024.
17. Kumar, R., Mohan, G., Narayanappa, N., Ranganath, G., Boraiah, B., Digital literacy in Indian farming: Opportunities, challenges and lessons learnt during Covid-19 pandemic, 2022. Authors Contribution. Retrieved from https://www.researchgate.net/publication/364223964_Digital_Literacy_in_Indian_Farming_Opportunities_Challenges_and_Lessons_Learnt_during_Covid-19_Pandemic_AUTHORS_CONTRIBUTION#fullTextFileContent. Accessed September 5, 2024.
18. Žalik, K.R. and Žalik, M., A review of federated learning in agriculture. *Sens. (Basel Switzerland)*, 23, 23, 9566, 2023.
19. Kamal, M. and Bablu, T.A., Mobile applications empowering smallholder farmers: An analysis of the impact on agricultural development. *Int. J. Soc. Anal.*, 8, 6, 36–52, 2023.
20. Gaur, A.K., *et al.*, Federated learning methods for collaborative multiclass classification of dry beans. *Genet. Resour. Crop Evol.*, 72, 1421–1439, 2025. https://doi.org/10.1007/s10722-024-02083-3.
21. Kabala, Mamba D., *et al.*, Image-based crop disease detection with federated learning. *Sci. Rep.*, 13, 19220, 2023. , https://doi.org/10.1038/s41598-023-46218-5.
22. Le, D.-D., Dao, M.-S., Tran, A.-K., Nguyen, T.-B., Le-Thi, H.-G., Federated Learning in Smart Agriculture: An Overview. *2023 15th International Conference on Knowledge and Systems Engineering (KSE)*, Hanoi, Vietnam, pp. 1–4, 2023.
23. Kumar, M., Aeri, M., Chandel, R., Kukreja, V., Mehta, S., Federated Learning CNN for Smart Agriculture: A Modeling for Soybean Disease Detection. *2024 IEEE International Conference on Interdisciplinary Approaches in Technology and Management for Social Innovation (IATMSI)*, Gwalior, India, pp. 1–6, 2024.

24. Fendor, Z., van der Velden, B.H.M., Wang, X., Carnoli, A.J., Mutlu, O., Hürriyetoğlu, A., Federated learning in food research. Retrieved from https://arxiv.org/abs/2406.06202. Accessed September 8, 2024, 2024.
25. Rai, M., *et al.*, Advancement in Agricultural Techniques With the Introduction of Artificial Intelligence and Image Processing, in: *Smart Village Infrastructure and Sustainable Rural Communities,* edited by Mohammad Ayoub Khan, et al., pp. 47–68. Hershey, PA: IGI Global Scientific Publishing, 2023. https://doi.org/10.4018/978-1-6684-6418-2.ch004.
26. Tripathi, P., Kumar, N., Rai, M., Khan, A., Applications of Deep Learning in Agriculture, in: *Artificial Intelligence Applications in Agriculture and Food Quality Improvement*, M. Khan, R. Khan, P. Praveen (Eds.), pp. 17–28, IGI Global Scientific Publishing, USA, 2022.
27. Tripathi, P., Kumar, N., Rai, M., Shukla, P.K., Verma, K.N., Applications of Machine Learning in Agriculture, in: *Smart Village Infrastructure and Sustainable Rural Communities*, M. Khan, B. Gupta, A. Verma, P. Praveen, C. Peoples (Eds.), pp. 99–118, IGI Global Scientific Publishing, USA, 2023.
28. Nath, B., Chen, G., O'Sullivan, C.M., Zare, D., Research and technologies to reduce grain postharvest losses: A review. *Foods*, 13, 12, 1875, 2024.

7

Federated Learning for Plant Disease Detection

Siddhartha Das[1]*, Sudipta Jana[1], Sudeepta Pattanayak[2], Pradipta Banerjee[3] and Sweety Maity[4]

[1]Department of Plant Pathology, M.S. Swaminathan School of Agriculture, Centurion University of Technology and Management, Paralakhemundi, Odisha, India
[2]Division of Plant Pathology, ICAR- Indian Agricultural Research Institute, New Delhi, India
[3]McGowan Institute for Regenerative Medicine, Department of Surgery, University of Pittsburgh, Pittsburgh, PA, USA
[4]Department of Molecular Biology and Biotechnology, Odisha University of Agriculture and Technology, Bhubaneswar, Odisha, India

Abstract
Federated learning represents an innovative approach in the field of machine learning, making it possible for distributed devices to train models together while safeguarding for privacy of data. This chapter explores the application of federated learning in the detection of diseases of plants, a vital responsibility for maintaining agricultural output and guaranteeing food security. Conventional machine learning techniques for identifying plant diseases necessitate extensive datasets, which are typically centralized. Creation of these datasets can be very difficult because of the diverse character of the agricultural environment and data privacy concerns. Federated learning tackles these issues by enabling individual devices, like smartphones used by farmers or edge sensors placed in the field, to conduct local training of models using their data, sharing only the updates of those models with a central server. Consequently, this server compiles these updates to develop a global model while never accessing the original data. Implementing federated learning for plant disease detection offers numerous benefits, such as enhancing data privacy, minimizing the necessity for data transmission, and utilizing diverse

**Corresponding author*: siddhartha.das10@gmail.com

Padmesh Tripathi, Bhanumati Panda, Shanthi Makka, Reeta Mishra, S. Balamurugan and Sheng-Lung Peng (eds.) Federated Learning for Smart Agriculture and Food Quality Enhancement, (165–192) © 2026 Scrivener Publishing LLC

datasets from different geographic areas, which strengthens the model's robustness and generalization capabilities. It additionally provides scalable, real-time monitoring for diseases, supporting prompt and precise identification for effective intervention and management. Challenges such as communication efficiency, model convergence, and handling non-independent and identically distributed data are discussed, with proposed solutions to optimize the federated learning framework for agricultural applications. The chapter discusses the potential of federated learning draft to revolutionize plant disease detection, fostering smarter and more resilient agricultural practices.

Keywords: Federated learning, agricultural productivity, plant disease, data transmission

7.1 Introduction

The cutting-edge method in machine learning (ML) called federated learning (FL) allows models to be trained across several dispersed devices or servers, each containing local data samples while maintaining data privacy by refraining from data exchange. This approach has recently emerged as a popular method for detecting plant diseases, providing a promising alternative to the limitations of conventional centralized ML techniques. Identifying plant diseases is essential to increase agricultural output and ensuring safeguards against hunger. Rapid intervention is made possible by accurate and timely identification of plant pathogens and diseases, lowering crop losses and decreasing dependence on harmful pesticides. Conventional techniques for detecting plant diseases using ML typically require large datasets to be aggregated in a central location, posing substantial threats to the security and privacy of data. FL tackles these issues by enabling individual farms or research institutions to jointly train a common model while retaining their data locally. This decentralized strategy improves data privacy while also utilizing a variety of datasets from various geographic areas, resulting in stronger and more generalized models. Furthermore, FL mitigates the issue of data scarcity, particularly in regions where data collection is challenging [1]. By facilitating the use of locally generated data, FL can empower farmers in remote or resource-limited areas to benefit from advanced plant disease detection technologies. The combination of FL and plant disease detection employs sophisticated methods like convolutional neural networks (CNNs) for classifying images, alongside secure aggregation protocols to maintain data privacy. This collaboration facilitates the creation of precise and effective disease detection models suitable for real-time applications, helping farmers make smart choices on crop management. The adoption of FL in this domain also opens up opportunities for continuous model improvement, as new data from different sources can be incorporated without compromising privacy. To summarize, employing FL in order to identify and characterize of phytopathogens and allied diseases marks a major advancement in the integration of artificial intelligence (AI) within agriculture, which promotes a food production system that is both sustainable and resilient. Agriculture is vital for human survival, and plant diseases present a major threat to crop yields and food security. Timely interventions and minimizing crop losses rely heavily on the early detection and precise diagnosis of plant diseases. Conventional approaches for identifying plant diseases, including surveys and lab analyses, have often been slow and relied heavily on specialists. The rise of digital agriculture has showcased how AI and ML can be utilized to enhance and streamline disease detection processes. Nonetheless, concerning centralized ML, it does consolidate significant amounts of data, leading to worries about the privacy and security of that information. One of the most promising solutions for these challenges would be FL, which has enabled decentralized training of ML models while preserving data privacy. This draft is precise informative content which focuses on utilizing FL to identify and characterize plant diseases, delving into its architecture, advantages, and challenges in agriculture.

7.2 Federated Learning

FL is a crucial strategy for collaboratively training a shared model by utilizing clients at the edge, while keeping data stored locally. This is done by training local models based on the data of each client, then aggregating model parameters—updates—from every client to form a global model on a central server. A number of advantages belong to such a paradigm: stronger privacy, lower costs of communication, and the possibility of using very diverse data sources without needing to store all the data centrally.

7.2.1 Key Components of Federated Learning

7.2.1.1 Clients

It refers to devices or entities that hold local data and are involved in the training process. In the realm of plant disease detection, potential clients could include particular farms, agricultural organizations, or research institutions.

7.2.1.2 Server

A globally central server entity for aggregating model updates from clients, maintaining the global model.

7.2.1.3 Local Training

Clients train models on their local data, generating local updates (gradients or model weights).

7.2.1.4 Model Aggregation

These updated models at the local level are then aggregated at the server into a new global model. Federated averaging (FedAvg) could be one of the most frequent techniques; that is, FedAvg.

7.2.1.5 Communication Protocol

Mechanisms for securely transmitting model updates between clients and the server.

7.3 Various Crop Diseases and Their Identification Strategies

FL was initially presented as a ML approach that operates over a network. Its techniques are formulated to educate algorithms on distributed devices while ensuring the local privacy of data. This approach represents the conventional, center-based, server-oriented methodology in FL. In essence, the process mainly entails collecting multiple local models and integrating them into a single global model, which subsequently intermingled each transaction cycle. The assemble process utilizes an aggregation algorithm which served foundation for uplifting updates from local models involved through centralized server. A way to aggregation that is frequently employed is FedAvg. The list is by no means comprehensive, and among the more recognized aggregation algorithms.

7.4 Tools Used in the Federated Learning

7.4.1 Federated Averaging (FedAvg)

A widely utilized technique in FL encompasses with FedAvg algorithm, which modify among multiple stratification and a model averaging streamline update through the server. FedAvg consists of combining local models into a global model by calculating a weighted average of each model's weights. This method does not share the actual data; only the model parameters are. This is among the most often used algorithms for FL. However, several empirical studies have revealed that only a few fine-tuning steps are required on the model output from FedAvg to achieve high test accuracy on new, unseen problems.

7.4.2 Federated Proximal (FedProx)

FedProx is another improved form of FedAvg, designed for dealing with some of the issues of heterogeneity in data and nonidentically distributed data [2]. This approach lessens the effect of devices that have uneven or very different data distributions.

7.4.3 Federated Stochastic Gradient Descent (Federated SGD)

In federated SGD, the aggregation process involves weighted averaging, where the weights are determined by the performance of each local model in relation to the correlation. In contrast to FedAvg, devices employing federated SGD do not have to divulge their new models after completing several local training cycles.

Utilization of Secure Multi-Party Computation Averaging (SMC-Avg)
SMC-Avg or Secure Multi-Party Computation Averaging: This aggregation method, introduced by Andrew Yao, is founded the principles of SMC-Avg. SMC simply means the process of keeping the inputs private while allowing several parties to compute a function. SMC-Avg is used in FL to safely aggregate model changes while keeping private individual device updates.

7.5 Advantages of Federated Learning to Identify Plant Diseases

Numerous methods of diagnosis and detection have been developed for crop diseases, which are regarded as abnormal conditions that impact a plant's growth, productivity, and quality [3, 4]. Recently, methods for determining disease outbreaks of plants utilizing deep learning (DL) and ML have been established [5, 6]. CNNs are typically used for plant disease characterization and detection [7, 8]. Specifically, VGG-16 and GoogLeNet have demonstrated accuracy rates of 91.04% and 91.8%, respectively, when applied to rice plants following data augmentation [9]. Additionally, VGG-16 and AlexNet have been utilized for analyzing tomato leaves; after hyperparameter adjustment, the results show 97.19% and 97.69% accuracy. A novel CNN architecture that has been tested on multiple plants and outperforms a number of pre-established models with up to 98.66% accuracy and less parameters was presented [10]. The vision transformer (ViT) addresses this concern. It achieves this by concentrating on all significant aspects of an image through the implementation of attention mechanisms [11]. This has produced impressive results in the classification of leaf diseases because it can handle photos at high resolution and understands the global context. It provides a novel approach for the detection of late blight in vineyards by fusing data and Transformer networks [12, 13]. It does this by utilizing three transformers to combine disparate data sources, such as weather and satellite photos. This creates a solid basis for promising results for the identification and categorization of foliar diseases through methods based on CNN and ViT, as evidenced by the numerous publications previously discussed. Nevertheless, none of these studies discusses the solution that is currently thought to be effective for creating more reliable general models: a distributed approach utilizing FL.

7.6 Data Collection and Preprocessing

In a FL setup for plant disease detection, data collection is decentralized. Various clients (e.g., farms and agricultural research centers) collect image data of plants showing symptoms of diseases. These images are often labeled with the type of disease or classified as healthy. Preprocessing steps include image augmentation, normalization, and potentially labeling with metadata (e.g., location and crop type) to enhance the learning process.

7.7 Model Training and Aggregation

7.7.1 Local Model Training

Each client will create a localized model utilizing their own dataset. In the present scenario, CNNs are often advantageous, having proven to be highly effective in identifying plant diseases through image classification. Clients can utilize pre-trained models (transfer learning) to improve performance, especially when dealing with small datasets.

7.7.2 Updating of Data, Model Construction, and Transmission

Once local training is completed, clients transmit model updates (such as weights) to the central server. To maintain privacy, methods such as differential privacy can be utilized to introduce noise into the updates.

7.7.3 Universal Model Aggregation

7.7.3.1 Federated Averaging (FedAvg)

The central server combines the incoming updates through techniques such as FedAvg. Following this, the global model is updated and sent back to the clients. This cycle repeats for several rounds until the model converges or reaches an acceptable level of performance.

7.7.3.2 Res Net 50

ResNet50, a CNN consisting of 50 layers and according to the VGG design (including VGG-16 and VGG-18) constructed. It stacks leftover blocks to create networks. Subsequent layers can, therefore, be added because there is already residual bottleneck block in the initial design, which can solve the problem of gradient leakage.

7.7.3.3 MobileNet-V2

Google-developed CNNs like MobileNet-V2 were created especially for mobile and integrated applications. This network employs depth-separable convolutions, which enhance speed and efficiency while reducing the key parameters, resulting in lightweight and low-latency models. Consequently, two new DL variants were introduced: the reduced MobileNet and the modified MobileNet. Compared to the VGG models, the reduced MobileNet models had an estimated classification efficiency of 98.30% using less parameters, showcasing its time efficiency, whereas the fine-tuned MobileNet reached accuracy levels nearly comparable to those of the VGG model [14].

7.7.3.4 *Vision Transformer (ViT)*

In 2021, the ViT was introduced [11]. Initially designed for processing of spontaneous language, the transformer architecture underpins this model. To accurately predict class labels for images, ViT examines sequences of image patches.

7.7.3.5 *DenseNet-121*

DenseNet-121 CNNs with dense connections between their layers are called DenseNet networks. This design preserves the spatial resolution by stacking dense blocks that combine inputs from previous layers.

7.7.3.6 *Inception V3*

Initially, Inception-V3, an advanced deep CNN complex architectural design, was developed to enhance the Inception-V1 model formulated by Google. It has numerous advances like factorized convolutions, auxiliary classifiers, and label smoothing.

7.7.3.7 *VGG-16*

The DL model VG-16, created by Oxford's Visual Geometry Group, is built upon the architecture of AlexNet. Unlike AlexNet, VGG-16 utilizes a greater number of compact 3 × 3 convolution filters, resulting in a reduced number of parameters.

Diverse Deep Learning (DL) Models are Employed for Detecting Plant Diseases

Diverse neural networks are used to analyze data and extract features (Table 7.1). Multiple hidden layers (perceptrons) abstract the features. After extracting low-level features, the perceptron combines them to create high-level features. In the end, plant diseases are categorized and found using these high-level characteristics. The process of creating an image dataset is the first simple step in a DL-based system. A trained DL model is then used to carry out classification and detection. Preparing an image dataset entails gathering images from publicly accessible datasets or from actual field observations. After this stage, the dataset is separated into validation, training, and test builds, and a number of picture pre-processing procedures are completed. Models are trained and validated using the training dataset. The typical technique used by DL algorithms to recognize and diagnose plant diseases is shown in Figure 7.1.

Table 7.1 Contemporary deep learning model comparison [30].

Sl. no.	Deep learning models	Principal attributes and advantages and disadvantages	Parameters
1.	VGG-Inception	An inception module and VGG cascaded version. Two 3 × 3 layers were used in place of five 5 × 5 to reduce the amount of parameters, use convolution layers. The accuracy of the tests showed improvement when compared to several established deep algorithm models, like as GoogLeNet, VGG-16, Inception-v3, ResNet, and AlexNet.	132 M
2.	LeNet	First CNN model. Fewer parameters than other CNN models. Limited computational capacity	60 K
3.	AlexNet	Regarded as the original CNN of today. Best image recognition capability available at the time. ReLU was applied in order to boost efficiency. The dropout technique was used to stop overfitting.	60 M
4.	Modified/ Reduced MobileNet	Fewer parameters in comparison to MobileNet. Comparable accuracy to that of MobileNet	0.5/0.54 M
5.	OverFeat	The first model to do object detection, localization, and classification with just one CNN. more parameters compared to AlexNet	145 M

(Continued)

Table 7.1 Contemporary deep learning model comparison [30]. (*Continued*)

Sl. no.	**Deep learning models**	**Principal attributes and advantages and disadvantages**	**Parameters**
6.	ZFNet	Increased accuracy and decreased weights (in comparison to AlexNet) by taking into account 7 × 7 kernels	42 M
7.	MobileNet	Examined the idea of depth-wise separable convolution. Significantly lowered parameters. Accuracy was achieved close to VGG and GoogLeNet.	4.2 M
8.	VGG	Decision function discrimination was thought to be enhanced by the inclusion of more non-linearity functions in 3 × 3 receptive fields. Model with a high computational cost because of its many parameters	133-144 M
9.	GoogLeNet	Less parameters in comparison to the AlexNet model. increased precision at the appropriate time	7 M
10.	ResNet	The problem of disappearing gradient was tackled. Greater accuracy compared to GoogleNet and VGG models	25.5 M
11.	DenseNet	Dense interconnections among the strata. fewer parameters with improved precision	7.1 M

(*Continued*)

Table 7.1 Contemporary deep learning model comparison [30]. (*Continued*)

Sl. no.	Deep learning models	Principal attributes and advantages and disadvantages	Parameters
12.	SqueezeNet	Accuracy comparable to AlexNet using 50 times less parameters. Consider 1 × 1 filters as an alternative to 3 × 3 filters. There were fewer input channels. extensive convolution layer activation maps.	1.25 M
13.	Xception	An approach to depth-wise separable convolution. Outperformed the Inception-v3 model, ResNet, and VGG models.	22.8 M
14.	VGG-Inception	An inception module was developed alongside a cascaded variant of VGG. To minimize the number of parameters, two 3 × 3 convolutional layers replaced five 5 × 5 convolutional layers. The testing accuracy showed improvement when compared to several prominent deep learning models like as VGG-16, GoogLeNet, Inception-v3, ResNet, and AlexNet.	132

DL, a subset of ML, was first utilized in 1943 to create a computer model that mimicked biological processes. There are two distinct eras in this subject of study's history: 1943–2006 and 2012–present. The field is still developing today. As illustrated in Figure 7.2, several significant developments occurred in the initial phase, such as handwritten text recognition through

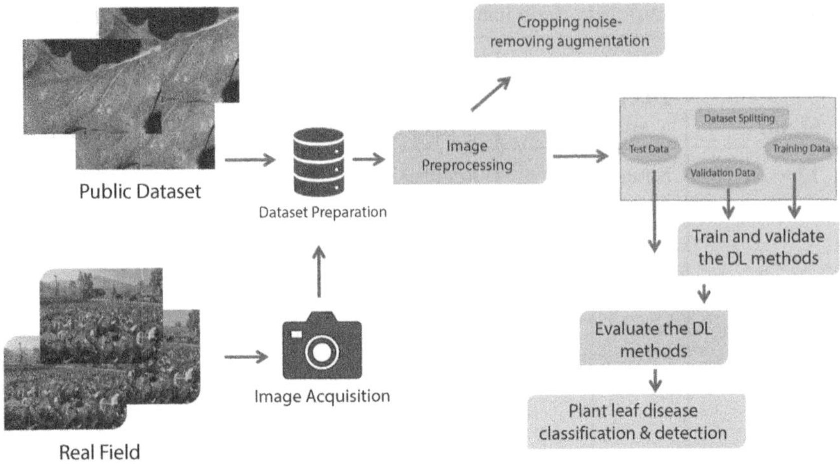

Figure 7.1 Classification and plant foliar disease diagnosis with deep learning: established procedures.

the LeNet architecture, the implementation of backpropagation, the application of the chain rule, the introduction of the Neocognitron, and the resolution of the training problem [15].

7.8 Other Associative Models

7.8.1 AlexNet

Many sophisticated DL models and architectures for visual detection, stratification, and categorization have been created since the advent of AlexNet. The goal of this CNN architecture, which was introduced in 2012, was to build a convolutional layer that was stacked consecutively. Graphical processing units (GPUs) are used by this trained network. Three of its eight layers are fully connected, and the other five are convolutional layers. AlexNet is unique in that it uses overlapping pooling, multiple GPUs, and rectified linear unit (ReLU) nonlinearity. One GPU utilizes a multi-GPU system to manage fifty percent of the model's neurons, with another GPU handling the rest fifty percent. Additionally, it demands less training time compared to other models. Similar to CNN's use of the Tanh function, this model is six times faster and requires less time thanks to its use of ReLU, a standard function. Overlapping pooling can make it more difficult to overfit and reduce error by 0.5%.

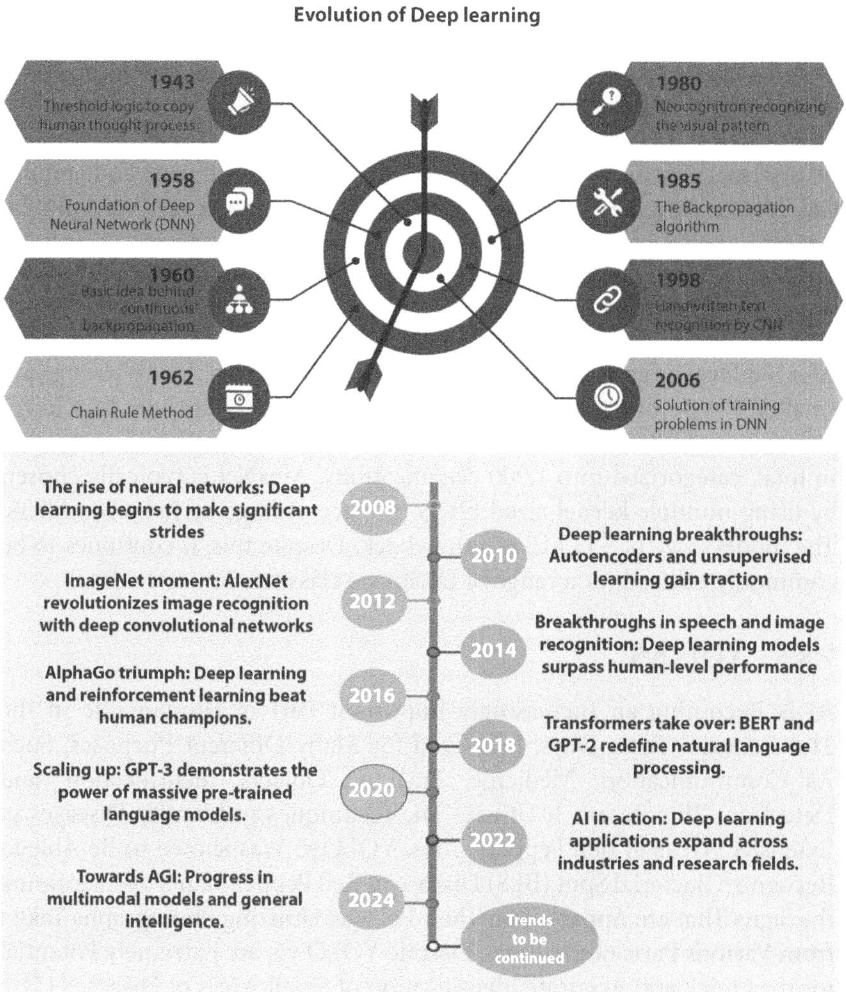

Figure 7.2 Overview of the development of deep learning from 1943 to 2024.

7.8.2 OverFeat

OverFeat is a classic convolutional neural network (CNN) architecture developed for integrated image classification, localization, and detection tasks, introduced by Pierre Sermanet, David Eigen, Xiang Zhang, Michael Mathieu, Rob Fergus, and Yann LeCun in 2013. It was notable for its multi-scale and sliding window approach and was a winner in the 2013 ImageNet Large Scale Visual Recognition Challenge. Initially, a single CNN was utilized for object detection, localization, and classification.

7.8.3 U-Net

Promising results were obtained in the classification of plant diseases using the U-Net architecture, which has demonstrated effectiveness with comparatively little medical imaging data. The research utilized sixty images displaying common diseases of bean plants, such as bean rust and angular leaf spot. Following the application of U-Net for segmentation, only the images depicting the disorders were initially classified [16].

7.8.4 VGG-16

AlexNet lays the groundwork for the VGG-16 DL model. VGG-16 employs smaller 3 × 3 convolution filters than AlexNet, which results in a lower parameter count. The collection consists of more than 14 million photos in total, categorized into 1,000 classifications. AlexNet is typically chosen by using multiple kernel-sized filters in place of large kernel-sized filters. The model's size of 533 MB is a drawback. Despite this, it continues to be commonly utilized for a range of DL image classification activities.

7.8.5 YOLOv5

AI is Becoming an Increasingly Important Part of Human Life in the 21st Century. These Days, It is Used for Many Different Purposes, Such As Communication, Medicine, Tracking Objects, Identification, and Detection. This Research Utilizes DL Techniques to Identify Diseases in Extensive Areas of Bell Pepper Crops. YOLOv5 Was Shown to Be Able to Recognize Bacterial Spot (BLS) Disease in Bell Pepper Plants by Examining the Signs That are Apparent on Their Foliage, Utilizing Photographs Taken from Various Parts of the Farm. Despite YOLO v5, are Extremely Potential for the Quick and Accurate Identification of Small Areas of Disease [17].

7.8.6 GoogLeNet

Computer vision tasks like object detection and image classification can be handled by the 22–successive layer DL network design, which achieved with 2014 ImageNet Large-Scale Visual Recognition Challenge; nowadays, it is employed in various computer-aided vision applications, including facial recognition, object detection, and adversarial training. This architecture's input layer receives an image with dimensions of 224 × 224. It accomplishes efficiency by keeping crucial spatial information while concurrently decreasing the input image. This architecture has excellent computational efficiency and is a powerful machine. Two max-pooling layers are positioned between nine inception modules that make up the GoogLeNet architecture. The main purpose of these optimal pooling layers is to trim the input by making it smaller in both width and height. To appreciate the advancement of deep convolutional networks in the realm of DL, one must comprehend the GoogLeNet design.

7.8.7 RCNN

Residual CNN is a faster Image taken in real field ResNet-50's precision (85 %) with region-based. The prediction percentage was calibrated using a variety of detectors, including RFCN, SSD, and Faster-RCNN.

7.8.8 PSPNet

Pyramid Scene Parsing Network (PSPNet) is one of the popular designs. Semantic segmentation networks, such as PSPNet, are specifically engineered to segment intricate situations that necessitate a comprehensive understanding of global context to accurately identify interconnected elements. To effectively segment leaf diseases, the Pyramid Scene Parsing Network (PSPNet) employs a multi-step approach [18, 19]. To extract fine details from an illustration of a maize leaf, it uses many convolutional layers. The leaf disease segmentation method of PSPNet allows for the diagnosis and management of agricultural ailment.

7.8.9 CapsuleNet

CNNs are the basis for image processing from the perspective of DL. Utilizing a novel CNN design called CapsuleNet, traditional neural networks can perform better and have fewer drawbacks. The proposed architecture utilizing CapsNet achieves an accuracy of roughly 94% [20]. The usage of mobile phones has surged significantly over the past few years. In this scenario, both models extract illustration of infected leaves utilizing smart mobile apps. Consequently, this approach can be readily adopted and expanded to include various species.

7.8.10 ZFNet

ZFNet is improved accuracy and decreased weights (in comparison to AlexNet) by considering 7×7 kernels.

7.8.11 ResNet-20

The residual network model was put forth by Microsoft researchers in 2015. It connects straight to the output by omitting a few layers' worth of training.

Skipping connections is necessary. This connection type's main benefit is that any layer that interferes with performance will be skipped *via* regularization. When it comes to accuracy, ResNet outperforms other systems, like highway networks and long short term memory (LSTM). This network fits the residual mapping by letting the network do the fitting [21].

7.8.12 NIN

The model is made up of a thermographic camera and two linear hyperspectral scanners. For classification, wavelengths between 430 and 2,376 nm were used, with a pixel-by-pixel increment of about 32 nm. It is necessary to practice the second derivative, which calculates the curve's slope change.

7.8.13 RFNet

The suggested model was measured using an infrared (IR) thermometer, a chlorophyll fluorescence detector, and a spectroradiometer. This model evaluated the photosystem II, quantum amount, and leaf surface area of 119 distinct vegetable varieties [22].

7.8.14 IRRCNN

The Inception Recurrent Residual CNN (IRRCNN) is a DL architecture designed to achieve cutting-edge performance in object recognition tasks by integrating the benefits of residual connections, recurrent convolutional layers, and Inception networks. The IRRCNN combines inception units, residual connections, and recurrent convolutional layers to more efficiently capture contextual information compared to comparable inception networks, residual networks, and recurrent CNNs (RCNNs) evaluated on standard datasets like CIFAR-100, TinyImageNet-200, and CU3D-100. In comparison to these models, it shows an improvement in accuracy of the categorization around 4.53%, 4.49%, and 3.56%, all achieved with the same number of network parameters.

7.8.15 IRCNN

The Inception Recurrent CNN (IRCNN) is an intense DL architecture that merges the advantages of recurrent connections in convolutional layers with the strengths of Inception networks. This architecture successfully models contextual information with the use of recurrent connections and the potent Inception network design, which improves object recognition performance. Empirical assessments of IRCNN using standard prototype data like datasets like SVHN, CIFAR-10, CIFAR-100, and MNIST have shown that it surpasses several well-known deep CNNs, including RCNNs, regarding recognition accuracy. IRCNN achieved better classification accuracy than RCNN. The evaluation was conducted on Equivalent Inception Networks and Equivalent Inception-Residual Networks using the improved CIFAR-100 dataset. These improvements were around 3.5%, 3.47%, and 2.54%, respectively. The Tiny ImageNet-200 dataset has shown that IRCNN outperforms other models like DenseNet, highlighting its advanced capabilities in object detection tasks.

7.8.16 SSD

The SSD model saves a substantial amount of time by identifying objects with remarkable detection accuracy in a single pass [23, 24]. By splitting predictions by aspect ratio directly, it obtains high detection accuracy and produces accurate predictions at various levels from feature maps [25]. The SSD employs neural networks to generate a vast array of bounding boxes in different fixed sizes and subsequently assesses whether each box contains an object class instance.

7.8.17 DenseNet

DenseNet, a deep convolutional neural network architecture known for its dense connectivity and efficient feature propagation, has shown remarkable performance in plant disease detection tasks, especially when combined with federated learning approaches. By leveraging DenseNet within a federated learning framework, plant disease detection models can benefit from the robust feature extraction and gradient flow of DenseNet, while also utilizing diverse, decentralized datasets collected from various geographic and environmental conditions. When integrated into federated learning, these models maintain strong performance across different clients, supporting early and accurate disease diagnosis while ensuring data security and adaptability to local variations in plant pathology.

7.9 Benefits of Federated Learning for Plant Disease Detection

7.9.1 Data Protection and Privacy Measures

FL ensures compliance with data protection regulations and mitigates the possibility of data leaks by keeping crude data on local gadgets, which lessens privacy concerns.

7.9.2 Scalability

FL's decentralized structure allows it to scale efficiently across numerous clients, making it suitable for widespread agricultural applications.

7.9.3 Utilization of Diverse Data

By leveraging data from various sources, FL can build robust models that generalize well across different environmental conditions, crop varieties, and disease manifestations.

Reduced Communication Expenses: Transmitting model updates instead of raw data results in smaller data transfers over the network, decreasing communication costs and accelerating the training process.

7.10 Implementation of DL Models

7.10.1 Other than Visualization Technique

Histogram approaches were utilized to demonstrate the model's significance, and CNNs were utilized to categorize diseases affecting maize plants [26]. To assess tomato leaves for disease identification, several CNN architectures were used, which comprises with GoogLeNet, ResNet, and AlexNet [27]. Plotting training/validation accuracy revealed the model's performance; among all CNN architectures, ResNet was thought to be the best. In order to identify illnesses in banana leaves, LeNet architecture was employed. The model was assessed using the CA and F1-score metrics in both color and grayscale formats. It was found that five CNN architectures were employed: the VGG architecture, which surpassed all other models, as well as AlexNet, AlexNetOWTbn, GoogLeNet, and Overfeat [28]. In addition to sophisticated DL models like GoogLeNet, ResNet-50, ResNet-101, Inception-v3, InceptionResNetv2, and SqueezeNet, three classifiers—support vector machines (SVMs), extreme learning machines (ELMs), and K-nearest neighbors (KNNs)—were used to identify eight distinct types of plant diseases [29]. Metrics including sensitiveness, specificity, and the F1 rating were used to assess the models' ideal performance; the combination of ResNet-50 and the SVM classifier produced the best outcomes. According to a report, a novel DL model called Inception-v3 has been applied to the identification of cassava-related ailments. Additionally, a comprehensive comparison was conducted by assessing classification accuracy after categorizing six diseases affecting tomato plants, utilizing the AlexNet and VGG-16 DL models [9]. It is important to note that no visual aids were employed in the processes described to detect signs of plant diseases.

7.10.2 Through Visualization Technique

By using this technique, we can more precisely categorize and recognize the diseases by seeing any signs on the foliage's of the affected plants (Table 7.2) [31]. Leveraging the CNN architecture of CaffeNet, a saliency map was utilized for visualization, and it was able to recognize and diagnose 13 distinct diseases. Its classification accuracy was substantially greater than the SVM model, reaching up to 96.30%. To detect disease symptoms, CNN architectures like AlexNet and GoogLeNet employ multiple filters alongside the publicly accessible PlantVillage dataset.

Table 7.2 Visualization techniques and mapping across multiple approaches.

• Display of feature with first to last layer
• Observe the first convolutional layer activations [32]
• Visualization of a saliency map [31]
• Disease categorization and localization using bounding boxes [42]
• Heatmaps were employed to pinpoint the disease's locations [34].
• Featured map applied to sick rice plants
• Method for visualizing symptoms [31]
• Core maps that combine features and spatial information
• HSV and K-color spaces indicate clustering.
• Disease detection by feature map [36].
• Methodology applied for image segregation [43]
• Heat map creation, binary threshold theory image segmentation, and image reconstruction on discriminant regions [44]
• Visualization of saliency maps [45]
• 2D and 3D saliency contour maps, and mesh graph images [40]
• Visibility of activation [46]
• Edge gap and segmentation map

Using indicators, including the F1 rating of precision (P), recall (R), and total accuracy, was estimated [32]. In order to evaluate performance and facilitate a clear comparison between GoogLeNet and AlexNet, categorization, color, and grayscale techniques were applied. GoogLeNet proved to be better. When the visualization of the initial levels was activated, the areas affected by the diseases became distinctly evident. We used a modified LeNet model that uses edge detection and segmentation maps to identify diseased areas in order to detect and properly diagnose diseases in olive trees. We evaluated how well AlexNet models, random forests, and SVMs recognized and classified four distinct diseases in cucumbers. These models used image analysis to reliably detect disease signs. Additionally, a cutting-edge DL approach involving a teacher-student network was developed, introducing an innovative method for visualizing symptoms of plant diseases. Various detectors, such as Faster-RCNN, SSD, and RFCN, were employed to fine-tune the prediction accuracy. Various well-known relative learning architectures, which encompasses with GoogLeNet, ZFNet, VGG, ResNet-50, AlexNet, and ResNetXt101, were employed for this study. To detect pests and diseases that impact banana plants, we utilized three CNN models—ResNet-50, InceptionV2, and MobileNet-V1—alongside two detection frameworks: Faster-RCNN and SSD. The findings indicated that ResNet-50 outperformed all the other architectures [33]. To forecast where a specific disease may manifest, various CNN model combinations were fed images of sick plants, such as heat maps [34]. Standardized LeNet architecture has been created to recognize and classify diseases affecting soybean plants [35]. The GoogLeNet-developed impingement technique makes it possible to identify disease-symptomatic regions. GoogLeNet performed better than the AlexNet model in attempts to identify tomato plant diseases. To visualize every characteristic within each block, two models—VGG-FCN and VGG-CNN—were utilized to identify and categorize various wheat diseases. Fusarium wilt in radish was detected using the VGG-CNN model's clustering technique. To detect certain illness signs or indicators, a DL model was used; in particular, a deep CNN network was successfully used to identify, categorize, and quantify eight stressors that damage soybeans [36, 37]. CNN was used to identify symptomatic patches on rice plants, allowing for the implementation of future maps. It was reported that a novel mobile application utilizing a deep residual neural network facilitated prompt disease identification by means of a hot spot [38]. In order to achieve color constancy, a customized hot spot technique was created using an algorithm that pools all of the hotspots that are present in the segmented image through alteration. Additionally, each identified specific region was characterized by two descriptors that could differentiate between the texture of the targeted spot along with the alteration of color which simultaneously noticed [39]. The DL algorithms VGG-16 and AlexNet, along with a correlation coefficient, were used to develop the sophisticated visualization method [40]. The dilation feature of the CNN model can be used to identify and classify cucumber diseases [41]. LeNet, a CNN model that uses different vegetation indexes and color space, shown remarkable efficacy in diagnosing grape illnesses [6]. A more accurate DL model cannot be practically implemented without a true/actual backdrop and environment. Unfortunately, most of the previously proposed techniques used a simple background dataset, which is not a realistic or naturalistic environment to identify and categorize plant diseases.

7.11 Challenges and Solutions in Federated Learning for Plant Disease Detection

7.11.1 Data Heterogeneity

Regional differences and farming practices can lead to significant discrepancies in agricultural data, resulting in non-independent and identically distributed distributions.

7.11.1.1 Solution

Techniques such as personalized FL, where each client maintains a personalized model that incorporates global knowledge while adapting to local data, can address data heterogeneity. Additionally, federated meta-learning can help models quickly adapt to new clients with minimal data.

7.11.2 Communication Overhead

Although FL minimizes the necessity of transferring raw data, the transmission of model updates, particularly for large neural networks, can still be significant.

7.11.2.1 Solution

Model updates can be made smaller by using model compression techniques like quantization and pruning. Asynchronous FL can also help by allowing clients to communicate updates at different times, alleviating network congestion.

7.11.3 Privacy and Security

Model inversion attacks can potentially cause sensitive information to leak during the transmission of model updates, despite the fact that FL improves privacy by keeping data local.

7.11.3.1 Solution

Model updates can use differential privacy to secure individual data points and introduce noise. Secure aggregation protocols ensure that the server only receives aggregated updates, preventing the reconstruction of individual updates.

7.11.4 Model Performance and Convergence

Ensuring that the federated model converges to a high-performance state across diverse data sources can be challenging.

7.11.4.1 Solution

FL algorithms can be enhanced with techniques like adaptive learning rates, client selection strategies to ensure diverse participation, and regularization methods to stabilize training.

7.12 Case Studies and Applications

7.12.1 Wheat Disease Detection

A FL framework was deployed across multiple wheat farms to detect common diseases like rust and blight. Each farm collected images using drones and smartphones. Local models were trained on these images and periodically updated to a central server. The global model achieved high accuracy in detecting diseases, and farmers received real-time alerts to take corrective actions, significantly reducing crop losses.

7.12.2 Tomato Plant Disease Classification

In a collaborative project involving agricultural research institutions, FL was used to classify diseases in tomato plants. Data collected from greenhouses and open fields were utilized to train localized models. The federated approach not only safeguarded data privacy but also enhanced the model's generalizability across various disease presentations and changing environmental conditions.

7.12.3 Future Directions

Federated learning for plant disease detection is an emerging approach that enables the collaborative training of machine learning models across multiple decentralized data sources—such as farms or research centers—without requiring raw image data to be shared or centralized, thereby preserving data privacy and reducing transfer costs. This method leverages advanced deep learning architectures like convolutional neural networks (CNNs) and vision transformers (ViTs) to classify plant diseases from leaf images, as demonstrated using datasets such as PlantVillage. Studies show that federated learning can achieve competitive accuracy compared to centralized methods, with performance influenced by factors like the number of participating clients, communication rounds, local training iterations, and data quality. Notably, models like ResNet50 have proven particularly effective in federated scenarios due to their balance of accuracy and computational efficiency, while larger vision transformers may be less suitable due to higher resource demands. Overall, federated learning addresses key challenges in plant disease detection by facilitating scalable, privacy-preserving, and efficient model development, which is especially valuable for sensitive agricultural data and diverse, geographically distributed stakeholders.

7.13 Various Kind of Integration through Edge, Multi-Modal and Reinforcement Learning

7.13.1 Integration with IoT and Edge Computing

Combining FL with IoT devices and edge computing can further enhance real-time disease detection [46] and intervention. Edge devices can preprocess data and train local models, reducing latency and reliance on centralized servers.

7.13.2 Multi-Modal Federated Learning

The precision and stability of disease identification algorithms can be improved by combining data from multiple sources, including soil sensors, meteorological data, and illustrations.

7.13.3 Federated Reinforcement Learning

Exploring federated reinforcement learning for optimizing agricultural practices, such as pest management and irrigation, can lead to more sustainable farming methods.

7.13.4 Federated Transfer Learning

Adapting previously trained models from various domains by utilizing federated transfer learning (e.g., medical imaging) can boost performance, especially in regions with limited agricultural data.

7.14 Conclusion

FL presents an innovative, unique, and revolutionary method for detecting plant diseases, providing an effective alternative to the restrictions imposed by traditional centralized models. By facilitating collaborative model training on various devices while keeping raw data private, FL guarantees the privacy and security of data, both of which are essential in agricultural sectors that handle sensitive information. This decentralized learning approach utilizes data from different geographical regions, improving the model's adaptability and resilience to a variety of plant diseases and environmental factors. The adaptability of FL allows continuous learning from real-time data, making it responsive to emerging plant diseases. Moreover, the reduced need for data transfer lowers communication costs and conserves bandwidth, making FL a cost-effective and scalable solution. Integrating FL with advanced technologies such as IoT sensors and edge computing further augments its efficiency, providing timely and accurate disease diagnostics. However, challenges remain, including addressing the heterogeneity of data sources, ensuring model convergence, and managing the computational overhead on local devices. Advances in optimization algorithms, hardware acceleration, and collaborative frameworks are pivotal to overcoming these hurdles. The future of FL in plant disease detection lies in interdisciplinary collaboration, bridging expertise from ML, plant pathology, and agronomy. As the agricultural sector increasingly adopts digital transformation, FL emerges as a key technology that promotes sustainable farming practices, minimizes crop losses, and improves global food security. Its potential to democratize access to high-quality disease detection tools empowers farmers worldwide, fostering a more resilient and productive agricultural ecosystem.

AI disclosure statement: The author acknowledges the use of AI-based tools for minor language editing or grammatical correction. However, all intellectual contributions, analyses, and conclusions presented in this chapter are entirely the author's own.

References

1. Li, T., Sahu, A.K., Zaheer, M., Sanjabi, M., Talwalkar, A., Smith, V., Federated optimization in heterogeneous networks. *Proc. Mach. Learn. Syst.*, 2, 429–450, 2020, http://arxiv.org/abs/1812.06127.
2. Lu, J., Hu, J., Zhao, G., Mei, F., Zhang, C., An in-field automatic wheat disease diagnosis system. *Comput. Electron. Agric.*, *142*, 369–379, 2017, DOI: https://doi. org/10.1016/j.compag.2017.09.012.
3. Borhani, Y., Khoramdel, J., Najaf, E., A deep learning based approach for automated plant disease classification using vision transformer. *Sci. Rep.*, 12, 1, 11554, 2022, https://doi.org/10.1038/s41598-022-15163-0.
4. Maillet, W., Ouhami, M., Hafane, A., Fusion of satellite images and weather data with transformer networks for downy mildew disease detection. *IEEE Access*, 11, 5406–5416, 2023, https://doi.org/10.1109/access.2023.3237082.
5. Yakkundimath, R., Saunshi, G., Anami, B., Palaiah, S., Classification of rice diseases using convolutional neural network models. *J. Inst. Eng. India Ser. B*, 103, 4, 1047–1059, 2022, https://doi.org/10.1007/s40031-021-00704-4.
6. Kerkech, M., Hafiane, A., Canals, R., Deep leaning approach with colorimetric spaces and vegetation indices for vine diseases detection in UAV images. *Comput. Electron. Agric.*, 155, 237–243, 2018, https://doi.org/10.1016/j.compag. 2018.10.006.
7. Tugrul, B., Elfatimi, E., Eryigit, R., Convolutional neural networks in detection of plant leaf diseases: A review. *Agriculture*, 12, 8, 1192, 2022, https://doi. org/10.3390/agriculture12081192.
8. Li, L., Zhang, S., Wang, B., Plant disease detection and classification by deep learning—A review. *IEEE Access*, 9, 56683–56698, 2021, DOI: 10.1109/ ACCESS.2021.3069646.
9. Rangarajan, A.K., Purushothaman, R., Ramesh, A., Tomato crop disease classification using pre-trained deep learning algorithm. *Procedia Comput. Sci.*, 133, 1040–1047, 2018, https://doi.org/10.1016/j.procs.2018.07.070.
10. Geetharamani, G. and Pandian, A., Identification of plant leaf diseases using a nine-layer deep convolutional neural network. *Comput. Electr. Eng.*, *76*, 323–338, 2019, https://doi.org/10.1016/j.compeleceng.2019.04.011.
11. Dosovitskiy, A., Beyer, L., Kolesnikov, A., Weissenborn, D., Zhai, X., Unterthiner, T., Dehghani, M., Minderer, M., Heigold, G., Gelly, S., Uszkoreit, J., Houlsby, N., An image is worth 16x16 words: Transformers for image recognition at scale, 2020, http:// arxiv.org/abs/2010.11929.

12. Chen, R., Qi, H., Liang, Y., Yang, M., Identification of plant leaf diseases by deep learning based on channel attention and channel pruning. *Front. Plant Sci.*, 13, 1023515, 2022, https://doi.org/10.3389/fpls.2022.1023515.
13. Wang, Y., Chen, Y., Wang, D., Convolution network enlightened transformer for regional crop disease classification. *Electronics*, 11, 19, 3174, 2022, https://doi.org/10.3390/electronics11193174.
14. Kamal, K.C., Yin, Z., Wu, M., Wu, Z., Depthwise separable convolution architectures for plant disease classification. *Comput. Electron. Agric.*, 165, 104948, 2019, DOI: https://doi.org/10.1016/j.compag.2019.104948.
15. Hinton, G.E. and Salakhutdinov, R.R., Reducing the dimensionality of data with neural networks. *Science*, 313, 504–507, 2006, DOI: 10.1126/science.1127647.
16. Kursun, R., Bastas, K.K., Koklu,.M., Segmentation of dry bean (Phaseolus vulgaris L.) leaf disease images with U-Net and classification using deep learning algorithms. *Eur. Food Res. Technol.*, 249, 10, 2543–2558, 2023, https://doi.org/10.1007/s00217-023-04319-5.
17. Mathew, M.P. and Mahesh, T.Y., Leaf-based disease detection in bell pepper plant using YOLO v5. SIViP, 841–847, 2022, https://doi.org/10.1007/s11760-021-02024-y, https://doi.org/10.1007/s11760-021-02024-y.
18. Sakshi, and Kukreja, V., Image segmentation techniques: statistical, comprehensive, semi-automated analysis and an application perspective analysis of mathematical expressions. *Arch. Comput. Methods Eng.*, 30, 1, 457–495, 2023, https://doi.org/10.1007/s11831-022-09805-9.
19. Sharma, R. and Kukreja, V., Segmentation and multi-layer perceptron: an intelligent multi-classification model for sugarcane disease detection, in: *2022 International Conference on Decision Aid Sciences and Applications (DASA)*, IEEE, pp. 1265–1269, 2022, DOI: 10.1109/DASA54658.2022.9765191.
20. Kurup, R.V., Anupama, M.A., Vinayakumar, R., Sowmya, V., Soman, K.P., Capsule Network for Plant Disease and Plant Species Classification, in: *Computational Vision and Bio-Inspired Computing. ICCVBIC 2019. Advances in Intelligent Systems and Computing*, vol. 1108, S. Smys, J. Tavares, V. Balas, A. Iliyasu, (Eds.), Springer, Cham, 2020, https://doi.org/10.1007/978-3-030-37218-7_47.
21. Das, S., Pattanayak, S., Behera, P.R., Application of machine learning: a recent advancement in plant diseases detection. *J. Plant Prot. Res.*, 62, 2, 122–135, 2022, DOI: 10.24425/jppr.2022.141360.
22. Odilbekov, F., Armonienė, R., Henriksson, T., Chawade, A., Proximal phenotyping and machine learning methods to identify Septoria tritici blotch disease symptoms in wheat. *Front. Plant Sci.*, 9, 685, 2018, DOI: https://doi.org/10.3389/fpls.2018.00685.
23. Li, W., Zhang, H., Wang, G., Xiong, G., Zhao, M., Li, G., Li, R., Deep learning based online metallic surface defect detection method for wire and arc additive manufacturing. *Rob. Comput. Integr. Manuf.*, 80, 102470, 2023, DOI: https://doi.org/10.1016/j.rcim.2022.102470.

24. Vig, S., Arora, A., Arya, G., Automated license plate detection and recognition using deep learning, in: *International Conference on Advancements in Interdisciplinary Research*, Springer Nature Switzerland, Cham, pp. 419–431, 2022, DOI: https://doi.org/10.1007/978-3- 031-23724-9_39.
25. Guravaiah, K., Bhavadeesh, Y.S., Shwejan, P., Vardhan, A.H., & Lavanya, S. Third eye: object recognition and speech generation for visually impaired. *Procedia Comput. Sci.*, *218*, 1144–1155, 2023, DOI: https://doi.org/10.1016/j.procs.2023.01.093.
26. Sibiya, M. and Sumbwanyambe, M., A computational procedure for the recognition and classification of maize leaf diseases out of healthy leaves using convolutional neural networks. *AgriEngineering*, *1*, 1, 119–131, 2019.
27. Zhang, K., Wu, Q., Liu, A., Meng, X., Can deep learning identify tomato leaf disease? *Adv. Multimedia*, *2018*, 1, 6710865, 2018, https://doi.org/10.1155/2018/6710865.
28. Ferentinos, K.P., Deep learning models for plant disease detection and diagnosis. *Comput. Electron. Agric.*, 145, 311–318, 2018, https://doi.org/10.1016/j.compag.2018.01.009.
29. Türkoğlu, M. and Hanbay, D., Plant disease and pest detection using deep learning-based features. *Turk. J. Electr. Eng. Comput. Sci.*, *27*, 3, 1636–1651, 2019, https://doi.org/10.3906/elk-1809-181.
30. Saleem, M.H., Potgieter, J., Arif, K.M., Plant disease detection and classification by deep learning. *Plants*, *8*, 11, 468, 2019, doi: 10.3390/plants8110468.
31. Brahimi, M., Arsenovic, M., Laraba, S., Sladojevic, S., Boukhalfa, K., Moussaoui, A., Deep learning for plant diseases: Detection and saliency map visualisation, in: *Human and Machine Learning*, pp. 93–117, Springer, Berlin, Germany, 2018, https://doi.org/10.1007/978-3-319-90403-0_6.
32. Mohanty, S.P., Hughes, D.P., Salathe, M., Using deep learning for image-based plant disease detection. *Front. Plant Sci.*, 7, 1419, 2016, DOI: https://doi.org/10.3389/fpls.
33. Selvaraj, J., Umapathy, S., Rajesh, N.A., Artificial intelligence based real time colorectal cancer screening study: Polyp segmentation and classification using multi-house database. *Biomed. Signal Process. Control*, 99, 106928, 2025, https://doi.org/10.1016/j.bspc.2024.106928.
34. De Chant, C., Wiesner-Hanks, T., Chen, S., Stewart, E.L., Yosinski, J., Gore, M.A., Nelson, R.J., Lipson, H., Automated identification of northern leaf blight-infected maize plants from field imagery using deep learning. *Phytopathology*, 107, 11, 1426–1432, 2017, DOI: https://doi.org/10.1094/PHYTO-11-16-0417-R.
35. Wallelign, S., Polceanu, M., Buche, C., Soybean Plant Disease Identification using convolutional neural network, in: *Proceedings of the 31st International Flairs Conference*, 2018.
36. Ghosal, S., Blystone, D., Singh, A.K., Ganapathysubramanian, B., Singh, A., Sarkar, S., An explainable deep machine vision framework for plant stress

phenotyping. *Proc. Natl. Acad. Sci.*, *115*, 18, 4613–4618, 2018, https://doi.org/10.1073/pnas.1716999115.

37. Barbedo, J.G.A., Plant disease identification from individual lesions and spots using deep learning. *Biosyst. Eng.*, 180, 96–107, 2019, DOI: https://doi.org/10.1016/j. biosystemseng.2019.02.002.
38. Picon, A., Alvarez-Gila, A., Seitz, M., Ortiz-Barredo, A., Echazarra, J., Johannes, A., Deep convolutional neural networks for mobile capture device-based crop disease classification in the wild. *Comput. Electron. Agric.*, 161, 280–290, 2019, DOI: https://doi.org/10.1016/j.compag.2018.04.002.
39. Johannes, A., Picon, A., Alvarez-Gila, A., Echazarra, J., Rodriguez-Vaamonde, S., Navajas, A.D., Ortiz-Barredo, A., Automatic plant disease diagnosis using mobile capture devices, applied on a wheat use case. *Comput. Electron. Agric.*, 138, 200–209, 2017, DOI: https://doi.org/10.1016/j.compag.2017.04.013.
40. Khan, M.A., Akram, T., Sharif, M., Awais, M., Javed, K., Ali, H., Saba, T., CCDF: Automatic system for segmentation and recognition of fruit crops diseases based on correlation coefficient and deep CNN features. *Comput. Electron. Agric.*, 155, 220–236, 2018, DOI: https://doi.org/10.1016/j.compag.2018.10.013.
41. Zhang, S., Zhang, S., Zhang, C., Wang, X., Shi, Y., Cucumber leaf disease identification with global pooling dilated convolutional neural network. *Comput. Electron. Agric.*, *162*, 422–430, 2019, DOI: https://doi.org/10.1016/j.compag.2019.03.012.
42. Fuentes, A., Yoon, S., Kim, S.C., Park, D.S., A robust deep-learning-based detector for real-time tomato plant diseases and pests' recognition. *Sensors*, *17*, 9, 2022, 2017, https://doi.org/10.3390/s17092022.
43. Ma, J., Du, K., Zheng, F., Zhang, L., Gong, Z., Sun, Z., A recognition method for cucumber diseases using leaf symptom images based on deep convolutional neural network. *Comput. Electron. Agric.*, *154*, 18–24, 2018, https://doi.org/10.1016/j.compag.2018.08.048.
44. Brahimi, M., Mahmoudi, S., Boukhalfa, K., Moussaoui, A., Deep interpretable architecture for plant diseases classification, in: *2019 signal processing: Algorithms, architectures, arrangements, and applications (SPA) IEEE*, pp. 111–116, 2019, doi: 10.23919/SPA.2019.8936759.
45. Nagasubramanian, K., Jones, S., Singh, A.K., Singh, A., Ganapathysubramanian, B., Sarkar, S., Explaining hyperspectral imaging-based plant disease identification: 3D CNN and saliency maps. *arXiv preprint arXiv:1804.08831, https://* doi.org/10.48550/arXiv.1804.08831.
46. Jiang, P., Chen, Y., Liu, B., He, D., Liang, C., Real-Time Detection of Apple Leaf Diseases Using Deep Learning Approach Based on Improved Convolutional Neural Networks. *IEEE Access*, 7, 59069–59080, 2019, doi: 10. 1109/ACCESS.2019.2914929.

8

Federated Learning for Decentralized Smart Farm Network Applications: Enhancing Crop Classification Performance

Mukesh Kumar Tripathi[1]*, Praveen Kumar Reddy[2], Vangara Nikitha[3], Nakshatra Reddy[3], Akshaya Gourisetty[3] and Kapil Misal[4]

[1]Department of Computer Science and Engineering (AIML), KIET Group of Institution, Delhi-NCR, Ghaziabad, India
[2]Department of Electronics and Communication Engineering, Guru Nanak Dev Engineering College, Bidar, Karnataka, India
[3]Department of Computer Science and Engineering, Vardhaman College of Engineering, Hyderabad, India
[4]Department of Master of Computer Application, Trinity Academy of Engineering, Pune, India

Abstract

This paper explores the use of federated learning (FL) in intelligent farming, focusing on crop classification. The results show better accuracy when training samples are divided into smaller subsets before feeding them into the model than when all samples are processed simultaneously during training. These findings indicate that FL can be applied successfully to innovative farming applications by improving crop classification accuracy using decentralized machine learning models trained with limited data resources while maintaining privacy and security concerns associated with sharing sensitive agricultural information across multiple organizations or locations. Future research could further explore more complex models and additional weather factors to improve prediction performance in real-world scenarios. Furthermore, integrating deep learning techniques alongside traditional statistical approaches might yield even higher accuracy and adaptability within FL frameworks utilized in agriculture. By employing FL, the study

**Corresponding author*: Mukeshtripathi016@gmail.com

Padmesh Tripathi, Bhanumati Panda, Shanthi Makka, Reeta Mishra, S. Balamurugan and Sheng-Lung Peng (eds.) Federated Learning for Smart Agriculture and Food Quality Enhancement, (193–216) © 2026 Scrivener Publishing LLC

successfully addresses two significant challenges in smart farming: data scarcity and privacy concerns. Agricultural datasets are often limited and scattered across various stakeholders, making centralized training approaches impractical. FL leverages these scattered datasets effectively, enabling the development of robust crop classification models without compromising data security. The decentralized nature of FL ensures that sensitive agricultural information, such as crop health, yield predictions, and weather patterns, remains protected. This aspect is particularly significant in modern farming, where data privacy is a critical concern due to the involvement of multiple stakeholders, including farmers, cooperatives, and government agencies. Finally, exploring ways to enhance communication protocols between participating devices would allow faster transfer of updated weights and improved overall collaboration efficiency among farms involved in FL networks focused on intelligent farming practices.

***Keywords*:** Federated learning, decentralized networks, smart farming, crop classification, edge computing

8.1 Introduction

The research adopts the federated learning (FL) approach to evaluate how closely predicted crop types match the original crop types in a given dataset. FL operates as a decentralized network where local edge nodes update their model weights and send these updates to a central server. The research focuses on hyper-tuning these models to optimize the federated averaging model for the intelligent farming dataset, ensuring effective aggregation and convergence within the decentralized network. Using satellite images, as discussed in [1, 2], has significantly contributed to analyzing soil and crop conditions in farmlands. This analysis helps determine the state of crops and soil, addressing various challenges by providing data-driven solutions. These solutions have improved farming by enabling predictions about crop harvesting times, which aids in making decisions to prevent poor harvests.

The exploration examined in [3] features the robotization of shrewd ranches, explicitly through a water system framework constrained by a versatile application. This framework permits ranchers to screen information caught by Internet-of-Things (IoT) gadgets around the plants. Be that as it may, an eminent restriction of their work is the requirement for more examination of the gathered information. Research by [4, 5] brought lights on predicting vermin invasions using datasets obtained from smart residences. In any case, as [6], these examinations did not analyze the accumulated data. From studies [7], it was seen that many edge contraptions in astute residences carried out essentially no assessment of the got data. Additionally, the shortfall of cooperation in results, gauges, or assessments is attributed to including different stages and working systems in their investigation. On the other hand, this paper utilizes a multi-named dataset where climatic limits are free factors and gather types are marks. The joined learning stage predicts crop types considering climatic limits for an innovative residence association. This approach tends to avoid past assessments' hindrances by organizing data examination and estimates inside a united stage. Throughout the long haul, analysts have found bioactive artificial materials in many plants commonly used as helpful fixes.

FL is a revolutionary approach to machine learning that addresses the critical concern of data privacy. Unlike traditional methods that centralize data for training, FL enables the creation of robust machine learning models while ensuring that data remains decentralized. This approach aggregates insights from diverse datasets in geographically dispersed locations, ensuring data security and developing accurate and versatile models. By doing so, FL upholds privacy standards and facilitates collaboration among various entities without compromising sensitive information. The decentralized nature of this methodology makes it particularly suitable for applications in agriculture, healthcare, and industries requiring sensitive data handling.

One emerging application of FL is assessing mango quality [8, 9]. Mangoes, known for their nutritional richness, vibrant flavor, and versatility, are significant in global markets. With an estimated annual production of 55 million tons, mangoes are a cornerstone of the agricultural economy and human diet. Their rich nutrient profile includes essential vitamins, minerals, fiber, and proteins, making them a critical dietary component worldwide. Mangoes are integral to diverse cuisines and food products and are consumed in raw or ripe forms. However, despite their importance, the scientific evaluation of mango quality still needs to be explored [10]. Traditionally, mango quality assessment has relied on either destructive method, which involve damaging the fruit to evaluate internal characteristics, or non-destructive techniques, which depend on external features like color, texture, and size. While both approaches have merits, each comes with limitations. Destructive methods are invasive and impractical for large-scale grading, whereas non-destructive methods may need more comprehensive accuracy. Combining these frameworks into a hybrid system integrating machine learning for mango grading offers a promising solution. The proposed system enhances mango quality grading by evaluating external and internal characteristics [11]. Machine learning models can analyze external features such as skin texture, color uniformity, and shape while simultaneously predicting internal qualities like sugar content, firmness, and acidity through advanced imaging techniques and sensor data. FL is crucial in this system, enabling data integration from various agricultural regions without requiring data centralization. By learning from diverse datasets, the model can accommodate variations in mango characteristics caused by geographical and climatic factors, ensuring a more reliable and universal grading system.

Mango grading is vital for both producers and consumers. For producers, quality grading helps maintain consistency, optimize pricing, and reduce post-harvest losses. It ensures consumers have access to high-quality fruit, enhancing satisfaction and trust. The integration of machine learning into this process not only improves efficiency but also minimizes human error and subjectivity in quality assessment. Furthermore, FL-driven mango quality assessment has broader implications for agricultural innovation. It demonstrates how technology can bridge gaps between traditional practices and modern demands, promoting sustainability, reducing waste, and fostering economic growth in agriculture. By ensuring privacy-preserving collaboration among researchers, farmers, and industry stakeholders, FL creates a framework for scalable and ethical solutions in agricultural technology [12]. The FL's ability to ensure data privacy while leveraging diverse datasets has immense potential in applications like mango quality assessment. The proposed system can transform how mango quality is graded by combining destructive and non-destructive methods within a machine learning framework. This innovative approach not only addresses a significant gap in agricultural research but also underscores the role of technology in creating sustainable and efficient systems for the global food industry.

The global demand for high-quality mangoes is growing steadily due to their rich nutritional value and versatility in various food products. As mangoes gain prominence in the agricultural market, practical methods for assessing and grading their quality have become essential. Quality grading ensures customer satisfaction, optimizes market value, and minimizes post-harvest losses. This process requires a comprehensive evaluation of both internal and external attributes. Internal features, such as soluble solid content, titratable acidity, pH levels, weight, dryness, firmness, moisture content, and maturity, are vital determinants of mango quality. To accurately evaluate and classify mangoes, integrating both physical and biochemical parameters is essential for defining their overall quality [13, 14]. Traditionally, grading systems have relied heavily on external attributes such as size, shape, and color. While external assessments provide a preliminary indication of quality, they may not always reflect internal characteristics. Advances in technology have highlighted the potential of more sophisticated techniques, such as near-infrared spectroscopy (NIRS), for assessing internal qualities. NIRS is a non-destructive method that uses light absorption patterns to analyze the chemical composition of food. This technique holds great promise for mangoes in evaluating parameters like sugar content, acidity, and moisture without causing any damage to the fruit. Combining NIRS with modern machine learning frameworks allows for the precise and efficient grading of mangoes, ensuring consistency and reliability in quality evaluation.

Developing innovative frameworks for mango grading is crucial to meet the growing market demand. One such advancement involves employing a neuromorphic approach-based intelligent system. Neuromorphic computing mimics the structure and functioning of the human brain, enabling the system to process complex data efficiently. This approach is efficient in agricultural applications, where diverse and nonlinear data patterns, such as those in mango quality grading, require advanced processing capabilities [15]. A neuromorphic system can integrate multiple data sources, including external features like texture, color, shape, and internal attributes derived from advanced imaging techniques. Hyperspectral imaging is a cutting-edge technology employed in this framework to estimate internal attributes. This imaging technique captures information across a wide range of wavelengths, offering detailed insights into mangoes' chemical and physical composition. Using hyperspectral imaging, the system can detect variations in sugar levels, acidity, and moisture content, critical indicators of fruit quality.

When coupled with machine learning algorithms, hyperspectral data can be analyzed to classify mangoes into different quality grades. The machine learning model can identify patterns and correlations between hyperspectral data and quality attributes, enabling accurate and automated grading. Integrating neuromorphic systems, hyperspectral imaging, and machine learning offers several advantages, including non-destructive testing, high precision, and scalability. It ensures that mangoes are graded consistently and reliably, meeting the expectations of consumers and market standards. The proposed framework enhances the efficiency and accuracy of mango grading and supports the expansion of the mango market. Farmers and producers can reduce post-harvest losses by leveraging advanced technologies, optimizing pricing, and ensuring high-quality mangoes reach consumers. Moreover, such systems promote sustainable agricultural practices by minimizing waste and maximizing resource utilization.

This paper proposes using hyper-tuned federated averaging models to ensure data privacy while evaluating a smart farming multi-labeled dataset. Unlike traditional methods, where data is shared with a central server, this approach involves training the dataset locally at the edge nodes [16]. This local training preserves data privacy, as the raw data is not shared at the edge nodes. The research utilizes PySyft, PyTorch, and Syft libraries to emulate the FL environment and test the model [17]. The federated averaging algorithm accurately predicts crop types without accessing the raw data, ensuring data privacy within a decentralized network. This approach leverages FL to enhance smart farming practices by providing accurate predictions while maintaining data security [18–20]. The investigation into FL applications in smart farming aims to address privacy concerns, enabling data scientists to evaluate smart farm datasets without accessing the actual data, thereby protecting data privacy. This research explores the use of FL in smart farming, allowing data owners to retain control over their data while still contributing to developing predictive models.

8.2 Related Work

Ensuring high-quality fruit post-harvest relies on effective quality assessment methods. Agriculture fruits are typically graded based on external and internal attributes, significantly influencing their marketability. External characteristics, such as size, shape, and color, often dictate consumer preferences, whereas internal attributes, including sweetness, texture, and ripeness, are critical for accurate grading. Among these, internal attributes hold more weight in determining overall quality.

A study on mango quality assessment during the ripening process highlights a systematic methodology. Image processing techniques are employed to enhance image quality and isolate the mangoes from their background. This pre-processing step ensures accurate feature extraction by eliminating irrelevant elements. Subsequently, a ranked search method is applied to evaluate and prioritize image attributes based on their information gain. This approach identifies the most significant features contributing to mango quality. This methodology provides a robust framework for assessing fruit quality, combining computational precision with biological relevance. The approach ensures comprehensive grading of mangoes by focusing on visual and internal characteristics. Such advancements improve the accuracy of quality assessment and support market-oriented agricultural practices, ensuring consumer satisfaction and boosting economic value. This innovative method represents a significant step forward in post-harvest fruit evaluation.

FL has emerged as a promising approach for decentralized data processing, enabling collaborative model training without centralizing data. This is particularly beneficial in smart farming, where diverse data sources from various farms can improve crop classification models [21]. This literature survey reviews recent papers on the application of FL in decentralized smart farm networks, focusing on enhancing crop classification performance. The FL framework has demonstrated significant improvements in accuracy, mainly when dealing with non–independent and identically distributed data, which is common in real-world scenarios where data distribution varies widely across different sources.

The presence of pests, diseases, and beneficial organisms in the ecosystem can affect rice production. Effective pest and disease management practices are crucial for maintaining healthy crops [22]. The choice of rice variety, including its resistance to pests, diseases, and environmental stresses, influences yield and quality. Protecting rice crops from these diseases is essential for ensuring food security, especially as the prevalence of plant diseases continues to rise. Effective disease management practices, including resistant rice varieties, proper field sanitation, timely application of fungicides and bactericides, and integrated pest management strategies, are critical to safeguarding rice production and minimizing crop losses [22–24].

The fruit industry is fundamentally driven by customer satisfaction. However, public trust in the industry has been diminishing due to inconsistent quality and concerns over health and safety. With consumers becoming increasingly health-conscious and vigilant, the need for accurate, reliable, and transparent methods for fruit quality assessment has grown. The concept of "quality" is multifaceted, varying across fruit types, target audiences, and intended applications. This variability makes assessing and grading fruit quality a progressively intricate task, requiring a nuanced approach that balances efficiency, accuracy, and consumer expectations [20, 25]. Historically, determining agriculture fruits' internal and external attributes has been a challenging and time-intensive process. Traditional evaluation methods relied heavily on visual inspections and manual assessments, often subjective and prone to error. For example, the flavor of a mango was judged based on tone or smell, while the presence of yellow spots on the skin was commonly used to identify diseases or deficiencies. While these methods provided a rudimentary understanding of fruit quality, they needed to be more comprehensive. The heavy reliance on external characteristics often overlooked critical internal factors such as sugar content, acidity, and texture, leading to inaccuracies in grading and consumer dissatisfaction [26].

Manual assessments, although widely practiced, introduced significant variability into the grading process. Factors such as human error, fatigue, and subjective judgment impacted the reliability of traditional techniques. Moreover, these methods were labor-intensive and needed more scalability, making them efficient for modern agricultural needs. Despite these limitations, conventional techniques formed the foundation for fruit quality evaluation and were instrumental in shaping the early research on mango grading. Research in traditional approaches primarily focused on external characteristics, emphasizing visible deficiencies and abnormalities. While this provided a baseline for assessing fruit quality, it needed to account for the complexity of internal attributes that play a crucial role in determining overall quality. For example, sweetness, firmness, and moisture content are critical parameters for consumer satisfaction and marketability but should have been addressed in early grading systems. This gap in evaluation highlighted the need for more advanced and holistic methods. The reliance on conventional techniques also hampered the industry's ability to address emerging consumer demands for transparency and quality assurance. As health-conscious consumers began seeking agriculture fruits with specific nutritional profiles and minimal defects, the inadequacy of traditional methods became apparent.

The inefficiencies of manual grading systems called for a shift toward more scientific and technology-driven approaches. Modern technological advancements have solved these challenges, enabling a more comprehensive and accurate fruit quality evaluation. Techniques such as hyperspectral imaging, NIRS, and machine learning models have revolutionized the grading process by allowing simultaneous analysis of internal and external attributes. These technologies not only enhance accuracy but also reduce the time required for assessments, addressing the limitations of traditional methods. For instance, NIRS can non-destructively evaluate internal qualities like sugar content and acidity, while hyperspectral imaging provides detailed insights into agriculture fruits' physical and biochemical properties. Machine learning algorithms analyze the data generated by these technologies, identifying patterns and correlations that facilitate precise grading. By combining these modern techniques, the fruit industry can move beyond the limitations of conventional methods, delivering consistent and reliable results. The transformation from traditional to modern approaches is essential for rebuilding public trust in the fruit industry. By adopting advanced technologies, the industry can meet consumer expectations for quality and transparency while ensuring sustainability and efficiency in operations. Integrating traditional knowledge with innovative techniques also provides a balanced framework for addressing the complexities of fruit quality assessment. The paper situates it within the broader context of IoT-based predictive models for agriculture [22]. The paper discusses integrating IoT devices with intelligent farming practices to gather and use climate data for predictive modeling. This integration aims to enhance the efficiency and productivity of outdoor farms by providing real-time insights and actionable data. The authors develop a predictive model that leverages climate data collected from IoT devices. It identifies patterns and correlations between various climatic factors and their impact on crop health and yield. The study [24] finds significant correlations between climate factors, such as temperature and soil moisture, and crop yield. These correlations are used to enhance the predictive accuracy of the model.

The paper [25, 27] proposes an IoT-based innovative farming system designed to minimize the need for human labor and reduce the wastage of time and natural resources. This is achieved through the automation of various farming tasks. The system applies to both traditional rural farms and urban mega-farms, showcasing its versatility and scalability. This dual applicability highlights the system's potential to transform diverse agricultural settings. The intelligent farming system includes a decision support component that leverages data analytics to provide actionable insights and recommendations for farmers. This helps optimize resource usage and improve crop yield. The collected data is transmitted to a central server, where it is aggregated and analyzed using advanced data processing techniques. The system includes a user-friendly interface that allows farmers to monitor the status of their farms in real time. Alerts and notifications are sent to farmers to inform them of any issues or necessary actions. By continuously monitoring and optimizing farming conditions, the system helps maintain ideal crop-growing conditions. This leads to improved crop health and higher yields. The system is highly scalable and adaptable, making it suitable for various types and sizes of farms [28]. The adaptability of the system ensures its effectiveness in different farming contexts.

The paper thoroughly overviews FL systems, covering their theoretical foundations, practical implementations, and emerging trends. It delves into the privacy-preserving aspects of FL, examining how these systems protect sensitive data during collaborative training processes [29]. The survey identifies critical challenges in implementing FL systems, such as communication overhead, model accuracy, and data heterogeneity, while highlighting potential opportunities for future research. The paper performs a comparative analysis of different FL frameworks and algorithms, evaluating their privacy, efficiency, and scalability performance. One of the significant challenges in FL is the communication overhead caused by frequent model updates. The paper discusses strategies to mitigate this issue, such as model compression and asynchronous communication. The scalability of FL systems is crucial for their widespread adoption. The survey discusses various architectural designs and optimization techniques to enhance the scalability of FL frameworks [30].

The paper identifies and categorizes the primary challenges in FL, including privacy, communication, computation, and data heterogeneity [31]. It provides an extensive review of the methods and algorithms developed to address the challenges in FL, covering both theoretical advancements and practical implementations. The authors propose several directions for future research to further advance the field of FL, emphasizing the need for more robust, efficient, and scalable solutions. The paper classifies the challenges in FL into several categories: privacy and security, communication efficiency, computation efficiency, statistical heterogeneity, and system heterogeneity. The survey discusses strategies like model compression, quantization, and asynchronous updates to reduce communication overhead [32]. Expanding the application scope of FL to new domains and exploring interdisciplinary approaches can uncover novel use cases and benefits.

The paper introduces adaptive FL techniques that adjust to the varying resource constraints of edge devices, ensuring efficient and effective model training [33]. It addresses the challenges that resource-constrained edge computing systems pose. The authors propose optimization algorithms to balance the trade-offs between computational load, communication cost, and model accuracy in FL. The authors develop adaptive learning algorithms that dynamically adjust the training process based on the available resources of edge devices. These algorithms consider device capabilities, network conditions, and energy availability. An optimization framework incorporates model pruning, quantization, and resource-aware scheduling techniques. By optimizing the frequency and size of model updates, the adaptive learning techniques effectively reduce communication overhead, which is a critical challenge in edge computing systems. The experimental validation confirms the practical applicability and effectiveness of the proposed methods. Future research could explore further refinements to these algorithms, such as more advanced adaptive techniques and their application to a broader range of edge computing scenarios [34].

The paper discusses integrating intelligent, innovative systems in hydroponic farming, emphasizing the role of technology in improving crop productivity. The study investigates how using intelligent systems in hydroponics can significantly improve crop yield and efficiency [30]. The paper underscores the sustainability benefits of hydroponics and the efficiency gains from using intelligent systems to manage resources such as water and nutrients. This automation ensures optimal growing conditions for the crops. The study employs data analysis and machine learning algorithms to interpret the sensor data and predict the best interventions for maximizing crop productivity. These algorithms help identify patterns and make informed decisions [35].

They compare the performance of the innovative system with traditional hydroponic methods in terms of crop yield and resource usage [36]. Using smart, innovative systems in hydroponics leads to a significant increase in crop yield compared to traditional hydroponic methods. Precisely controlling environmental conditions and nutrient supply is critical to this improvement. The intelligent system enhances resource efficiency, particularly in water and nutrients. The automated control systems minimize waste and ensure crops receive the optimal resources. The real-time monitoring capabilities of the intelligent sensors allow for immediate adjustments to be made, preventing adverse conditions that could negatively impact crop growth. The data collected and analyzed by the intelligent systems provide valuable insights into the optimal conditions for different crops. These insights can further refine the hydroponic processes and improve productivity. The smart hydroponic system contributes to sustainable farming practices by reducing the reliance on soil, conserving water, and minimizing chemical fertilizers. The author in [37] utilized a covetous calculation and a two-greatness picture scientific arrangement where the edge hubs are vehicular. It very well may be found that the eager calculation improves model exactness and conglomeration effectiveness in a united learning vehicular organization. Their work rouses working on the exhibition of FL models in canny cultivating.

In this exploration, as portrayed in Figure 8.1 of the referred to paper, the design includes versatile edge processing computing (PCs) multi-access edge computing (MECs) getting information from IoT gadgets. The MECs perform nearby preparation on this information and just send their refreshed loads to the server. The MECs then utilize this new refreshed worldwide model for the following preparation cycle. This cycle rehashes until intermingling is accomplished. The author in [38] talked about their adjusted C-portion Unified Stochastic Angle Drop calculation, which thinks about the proportion of online members to the absolute number of members inside the united organization. Their altered calculation was accomplished somewhere in the range of 99.65% and 99.85% exactness during preparation to utilize various upsides of the C-division. In spite of the amazing outcomes from their trial and error, it was seen that a similar learning rate was utilized for each of the four different divisions.

The cycle is portrayed in Figure 8.2, showing the organization stream grouping beginning from sensors that catch information and send it to

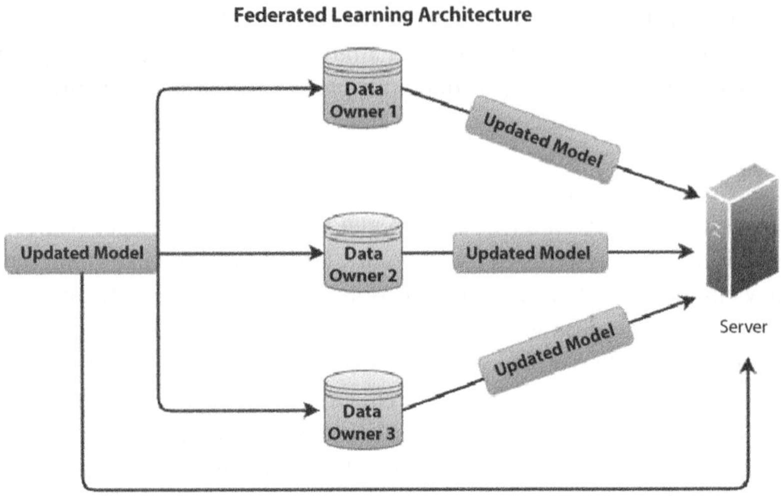

Figure 8.1 Federated learning architecture [38].

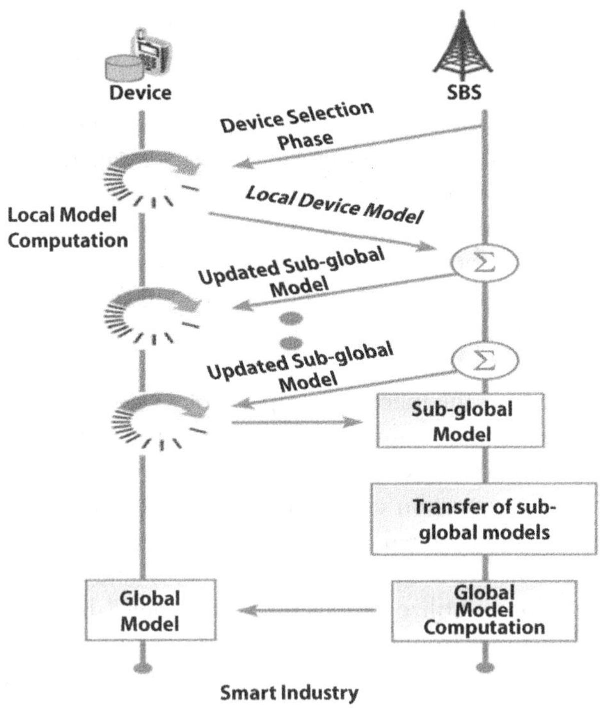

Figure 8.2 Federated learning sequence [39].

edge gadgets [39]. Dissimilar to traditional AI, where information is shipped off the cloud for preparation, unified learning adopts an alternate strategy. The server is situated at the edge hubs where the information dwells. Edge gadgets utilize the underlying worldwide model from the server to prepare their neighbourhood models and afterward send their refreshed loads back to the server. Remarkably, the total server never gets to the crude information from the edge gadgets, guaranteeing information protection and security all through the investigation. Table 8.1 presents the key features and associated challenges observed in existing systems utilizing Federated Learning for decentralized smart farm network applications.

8.3 Methodology and Experimental Setup

The research dataset comprises climatic features, including temperature, humidity, potential of hydrogen (pH), and rainfall, as independent variables, whereas the dependent variable includes equally distributed classes of rice, maize, and chickpea, indicating a balanced dataset. The data has been partitioned into 80% for training and 20% for testing using the Scikit-learn library. The research experiment is conducted in a FL environment, where each node is homogeneous, managing data with the same attributes and features, including climatic data such as temperature, humidity, pH, and rainfall. The aim is to predict a specific crop type using climatic data as independent features, with crop types serving as the labels. A modified federated averaging algorithm is employed to achieve this objective. The experiment leverages the Syft library on a decentralized platform, ensuring that data remains at the edge nodes while allowing the data scientist to remotely train the model without directly accessing the raw data, thus preserving privacy. This approach facilitates a distributed and privacy-preserving machine learning model, leveraging the advantages of FL for intelligent farming applications. For the experimental setup, simulated farm nodes are created to represent diverse environmental conditions and crop data distributions, using publicly available datasets such as crop data or other climatic datasets. In this setup, the data owner acts as the custodian of the data. The process begins with the data owner establishing a connection through the duet server and awaiting the data scientist's connection request. Once the connection is established, the data owner trains its local dataset and sends the locally updated weights to the aggregate server or the data scientist. The updated global model is then sent back to the edge devices, initiating another iteration of training. This iterative process continues until the model converges.

Table 8.1 Features and challenges found in existing systems.

S. no.	Application	Methods	Key findings	Limitations	Ref.
1	Federated learning for crop type classification in smart farming	Federated averaging and convolutional neural networks (CNNs)	Demonstrated improved accuracy in crop classification while maintaining data privacy	Limited to specific crop types; scalability issues	[28]
2	Privacy-preserving crop prediction using federated learning	Federated learning and decision trees	Maintained high prediction accuracy with data privacy; tested on multi-labeled datasets	Requires significant computational resources	[29]
3	Smart farming with federated learning: A case study	Federated learning and support vector machines (SVMs)	Improved data security and accuracy for smart farm applications	Limited edge node capabilities	[30]
4	Decentralized machine learning for crop yield prediction	Federated learning and random forests	Achieved high accuracy in crop yield prediction while preserving data privacy	Data heterogeneity among edge nodes	[31]
5	Enhancing smart agriculture with federated learning	Federated averaging and k-nearest neighbors (k-NNs)	Enhanced predictive performance and data privacy	Communication overhead between edge nodes	[32]

(Continued)

Table 8.1 Features and challenges found in existing systems. (*Continued*)

S. no.	Application	Methods	Key findings	Limitations	Ref.
6	Federated learning approaches for agricultural data	Federated learning and neural networks	Demonstrated the potential of FL in agricultural data analysis and crop classification	High energy consumption at edge nodes	[33]
7	Machine learning in decentralized smart farming	Federated learning and Gaussian naïve Bayes	Outperformed traditional ML models in accuracy and convergence	Complexity in model coordination	[34]
8	Crop classification using federated learning: A review	Federated learning and ensemble methods	Comprehensive review of FL applications in crop classification	Lack of real-world deployment cases	[30]
9	Privacy-aware crop prediction in smart farms	Federated learning and logistic regression	Maintained data privacy while achieving competitive accuracy	Dependency on network reliability	[35]
10	Leveraging federated learning for precision agriculture	Federated learning and gradient boosting machines	Improved accuracy in precision agriculture tasks; reduced data sharing risks	High initial setup costs	[36]

Gaussian Naïve Base (NB) classifiers

The authors in [39] discuss that the binary relevance (BR) method decomposes a multi-class dataset into several independent binary variables, each representing a separate label. Conversely, according to author, the classifier chain (CC) method using Gaussian Naive Bayes (NB) also breaks down the multi-class dataset into multiple binary variables. However, unlike the BR method, the CC approach identifies and leverages the correlations between the dependent variables, offering an enhancement over the BT Gaussian NB model. This improvement enables the CC method to achieve better predictive performance by considering the relationships between the labels. Furthermore, the authors in [40] explain that the label powerset (LP) method with Gaussian NB transforms a multi-label dataset into multiple multi-class single-label classification problems. This approach differs from methods that treat each label independently. The Gaussian NB algorithm, utilized in this context, is based on the NB theorem. This theorem assumes that the features are independent given the class label and follow a Gaussian (normal) distribution. By converting the problem into single-label classification problems, the LP method can leverage the Gaussian NB classifier to manage the multi-label nature of the data more effectively.

8.3.1 Federated Learning

FL can offer significant advantages in agriculture by enabling collaborative model training across different farms, agricultural research centers, and equipment without the need to share sensitive or proprietary data. Imagine a network of farms participating in a FL project to monitor crop health. Each farm collects data using drones and sensors, capturing images and environmental data. Locally, each farm trains a model to identify issues like pest infestations or nutrient deficiencies. The model updates are then sent to a central server, aggregating them to create a robust, generalized model. This model is shared back with all farms, enhancing their ability to promptly detect and respond to crop health issues.

A step-by-step description of the FL process, involved at each step:

- Each data owner iii has a local dataset, Di={(xi1,yi1),…,(xiNi ,yiNi)}
- A global model w is initialized. This can be a set of weights for a neural network or parameters for another machine learning model.
- The data owners connect to the central server (or aggregator) *via* a secure communication channel.

- This often involves performing several epochs of stochastic gradient descent (SGD) or another optimization method.
- Each data owner computes the model updates Δwi.
- The central server aggregates the received updates to form an updated global model.
- The central server sends the updated global model www back to all data owners.
- This iterative process continues until the global model converges, meaning the changes in model performance become negligible or meet predefined criteria.
- The final global model w is deployed for inference tasks.

8.4 Results and Discussion

These models were implemented within a Jupyter Notebook environment. The results obtained from these models are displayed in Figures 8.3, 8.4, and 8.5, respectively, showcasing the effectiveness and performance of each modeling approach in predicting the dependent variables based on the provided independent variables. FL can substantially benefit agriculture by facilitating collaborative model training across various farms, agricultural research centers, and equipment manufacturers without sharing sensitive or proprietary data. Imagine a network of farms engaged in a FL initiative to monitor crop health. Each farm gathers data using drones and sensors, capturing images and environmental data. Locally, each farm

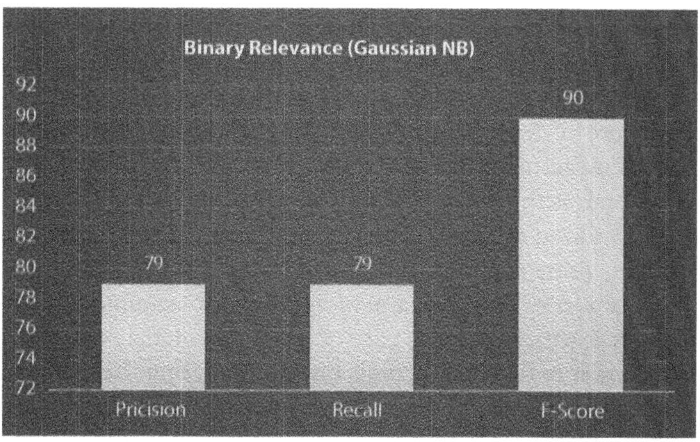

Figure 8.3 Binary relevance (Gaussian NB).

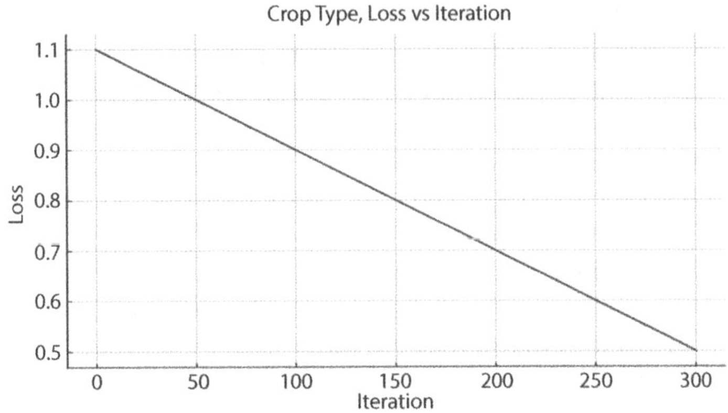

Figure 8.4 Loss using SGD optimizer, learning rate = 0.01.

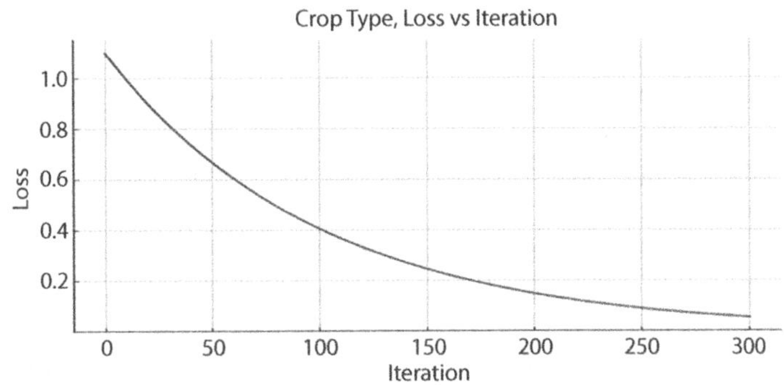

Figure 8.5 Loss using adam optimizer, learning rate = 0.001.

trains a model to identify issues such as pest infestations or nutrient deficiencies. The model updates are then sent to a central server, where they are aggregated to create a robust, generalized model.

The F1-score of 0.69 accomplished by the model connotes that the symphonious mean of accuracy and review is 0.69. This score mirrors the example normal, as the dataset is multi-named, and each occasion's example normal is utilized during assessment. Hyperparameter tuning included changing the gaining rate from 0 to 1, and trying different things with various analyzers like SGD and Adam, in light of past examination. Utilizing a SGD enhancer with a learning pace of 0.001, the model recorded a computational season of 0.00013 seconds during preparing. Notwithstanding, it possibly accomplished 23% exactness while anticipating rice, demonstrating huge mistakes and high misfortune values. This outcome proposes that the SGD enhancer at a learning pace of 0.001 performed ineffectively,

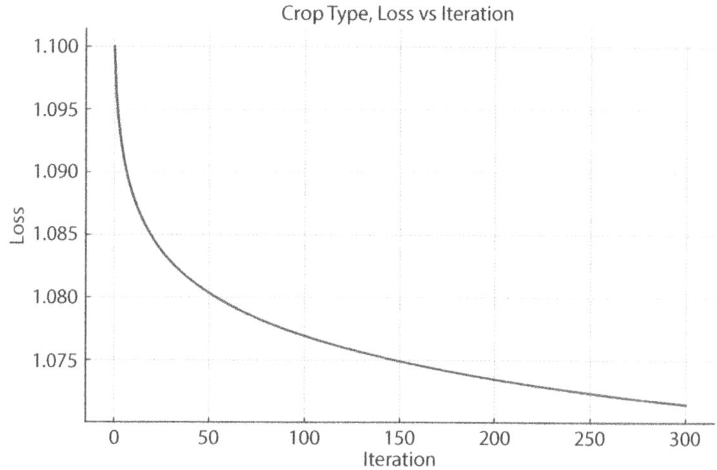

Figure 8.6 Loss using adam optimizer, learning rate = 0.01.

as just 23% of the anticipated names matched the real marks. Further hyper-tuning was performed with a united learning model, utilizing an SGD enhancer and a learning pace of 0.01. Albeit this approach prompted a few enhancements, the model actually experienced unfortunate combination due to overfitting, bringing about an exactness of 77%.

From the outcomes portrayed in Figure 8.4, the model assessment showed a base deficiency of 1.096. Figure 8.5 shows a further improvement with a base misfortune worth of 0.7, whereas Figure 8.6 delineates a far better least misfortune worth of 0.6, with observable union after 100 cycles. The misfortune kept on improving, settling at 0.1 after only 20 emphases, addressing a critical improvement over the underlying learning paces of 0.001 and 0.01 for the SGD enhancer, as well as the learning pace of 0.001 for the Adam analyzer. These discoveries highlight the productivity of the Adam enhancer, which exhibited better union contrasted with the SGD enhancer while hyper-tuning the united learning model boundaries for a shrewd ranch dataset. This affirms that unified learning models can arrive at their nearby minima at lower learning rates yet require longer preparation times to accomplish combination.

8.5 Conclusion

Climatic boundaries were treated as free elements, while crop types were replaced as reliant highlights. Preparing the dataset with the changed model uncovered that the Adam streamlining agent empowered the model to accomplish its neighborhood minima, successfully taking care of the valid and misleading positive anticipated mark classes and the valid and bogus negative anticipated subordinate factors. Different consonant mean qualities were assessed for the multi-marked dataset, which included temperature, stickiness, pH, and precipitation as free factors, as well as rice, maize, and chickpea as names. Accurate prediction of crop types is essential for intelligent farming and resource optimization. Traditional models for assessing agricultural data, such as paired significance Gaussian NB, CC Gaussian NB, Mark powerset Gaussian NB, and unified averaging models, have proven effective in specific contexts. However, exploring advanced architectures like the Swin Transformer for predicting crop types introduces a novel dimension to this field. This research proposes employing the Swin Transformer to evaluate climatic parameters as independent variables to predict crop types as dependent variables. Furthermore, a federated split learning model is integrated to ensure privacy and data security during training. This methodology enhances data utilization without compromising security by distributing the learning process across edge nodes with homogeneous datasets. The combination of Swin Transformer's robust feature extraction and Federated Split Learning's decentralized architecture aims to improve prediction accuracy and efficiency. This innovative approach could significantly advance intelligent farming technologies, setting a new benchmark for sustainable agriculture.

AI disclosure statement: The author acknowledges the use of AI-based tools for minor language editing or grammatical correction. However, all intellectual contributions, analyses, and conclusions presented in this chapter are entirely the author's own.

References

1. Manoj, T., Makkithaya, K., Narendra, V.G., A Blockchain-Assisted Trusted Federated Learning for Smart Agriculture. *SN Comput. Sci.*, 6, 3, 221, 2025.
2. Rizwan, M., Habib, Z., Sarwar, S., Iqbal, M., Safyan, M., Almakhles, D., An Exploratory Framework for Intelligent Labelling of Fault Datasets. *International Conference On Human-Centered Cognitive Systems (HCCS 2022)*, 2022, doi: 10.1109/hccs55241.2022.10090366.
3. Islam, B., Iqbal, M., Ubakanma, G., Skillbot: A Conversational Chatbot based Data Mining and Sentiment Analysis. *International Conference On*

Human-Centered Cognitive Systems (HCCS 2022), 2022, doi: 10.1109/HCCS55241.2022.10090342.

4. Khatri, S., Iqbal, M., Ubakanma, G., SkillBot: Towards Data Augmentation using Transformer language model and linguistic evaluation. *International Conference On Human-Centered Cognitive Systems (HCCS 2022)*, 2022, doi: 10.1109/HCCS55241.2022.10090376.
5. Konduru, P., Tripathi, M.K., Khairnar, D., Patil, P., Patil, M., Prashant, G., Emerging technology for smart farming in the agriculture domain: Application and future perspective, in: *2023 International Conference on Advances in Computation, Communication and Information Technology (ICAICCIT)*, IEEE, pp. 88–93, 2023.
6. Idoje, G., Mouroutoglou, C., Dagiuklas, A., Kotsiras, A., Muddesar, I., Alefragkis, P., Comparative Analysis of Data using Machine Learning Algorithms: A hydroponics system use case. *Smart Agric. Technol.*, 4, 100207, 2022, doi: 10.1016/j.atech.2023.100207.
7. Tripathi, M.K., Neelakantappa, M., Prashanthi, T., Sukte, C.D., Pandurang, D.D., Bhosle, N.P., Plant Leaf Disease Classification in Precision Farming With Hybrid Classifier: Colour, Deep and Pattern-Based Feature Descriptors. *J. Phytopathol.*, 173, 1, e70030, 2025.
8. Harakannanavar, S.S., Rudagi, J.M., Puranikmath, V., II, Siddiqua, A., Plant leaf disease detection using computer vision and machine learning algorithms. *Int. J. Intell. Syst. Appl. Eng.*, 3, April, 305–310, 2022, doi: 10.1016/j.gltp.2022.03.016.
9. Sharma, R., Artificial intelligence in agriculture: a review, in: *2021 5th International Conference on Intelligent Computing and Control Systems (ICICCS)*, IEEE, pp. 937–942, 2021.
10. Sarma, K.K., Das, K.K., Mishra, V., Bhuiya, S., Kaplun, D., Learning aided system for agriculture monitoring designed using image processing and IoT-CNN. *IEEE Access*, 10, 41525–41536, 2022.
11. Tripathi, M.K., Neelakantappa, M., Mahalle, P.N., Channapattana, S.V., Deshmukh, G., Prashant, G., Detection and Classification of Mango Fruit-Based on Feature Extraction Applying Optimized Hybrid LA-FF Algorithms, in: *Data-Centric Artificial Intelligence for Multidisciplinary Applications*, pp. 177–185.
12. Chandre, P., Mahalle, P., Shinde, G., Intrusion prevention system using convolutional neural network for wireless sensor network. *IAES Int. J. Artif. Intell.*, 11, 2, 504–515, 2022, doi: 10.11591/ijai.v11.i2.pp504-515.
13. Moorthy, C.VKNSN., Tripathi, M.K., Joshi, S., Shinde, A., Zope, T.K., Avachat, V.U., SEM and TEM images' dehazing using multiscale progressive feature fusion techniques. *Indones. J. Electr. Eng. Comput. Sci.*, 33, 3, 2007–2014, 2024.
14. Elbasi, E., Mostafa, N., AlArnaout, Z., Zreikat, A.I., Cina, E., Varghese, G., Shdefat, A. et al., Artificial intelligence technology in the agricultural sector: A systematic literature review. *IEEE Access*, 11, 171–202, 2022.

15. Applalanaidu, M.V. and Kumaravelan, G., A review of machine learning approaches in plant leaf disease detection and classification, in: *2021 Third International Conference on Intelligent Communication Technologies and Virtual Mobile Networks (ICICV)*, IEEE, pp. 716–724, 2021.
16. Goswami, T., Roy, U.B., Kalavala, D., Tripathi, M.K., Implementing generative adversarial networks for increasing performance of transmission fault classification. *Indones. J. Electr. Eng. Comput. Sci.*, 34, 2, 1024–1032, 2024.
17. Rodríguez-Lira, D.-C., Córdova-Esparza, D.-M., Álvarez-Alvarado, J.M., Terven, J., Romero-González, J.-A., Rodríguez-Reséndiz, J., Trends in Machine and Deep Learning Techniques for Plant Disease Identification: A Systematic Review. *Agriculture*, 14, 12, 2188, 2024.
18. Sarkar, C., Gupta, D., Gupta, U., Hazarika, B.B., Leaf disease detection using machine learning and deep learning: Review and challenges. *Appl. Soft Comput.*, 145, 110534, 2023, doi: 10.1016/j.asoc.2023.110534.
19. Tripathi, M.K., Hasini, S., Homamalini, M., Neelakantappa, M., Pothole Detection based on Machine Learning and Deep Learning Models, in: *2023 International Conference on Advanced Computing Technologies and Applications (ICACTA)*, IEEE, pp. 1–9, 2023.
20. Harakannanavar, S.S., Rudagi, J.M., Puranikmath, V., II, Siddiqua, A., Plant leaf disease detection using computer vision and machine learning algorithms. *Int. J. Intell. Syst. Appl. Eng.*, 3, April, 305–310, 2022, doi: 10.1016/j.gltp.2022.03.016.
21. Singh, P., Tripathi, M.K., Patil, M.B., Shivendra, Neelakantappa, M., Multimodal emotion recognition model via hybrid model with improved feature level fusion on facial and EEG feature set. *Multimed. Tools Appl.*, 84, 1, 1–36, 2025.
22. Chandre, P., Mahalle, P., Shinde, G., Intrusion prevention system using convolutional neural network for wireless sensor network. *IAES Int. J. Artif. Intell.*, 11, 2, 504–515, 2022, doi: 10.11591/ijai.v11.i2.pp504-515.
23. Tripathi, M.K. and Shivendra, Improved deep belief network for estimating mango quality indices and grading: A computer vision-based neutrosophic approach. *Netw. Comput. Neural Syst.*, 35, 3, 249–277, 2024.
24. Rajamohan, K., Image processing and artificial intelligence for precision agriculture, in: *2022 International Conference on Innovative Computing, Intelligent Communication and Smart Electrical Systems (ICSES)*, IEEE, pp. 1–8, 2022.
25. Sarkar, C., Gupta, D., Gupta, U., Hazarika, B.B., Leaf disease detection using machine learning and deep learning: Review and challenges. *Appl. Soft Comput.*, 145, 110534, 2023, doi: 10.1016/j.asoc.2023.110534.
26. Jackulin, C. and Murugavalli, S., A comprehensive review on detection of plant disease using machine learning and deep learning approaches. *Meas.: Sens.*, 24, September, 100441, 2022, doi: 10.1016/j.measen.2022.100441.
27. Kwabena, P., *et al.*, CCMT: Dataset for crop pest and disease detection. *Data Br.*, 49, 109306, 2023, doi: 10.1016/j.dib.2023.109306.

28. Federated learning: Applications, challenges and future directions. *Int. J. Hybrid Intellint Syst.*, 18, 1-2, 19–35, 2022, doi: 10.3233/HIS-220006.
29. Mahalle, P.N., Shinde, G.R., Ingle, Y.S., Wasatkar, N.N., *Data Centric Artificial Intelligence: A Beginner S Guide*, Springer Verlag, Singapore, 2023.
30. Naveen Kumar, K., *et al.*, Precision Guided Approach to Mitigate Data Poisoning Attacks in Federated Learning. *ACM CODASPY 2024*, 2024, doi: 10.48550/arXiv.2404.04139.
31. Shafik, W., Tufail, A., Namoun, A., De Silva, L.C., Apong, R.A., A systematic literature review on plant disease detection: Motivations, classification techniques, datasets, challenges, and future trends. *IEEE Access*, 11, 59174–59203, 2023.
32. Balafas, V., Karantoumanis, E., Louta, M., Ploskas, N., Machine learning and deep learning for plant disease classification and detection. *IEEE Access*, 11, 114352–114377, 2023.
33. Mehta, S., Kukreja, V., Gupta, R., Empowering Precision Agriculture: Detecting Apple Leaf Diseases and Severity Levels with Federated Learning CNN, in: *2023 3rd International Conference on Intelligent Technologies (CONIT)*, IEEE, pp. 1–6, 2023.
34. Zhou, H., *et al.*, Vertical Federated Learning: Concepts, Advances and Challenges. arXiv. doi: 10.48550/arXiv.2211.12814, 2023.
35. Kumar, M., Aeri, M., Chandel, R., Kukreja, V., Mehta, S., Federated Learning CNN for Smart Agriculture: A Modeling for Soybean Disease Detection, in: *2024 IEEE International Conference on Interdisciplinary Approaches in Technology and Management for Social Innovation (IATMSI)*, IEEE, vol. 2, pp. 1–6, 2024.
36. Rajput, K., Garg, A., Kukreja, V., Mehta, S., Smart Agriculture: Innovating Soybean Leaf Disease Detection with Federated Learning CNNs, in: *2024 3rd International Conference for Innovation in Technology (INOCON)*, IEEE, pp. 1–6, 2024.
37. Aledhari, M., Razzak, R., Parizi, Reza, Saeed, F., Federated Learning: A Survey on Enabling Technologies, Protocols, and Applications. *IEEE Access*, 8, 1–1, 2020. 10.1109/ACCESS.2020.3013541.
38. Khan, L.U., Alsenwi, M., Yaqoob, I., Imran, M., Han, Z., Hong, C.S., Resource optimized federated learning-enabled cognitive internet of things for smart industries. *IEEE Access*, 8, 168854–168864, 2020.
39. Vats, S., Khanna, M.S., Kukreja, V., Mehta, S., Federated Learning CNN Empowers Precision In Agriculture: A Case Study in Soybean Leaf Disease, in: *2024 IEEE International Conference on Interdisciplinary Approaches in Technology and Management for Social Innovation (IATMSI)*, IEEE, vol. 2, pp. 1–6, 2024.
40. Misra, N.N., Dixit, Y., Al-Mallahi, A., Bhullar, M.S., Upadhyay, R., Martynenko, A., IoT, big data, and artificial intelligence in agriculture and food industry. *IEEE Internet of Things J.*, 9, 9, 6305–6324, 2020.

9

Revolutionizing Agriculture Yields through Federated Learning

Ramit Sehgal, Nitendra Kumar* and Yash Dwivedi

Amity Business School, Amity University, Noida, Uttar Pradesh, India

Abstract

This chapter explores the potential of federated learning to transform crop yield prediction, which is such an important part of modern agriculture. The agricultural sector is still facing such serious problems as climate change and the increased demand for food. At the same time, traditional methods used for crop yield prediction cannot cope with even basic requirements, including scalability and data privacy, due to their centralization. In this chapter, we start by introducing a precursor to the role of federated learning in agriculture, hinting at the current applications and success stories that demonstrate its efficacy. The background develops an overview regarding crop yield prediction under traditional methodologies and their disadvantages, which is then followed by developments in machine learning and its further role in big data. The paper elaborates on the cause-and-effect relationship and how other factors like soils, water, and pest management factors affect the yield. It then explores the challenges of using machine learning techniques in that arena. This paper illustrates practical applications of federated learning in agriculture settings through case studies. We outline some of the benefits while being candid about the challenges that federated learning faces. Ultimately, the chapter envisions a future in which federated learning will become a means to revolutionize crop yield predictions by contributing to sustainability and food security.

Keywords: Federated learning (FL), crop yield prediction, IoT integration, blockchain technology, agricultural innovation, regulatory frameworks

**Corresponding author*: nkshukla.kumar4@gmail.com

Padmesh Tripathi, Bhanumati Panda, Shanthi Makka, Reeta Mishra, S. Balamurugan and Sheng-Lung Peng (eds.) Federated Learning for Smart Agriculture and Food Quality Enhancement, (217–246) © 2026 Scrivener Publishing LLC

9.1 Introduction

Agriculture is one of the most essential industries for ensuring food security and economic stability. It requires ongoing innovation to meet the increasing demands of a growing global population. Crop yield prediction, an important aspect of agricultural management, has long been used by farmers, governments, and agribusinesses to plan and optimize resource allocation, limit risks, and enhance decision-making processes [1]. Traditionally, these forecasts have been generated using statistical models based on historical data such as climatic conditions, soil properties, and farming methods [2]. However, with the emergence of big data and machine learning (ML) technology, the landscape of agricultural yield prediction has changed considerably. In recent years, several ML and artificial intelligence (AI) approaches have developed as a breakthrough in developmental exploration models for more accurate and widely expanded prediction. Neural networks, decision trees, and ensemble methods are more effective than classical statistical methods when identifying a complex non-linear relationship in data [3, 4]. However, centralized ML models still face challenges: data privacy, logistical aspects of data gathering, or dealing with highly diverse data originating from multiple locations that can scarcely be considered homogeneous [5–7].

These problems are tackled by the so-called federated learning (FL) which is an emerging solution for the training of ML models where many data sources contribute without sharing the data [8]. FL is an approach where both data training and model refining are done in a distributed manner where for instance, farms or agricultural research centers can individually train models on local data and share only the update of the model weight gradients with the main server. This approach preserves the privacy of the data. It allows the accumulation of the results from different regions, which is perfect for improving AI techniques in agriculture which often involves the sensitive and scattered data and outcomes of different geographical locations [9].

9.1.1 The Need for Improved Crop Yield Prediction

It has never been as important to be able to predict yield as accurately as possible. It is predicted that food production will rise approximately by 70% by 2050 due to the increased global population [10]. Furthermore, the impacts of climate change have increased the risks of food production by reducing agricultural activities due to variability that results from frequent droughts, floods, and extreme climate conditions, which are contrary to climate trends for farming [11]. These challenges necessitate a finer, while timely, understanding of how micro-climate influences, soil conditions, and the specific crop variety being grown. Crop yield prediction may provide information to farmers to assist them make informed decisions regarding crop planting, water availability, nutrition support, and crop harvesting times.

In the past, crop yield prediction has been done by statistical models which include but are not limited to; Linear regression or time series analysis [2]. Unfortunately, when it comes to modelling these factors, there are commonly used models that, although helpful for simpler model systems, were originally based on the assumption of linear interactions between these factors and do not consider the multiple feedback and dynamic complex relationships that exist in agricultural ecosystems. For example, crop development is closely related to temperature or rainfall, but it is also relevant to the nature and properties of the soil and the variety of crops grown. Thus, the traditional approaches contain less accuracy in the parameter estimation and lower predictive capability, at least when the climate changes [12].

The difficulties outlined above have been effectively dealt with by one of the most rapidly expanding branches of ML—deep learning. Due to the envelopment of a big amount of data encompassing all the environmental factors and patterns in farming, ML models can determine multiple, non-linear interactions between inputs, such as climatic data or soil type, and outputs, such as crop yield [3, 4]. These models can be a work in progress as more data is brought into the process making them more responsive to existing conditions.

Nevertheless, the use of ML in crop yield prediction is equally faced with the following challenges. Another challenge test is the data issue: lack of data or poor quality of data is one of the most important problems that we face. Agricultural data is dispersed and dispersed in a very poor and uncoordinated manner across various organizations, regions, and even farms. What is more, most of this data is considered sensitive, as it contains information about the productivity of the farm, the condition of the soil, or diseases of crops, for example. This is data that farmers or agribusiness are very loath to share especially if this could be in the hands of competitors or lead to increased regulatory attention [13]. Furthermore, the nature of the data gathered from agriculture concerning weather and sensorial information from the soil as well as satellite creates technical barriers to data fusion and analysis in providing significant information.

9.1.2 The Role of Federated Learning in Agriculture

More so, FL exploits many of the problems central to the ML model by enabling different parties to build models together without having to share their data [8]. As in an FL framework, individual participants, including farms, research institutions, or agricultural cooperatives, learn local models on their data and then send only updates to a central server. These updates are pooled to the central server which forms a global model that benefits from the knowledge of all participants without violating their privacy.

Below are a few advantages of the FL approach in crop yield prediction: First, it means a possibility to involve several farms or regions, which will result in the development of more accurate and universal models. Thanks to the data aggregation from different geographical areas, FL models can adapt the information for climates, types of soil, or farming techniques that may be constant in one location and vary in another maximum bootstrap 76 [14]. Second, FL has an advantage over data privacy in that raw data about the individual participants are stored locally on the participants' devices. This is especially so in agriculture where data protection can be a major impendent to cooperation [9].

However, FL can bring solutions to some of the logistic issues that are inherent to data collection in agriculture. In most instances, farms are situated in remote or rural zones where the Internet connection is poor to permit the transfer of massive data to a central unit for analysis. FL prevents direct data transmission because only model updates require transmission between participants and the central server. This cuts down the data transfer needs and makes it possible for farms in areas where internet connection is poor to contribute to model training [15].

9.1.3 Current Applications and Success Stories

Although FL is a relatively new method, it has already been applied in a variety of sectors. Researchers and specialists have effectively applied FL to fields like healthcare, banking, and mobile technology, proving its ability to address real-world issues and stimulate innovation in these industries. In healthcare, for example, FL has been used to develop models for disease diagnosis with no expectation that hospitals will input patient data [13]. In finance, FL has enabled banks to jointly build fraud detection algorithms while keeping individual datasets secure from other banks. The following portion of this article will demonstrate that FL may effectively manage privacy and collaboration issues in sensitive sectors, showcasing these success stories with a variety of actual use cases.

At present, the agricultural sector encounters limited application of FL, yet researchers have identified its potential to predict crop yield and solve other precision agriculture problems. Li *et al.* [14], for instance, showed how FL is capable of yielding wheat based on data obtained from farms in different areas. The study found that the model was more accurate and generalized than typical centralized models while also protecting farm data privacy.

The second example is based on the FL application in climate modelling: The use of FL has helped aggregate regional climate data to improve weather forecasting and avoid intrusion into individual data [15]. Better weather forecasts can directly influence crop yield forecasts, as weather data is the most important input in crop performance.

9.1.4 The Future of Federated Learning in Agriculture

The benefits, which farmers and agriculture in general can reap from FL, are tremendous. Co-location is another complementary approach to increase the length of corresponding indexes as more farms use digital technologies, gather information on the state of soil, plants, and weather, and thus enhance the opportunities to use this data for a more accurate estimation of crop yield. FL is a viable solution for tapping into this data without squandering precious resources in privacy or data exchange, making it a more effective means of getting around some of the issues that come with the traditional ML models.

Several trends will influence the future of FL in agriculture as follows: First, a mobile and cloud era in FL and Internet-of-Things (IoT) devices will enhance data collection and processing at the farm level and allow farms to be part of the FL process without involving large data centers [15]. Second, researchers could combine FL with another innovative undertaking such as blockchain to augment data protection in agricultural partnerships [16, 17]. In conclusion, extending research on the use of FL in agriculture will lead to the development of better algorithms and methods as a more diverse and more volatile character of agricultural data is recognized.

Enhancing agricultural output estimates with FL may be an advantageous approach, distinctively when dealing with issues like dispersed data, privacy concerns, and differences in data sources. It offers a workable solution to safeguard sensitive data and enhance reliability. With this capacity to redraw the boundaries of complexity without relocating data, FL has the BPO potential that successfully addresses challenges in the agricultural sector and therefore enhance food security and sustainability.

9.2 Overview of Crop Yield Prediction

Crop yield forecast is an integral part of the entire crop production planning and control cycle. Appropriate forecasting relative to crop yield becomes important for resource allocation to minimize waste and achieve sustainability of food systems [1]. This information supports farmers, agribusiness companies, authorities, and scholars while deciding when to plant, how to build fertilization programs, when to harvest, and how to sell obtained crops. In earlier days, agriculturalists' empirical knowledge with simple statistical models was relied on for crop yield predictions, but developments of methods for data collection and advances in AI have now increased the accuracy of yield forecasting.

9.2.1 Traditional Methods of Crop Yield Prediction

Traditionally, crop yield prediction is a regression-based statistical model. Other models utilized historical data, including past crop yields, climate, and soil type, to predict future yields. Linear regression models were often used where it was presumed that variables under study held linear and non-correlated relationships [2]. However, these approaches did not analyze the dynamics of agricultural systems and relationships between climate factors, soil, and crop developmental stages [12].

Through statistical methods, crop simulation models have also been used to estimate crop growth based on specific input requirements of the environment at different growing stages. Testing models such as the Decision Support System for Agrotechnology Transfer and Agricultural Production Systems Simulator have been used to simulate crop production. These models encompass soil moisture, nutrient status, and pest control, but they are often sensitive to calibration and parameterization [18]. Also, they are constrained by the need for accurate input data and cannot easily make cross-regional or crop comparisons.

9.2.2 Limitations of Traditional Methods

However, many statistical and crop simulation models, which have been commonly used for many years for yield prediction, possess significant restrictions. First, these models are usually based on a fixed parametric relation between inputs and yield, and it is difficult to describe the non-linear and dynamic nature of agricultural systems using such models [19]. For instance, crop yields are affected by myriad factors, including temperature, precipitation, other features of soils, pests and diseases, irrigation, and fertilizer, among others [1]. These interactions are often reduced in traditional models, and, therefore, yield prediction is likely to be off when climate changes, as noted by Challinor [11].

The second limitation that Chu and Miao mentioned is that the system requires much calibration. Crop simulation models are expected to mimic the conditions of farms or regions, and, hence, the models need to be parameterized accurately. It can be very time-consuming, and the knowledge of agronomic practices, field characteristics, and local climate is inherent in the process. However, as the impacts of climate change increase and new information for recalibrating these models becomes irrelevant, this approach becomes inefficient as well [20].

9.2.3 Advancements in Machine Learning for Crop Yield Prediction

But with the emergence of big data and ML techniques, researchers moved to more complex models, capable of treating the essential non-stationary nature of agricultural systems. ML models can be especially useful for crop yield prediction as they usually involve determining a relationship in a large dataset where programming is not required [3, 4]. They also can include a mass of different data from satellites, weather stations, sensors of soil moisture, historical data on yields, etc.

The most frequently applied approach to ML for yield estimation belongs to the class of ANNs. ANNs are general-purpose models, which allows them to easily learn relationships that are not linearly defined between input and output. For instance, Shahhosseini *et al.* [21] applied deep learning models for estimating food yield, particularly maize, based on weather parameters with relatively high accuracy compared to other conventional models. In the same way, random forests and support vector machines have been used in crop yield prediction and have yielded better results than regression models [2].

A consequent promising sector is the application of activities of remote sensing data, like satellite imagery and drone sensors, for the evaluation of the cereal crops' status and yield estimation. Remote sensing gathers timely information on crop status, soil water content, and vegetation vigour that can be fused with a ML algorithm for enhanced yield forecasts [22]. Because farmers can now utilize the data generated to make better decisions, these advances have completely changed the way farmers manage their crops.

Table 9.1 Comparison of crop yield prediction methods.

Method	Advantages	Limitations
Statistical Models	Simple to implement, requires minimal computational power	Assumes linear relationships, struggles with non-linearities
Crop Simulation Model	Simulates detailed crop growth process	Requires extensive calibration, limited by data quality
Machine Learning Model	Can handle complex, non-linear interactions	Requires large datasets, potential for overfitting
Remote Sensing	Real-time data, highly accurate	Expensive, requires advanced technical skills

Table 9.1 provides a comparison of traditional crop yield methods with advanced ML techniques, highlighting the advantages of the latter in terms of accuracy, flexible, and scalability.

9.2.4 The Role of Big Data in Crop Yield Prediction

Big data has a central function in increasing the solidity of crop yield estimation models. The explosion of information arising from sensors deployed on farms, satellites, and aerial imagery, together with smart machines, has enhanced farmers' access to information such as growth progress, weather patterns, and the quality of the soil. When integrated with ML, such data helps to predict yield at a much finer level. For example, in the case of production rates, which have formerly been provided as yields, data can now be updated live in addition to figures of yield, helping farmers to make better decisions regarding their watering, feeding, and pest control.

Advanced strategies also allow increased early problem identification related to crops, enabling farmers to identify problems like nutrient deficiency or pest attacks [23]. Of the two aspects, the former offers an organization the predictive capability it needs to manage resources more effectively to increase productivity while minimizing wastage.

9.2.5 Challenges in Adopting Machine Learning for Crop Yield Prediction

Despite the huge potential in predicting crop yields, ML methods have some gaps that need to be addressed. Large, high-quality datasets are some of the most difficult things to come by when it comes to applying big data techniques. According to Kamilaris and Prenafeta-Boldú [3, 4], ML models work based on the use of large datasets that enable the model to learn about the correlation between input variables and the yield outcomes. However, the acquired agricultural data is frequently dispersed across several organizations and geographical areas, making the consolidation and accumulation of the data more difficult. Additionally, there may be expenses associated with gathering data in agriculture, particularly when sensor networks or remote sensing are used.

The other difficult problem is model interpretability. Some ML models, especially deep learning models, are usually referred to as black boxes; this is because most of the time it is hard to follow the movements of the model [21]. This lack of transparency can be especially concerning in agriculture, where the farmer and the agronomist need to understand what is driving the prediction if they are to act on it.

Accounting and data privacy is another issue with the use of big data in farming because some data owned by farmers is sensitive. Farmers may be averse to disclosing the details of practices and yields to others, including third parties, in a way that has the intention of getting improved crop yield predictions [13]. This presentation was previously introduced in the introduction; FL is becoming a solution for this problem by enabling farmers to work together on multiple learning models and yet never share the data the model works on.

9.2.6 Case Studies and Applications

Numerous case studies, such as the one being reviewed, demonstrate how the machine-learning method may enhance agricultural production prediction. For instance, Jeong *et al.* [24] used several ML algorithms on rice yield mechanics in South Korea, which included temperature, rainfall, and moisture contents of the soil. According to their findings, ML was more accurate than conventional regression models, especially under volatile weather conditions.

In another clause, van Klompenburg, Kassahun, and Catal [25] tried to compute the yield prediction of wheat in the United States utilizing ML calculations depending on yield records and remote sensing index. It emerged that the random forest models offered the best accuracy, and this reduced error rates when compared to conventional models.

In India, Shukla and Jharkharia [26, 27] used deep learning to assess potential sugarcane yields, showing ML methods can be used in many settings in agriculture. The authors of these case studies illustrate that ML has become an increasingly important factor in increasing agricultural productivity and climate change adaptation.

9.2.7 The Future of Crop Yield Prediction

In the future, crop yield prediction is expected to use multiple data sources, such as remote sensing imagery, IoT sensors, and genomics data, in order to deploy more precise models of crop yields. FL and edge computing are improving to encourage the involvement of farmers in the development of data-driven approaches to predict crop yields by sharing their private data [15].

Therefore, conventional approaches for crop yield forecasting have been heinous, but with the emergence of ML systems and big data, they have proved to be more accurate and can be scaled up than traditional methods. With the help of these technologies, farmers would be in a position to optimize the use of their inputs and work on the elimination of wasteful practices or outcomes within the constantly evolving environmental and production conditions in the agricultural profession.

9.3 The Importance of Crop Yield Prediction

Yield estimation for crops is a very important component of contemporary agriculture because it provides the base for agricultural decision-making vis-à-vis food security, resource utilization, and decision-making by stakeholders in the agriculture industry. Risk predictions make it easy to make the right choices in farming, policymaking, and agronomic practices aiming at impacting productivity and sustainability. This section looks into the causes of crop yield, ways used today to predict crop yield, and the difficulties involved in predicting accurate crop yield.

9.3.1 Factors Influencing Crop Yield

Understanding the factors that influence crop yield is essential for improving prediction accuracy. These factors can be categorized into climatic, biological, and management aspects.

9.3.1.1 Climate Factors

Adverse climatic factors are known to affect seed germination as well as the growth and production of crops. They include temperature, rainfall, humidity, and solar radiation, which play an incredibly vital role in the development of a plant. For instance, studies conducted by Lobell *et al.* [1, 12]. Undoubtedly, yield declines with rising temperature, especially for crops such as; wheat and maize. Because global temperatures are rising because of climate change, the chances of yield reductions also increase, and prediction becomes even more important.

i. **Temperature:** The temperature at which a grower of different crops is most appropriate in its growing process differs. High temperature affects and increases heat stress which reduces the yield. For instance, Lobell *et al.* [1, 12] have revealed that at an optimal temperature range of 30°C/35°C optimal temperature, wheat yield may reduce by around 6% for each degree Celsius above optimum temperature.
ii. **Precipitation:** It is also very important when this water is available and the quantity is the correct quantity to make correct moisture levels for soil to support crop growth. Water deficits during development stages cause stress from drought and reduce yields.
iii. **Humidity:** These diseases and pests make the yields of crops to be very low especially if they are grown using high humidity. For example, sometimes, levels of humidity can be high resulting in fungal diseases that harm yields.
iv. **Solar Radiation:** Lack of sufficient light can at times lead to slow photosynthesis, the process by which plants absorb light energy and transform it into chemical energy. For proper growth and crop development, the plants demand enough solar radiation.

9.3.1.2 Soil Conditions

Soil quality also plays a major role in the yield determination of crops. Soil nutrients, acidity, and aggregation of soils impact plants' health in one way or another. Several authors have provided evidence that better management practices of soil can enhance production outputs [1].

i. **Nutrient Availability:** Nitrogen phosphorus and potassium are some nutrients that crops are needed in large amounts before they can grow mature. Deficient soils deprive yield potential.
ii. **Soil Structure:** Organized soils yield good root structures and water-holding capacity which benefits crop production.

Fragmented structures, on the other hand, hinder root penetration and soil water uptake while compacted structures reduce the ability of the soil to penetrate roots and take up water.

iii. **Soil pH:** These include the nature of soil implying that the variation in pH will determine the extent to which nutrients are available. Ideal soils for the growth of most crops have a slightly acidic to neutral pH of between 6.0 and 7.0.

9.3.1.3 *Water Availability*

Agriculture is one of the leading consumers of water; it consumes about 70% of the freshwater used in the world (FAO, 2021). The tricky part is the water aspect; it is as important as irrigation water or rainwater to make crops healthy.

i. **Irrigation Systems:** Irrigation techniques, for instance, drip irrigation, are, however, efficient in their usage, and yields are likely to be improved. Research also revealed that optimized irrigation was probably to change from a desert area yield by 50%.
ii. **Natural Rainfall:** An attribute of climate change, changes in the patterns of rainfall put farm activities in a situation where they are not quite certain of water availability.

9.3.1.4 *Pest and Disease Management*

That is, control of pests and diseases influences high-yield production in profitable crop-production enterprises. Several authors agree with that pests and diseases play a negative role in crop management and are responsible for crop losses ranging from 20% to 40%. Reduction of these losses can be achieved through biological, cultural, and chemical methods using Integrated Pest Management.

i. **Pesticide Use:** Farmers use pesticides; they rely on them to protect their products. But with chemicals, they develop a resistance that facilitates the usage of integrated measures all over the world.
ii. **Monitoring Systems:** Measures: The incorporation of monitoring techniques to check for signs of pests and disease, which, if early spotted, helps curb and can also protect yields.

9.3.1.5 *Agricultural Practices*

Farmer's practices play a great role in the productivity of the crops grown. Conservation of practical measures of the social use of soil fertility and other natural resources available in countries can also increase the yields of crops.

1. **Crop Rotation:** Cropping in spare triennial cycles can help replenish nutrient content in the soil and minimize pests and diseases.
2. **Fertilizer Application:** This path highlights that the timing and choice of fertilizer could tremendously affect the yield. Scientists have it that more efficient use of fertilizers could help boost yield by between 30% and 50%.

9.3.2 Current Methodologies for Crop Yield Predictions

The classical methods of estimating yield have changed over the years using statistical and ML approaches.

9.3.2.1 *Statistical Methods*

In the past, crop yield prediction has mainly used statistical techniques, including regression analysis and time series forecasting. These methods involve the use of yield history information and associate this with the climatic factors and come with a forecast. However, they usually do not incorporate multi-scalar and multi-step dependence patterns that are inherent in agricultural data.

9.3.2.2 *Machine Learning Techniques*

Yield prediction in recent years has adopted ML techniques due to the improvement in technology. Techniques like deciding trees, random forests, or even the more complex neural networks can deal with relations that exist between more than one variable and thus welcome increased accuracy when making predictions.

i. **Decision Trees:** Decision trees as a tool can be easily interpreted by the people involved because they map the decision-making process well. They can always cope with non-linear relationships and make good predictions about the outcome.

ii. **Random Forests:** This sort of ensemble learning greatly enhances the planning of several models of the decision trees and reduces the amount of overfitting.
iii. **Neural Networks:** Neural networks are versatile models when it comes to accurately predicting the shapes of signals in data. It allows them to work with big data and relate variables in a complicated manner.

Despite the advantages of these methodologies, one significant challenge remains: the requirement of a significant and diverse sample size for training them aptly. Due to the fragmentation of data accumulated in the agricultural field, many challenges arise in creating a vast range of integrated predictive models.

9.3.3 Challenges in Crop Yield Prediction

The accurate prediction of crop yield is an exercise that is marred by several factors that ought to be overcome to improve on the forecasting mechanisms developed.

9.3.3.1 Data Silos

Farm data is often not fully stored, often distributed between various organizations, districts, and farms, and it often creates data blocks, which hinders comprehensive dataset access. These silos can hamper the creation of proper models that can be easily scaled to varying environments to generate successful outcomes [3, 4].

9.3.3.2 Privacy Concerns

The participants may not want to disclose other information that ought to be collected from the farmers, such as information about the soil and the health of the crops and other procedures that are undertaken on the farm due to privacy. This reluctance to share data can, therefore, lead to an increase in data silos, making cross-stakeholder contributions unable to publish on the identified date.

9.3.3.3 Environmental Variability

Agricultural systems are unambiguously complicated and are subjected to numerous environmental impacts in contrast to other systems.

This variability has made it difficult to build well-fitting general models that capture all the different environments in which crops are grown [1].

9.3.3.4 *Technological Limitations*

However, even with these progresses in available methodologies of prediction, the constraints posed by technology including by low computing capacity and infrequently sourced data may slow down the use of these methods in practical agriculture.

9.4 Federated Learning in Agriculture

FL is an innovative approach that can solve the issues created by data privacy and access to data in agricultural research. Hence, in FL, many entities for instance farms and agricultural research institutions, can collectively train the ML model even without the need to exchange sensitive data [8]. This section traces FL in agriculture, how it has been utilized, and the advantages therein, which exposes the potential to improve the accuracy of crop yield estimation.

9.4.1 Application of FL in Agriculture Settings

FL can dramatically transform the agriculture sector by solving major issues concerning data acquisition and protection. It allows a variety of farms to participate in a single model training process despite maintaining their data secrecy. Such an idea is most fitting in crop yield prediction, where using a plethora of data collected across different locations improves both the model accuracy and interpretability [13].

Table 9.2 Farm data contributions to model development.

Farm ID	Location	Data type	Model contribution
Farm 1	Midwest, USA	Climate and soil data	Provide model surplus updates on regional conditions
Farm 2	Southeast, USA	Irrigation and pest data	Updates model according to water availability
Farm 3	Northwest, USA	Soil moisture data	Contributes insights on soil quality

Data Contribution from Farms
In a FL system setup, none of the participating farms sends their raw data, but contributes to the formulated model. For example, consider three farms located in different regions.

In Table 9.2, each farm contributes valuable insights based on its unique environmental conditions, enhancing the model's ability to generate across diverse contexts.

9.4.2 Case Studies and Examples

Hence, FL has the promise and proven potential in many domains, and its use in agriculture is increasingly. Its ability to predict crop yields is exemplified in several studies.

Case Study: Predicting Disease Outbreaks
A similar example is a work by Sheller and his team, who employed FL to forecast diseases in crops [13]. Using the proposed approach of aggregating data from several farms but keeping privacy in mind, the model proved to outperform other conventional models in terms of accuracy of disease occurrence prediction.

Case Study: Improving Yield Forecasting Models
A similar case is the group of farms in a particular area where FL was implemented to improve the yield forecast models. Every farm estimates a local model on-farm data, and the combined model proved superior to individual models. From this cooperation, there was an enhanced appreciation of the factors that influence yield and therefore enhanced custodianship practices.

9.4.3 Benefits of Federated Learning

The adoption of FL in agriculture offers several key benefits:

i. **Data Privacy:** FL makes certain that the critical data of farms is contained within the local setting and, at the same time, overcomes the key barriers of privacy issues that affect shared learning environments. They will never misuse the data that farmers will provide them with during model training.
ii. **Improved Model Performance:** Thus, FL motivates participation from multiple areas for improving the model both in terms of efficiency and reliability. This improvement is especially useful for agriculture, as the variability of the

environment has a major influence on the number of crops justified.

iii. **Resource Efficiency:** FL is considered more efficient because it limits the centralization of data and their processing compared to the traditional approach. It keeps down the costs of data exchange and utilizes the local processing capacity for model building.

iv. **Scalability:** Because FL decentralizes the learning process, it is easy to scale it up. Indeed, as more farms are incorporated into the network, the model keeps on getting better because of increased data input without affecting privacy.

9.4.4 Challenges of Federated Learning in Agriculture

Despite its potential, several challenges must be addressed to effectively implement FL in agricultural settings:

i. **Communication Overhead:** In FL, the participating farms need to constantly exchange data with the central server to update the used model. This communication can bring in matters pertaining to latency and bandwidth, which are even worse in rural areas due to the internet connection.

ii. **Model Convergence:** There might be a tough time to reach a consensus in FL models because farm data distribution is quite diverse. External factors on the farms can vary, and this means that the data generated can be heterogeneous, making it harder to train a model.

iii. **Data Quality:** The quality of data provided by farms will determine the performance of the model applicable to the farms. Keeping data from farmers honest can pose some difficulty because not all the farmers have the same kind of technology used in data collection.

9.5 Implementation of Federated Learning for Crop Yield Prediction

FL is trending as a new model shift in the paradigm level of ML, especially within domains of sensitive data like agriculture. While acting as a decentralised system that empowers the training of ML models, FL provides multiple entities, including farms, with a means for model predictive training without having to share fundamental data. This section provides a description of how FL can be used in the prediction of crop yield, the process followed, the architectures that are possible, the challenges that may be encountered during the process, and, in view of this, the impact of FL on agricultural productivity.

9.5.1 Understanding Federated Learning

FL is a collaborative framework where many participants (in this case, farms) jointly train a model, but the local data are not aggregated. To achieve this, each participant does not upload its raw data but trains the model locally and uploads only the changes made to the model, for instance, weights or gradients, to a central server. This process also helps to protect the user's data privacy and uses various data sets collected from multiple places to make the model more reliable and accurate [8].

Figure 9.1 shows the FL setup, whereby multiple farms (nodes) work to train local models and share updates with a central server which further improves the global model based on the updates.

9.5.2 Steps for Implementing Federated Learning in Crop Yield Prediction

Implementing FL for crop yield prediction involves several key steps:

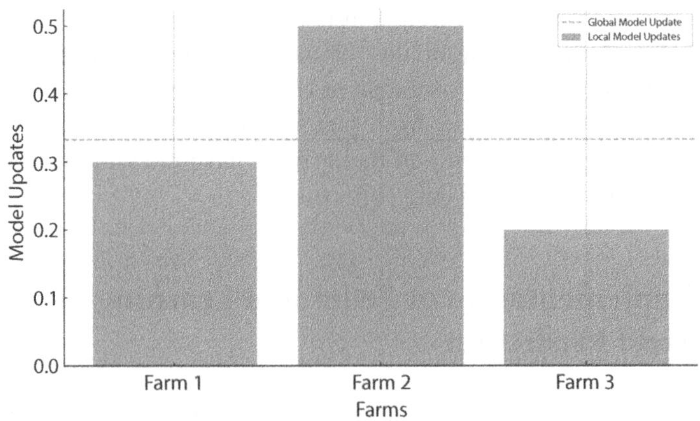

Figure 9.1 Federated learning framework.

9.5.2.1 *Step 1: Data Collection*

All the farms involved in the study record information that could affect crop yield estimation, such as climate conditions, soil characteristics, and past yields. This data collection should also consider issues like the quality of the data, the resolution of the data, and the kind of variables involved, as pointed out by Kamilaris and Prenafeta-Boldú [3, 4].

Data Types: Common Data Types Include:

i. Climate Data: climate factors, including temperature, precipitation, and humidity.
ii. Soil Data: such things as pH, nutrient, and moisture.
iii. Management Practices: area-specific fertilizer rates and irrigation techniques.

9.5.2.2 *Step 2: Local Model Training*

Each firm fits a sub-model based on its information. This process normally entails deciding on an ideal ML algorithm most relevant to the data nature besides the complexity of relationships in different variables, such as linear regression, decision trees, and neural networks, among others.

Model Selection: It is apparent that the kind of model used has an impact on the degree of accuracy of predictions. The literature also shows that the random forest type of ensemble method, for instance, performs better than simpler models in agricultural applications more often [28].

9.5.2.3 *Step 3: Model Update Sharing*

When training the local models, each farm transfers to the central server only the updates of models (e.g., gradients). This makes the approach economical regarding data transport and is very confidential. These updates are then accumulated to form a fresh global model by the server.

Aggregation Process: The most frequently used techniques of data aggregation are as follows:

i. **FedAvg:** A frequently applied algorithm that computes the weighted mean of local models [8].
ii. **Weighted Averaging:** This method factors in the number of data points that were used when training, where the farms with large volumes were afforded more importance.

9.5.2.4 *Step 4: Global Model Distribution*

From here, the global model is updated and sent back to the farms included in the process. Local predictions or new local data sets can then be incorporated to continue the training process for each farm for iterations through the contact step in this design.

Figure 9.2 can depict the model update process, illustrating the flow of model updates from farms to the central server and back.

9.5.3 Potential Architectures for Federated Learning in Agriculture

In the implementation of FL, the following architectures can be employed effectively in the agricultural domain. Below are some common architectures.

Centralized Federated Learning
In this architecture, a single master node or a single server looks after the entire training process. Every farm transmits some model updates to the server, which then combines the updates and distributes the global model. This approach is suitable for environments in which farms enjoy a strong internet connection.

The Proposal of Decentralized Federated Learning
Specifically, decentralized FL only requires farms to exchange model updates with each other instead of the need for a central server. This architecture can improve reliability and decrease the risk of communication

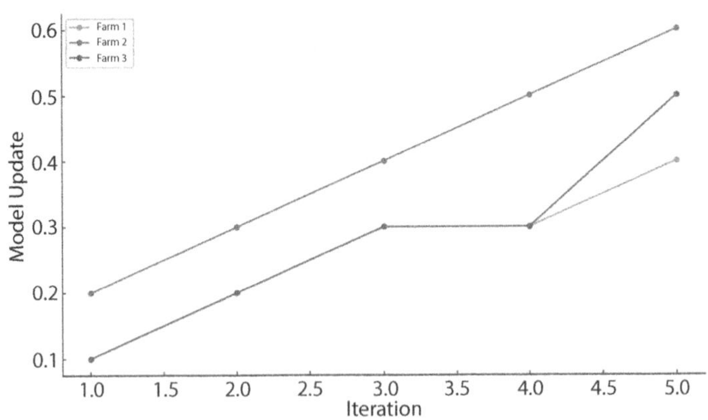

Figure 9.2 Model update process.

problems but at the same time makes the coordination task more challenging [29].

A novel approach to facilitate distributed learning is called hierarchical federated learning

Intermediate nodes are also included in this actualization to ensure that the principal nodes, which can be regional hubs or cooperatives, compile updates from various farms and then forward them to the main server. This approach can be helpful where connectivity is a problem, and it also scales the amount of work done by the central server.

Figure 9.3 illustrates different FL architectures (centralized, decentralized, and hierarchical) with a flow diagram showcasing the communication between farms and servers.

9.5.4 Challenges in Implementing Federated Learning

Although an enabling framework of FL as detailed above presents a host of benefits, it is noteworthy that several issues require consideration during the FL.

9.5.4.1 Communication Overhead

There is always a high chance of communication latency and even the passing of large amounts of bandwidth between farms and the central server. Hence, this challenge can be highly experienced in remote or connected

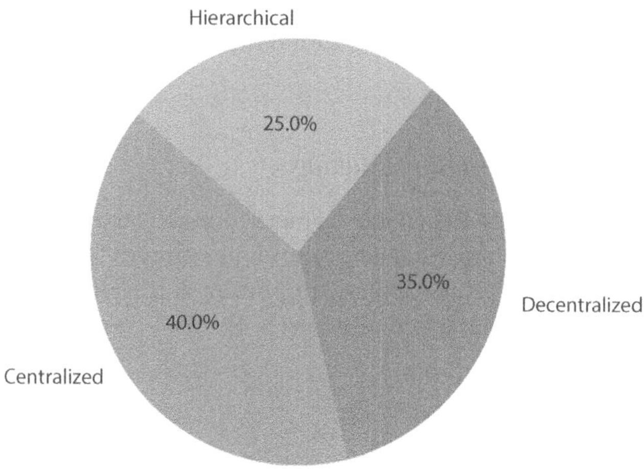

Figure 9.3 Federated learning architectures.

rural regions with weak and limited internet connectivity. Hard *et al.* [30] noted that these problems can be addressed by improving communication.

9.5.4.2 Data Heterogeneity

Due to the variation in users' needs, localization, dynamics, and size, data distributed across farms may differ in quality, quantity, and kind. Such variation results in the model convergence problem whereby the global model does not capture the characteristics of different farms across the regions. Managing data heterogeneity is a critical requirement to achieve the successful training of models.

9.5.4.3 Technical Expertise

For FL to work, there must be minimal technical skills to engage in farming, among Bragard's target groups. Capacity should be built to enable farmers to engage in meaningful FL processes [13].

The Swiss Regulated and Privacy Aspects of FinTechs
As FL improves data privacy, new questions arise for data ownership and usage rights. Readily understandable laws must be set in place to safeguard the farmers' rights about the distribution of benefits of modeling with the various farmers and other companies.

9.5.5 Implications for Agricultural Productivity

The application of FL to the field of crop yield prediction has a vast impact on boosting food production. By utilising a range of datasets more sensitively to privacy, FL allows the creation of more accurate models in terms of prediction.

9.5.5.1 Enhanced Decision-Making

Increased yield predictions enable farmers to make better decisions when it comes to input use, crop husbandry, and measures to take when disaster strikes. This capability is particularly important at this time of climate volatility and uncertain market conditions [3, 4].

9.5.5.2 Increased Collaboration

While using FL, farmers collaborate to discuss their findings or can simply share their experiences and ultimately learn from one another while their

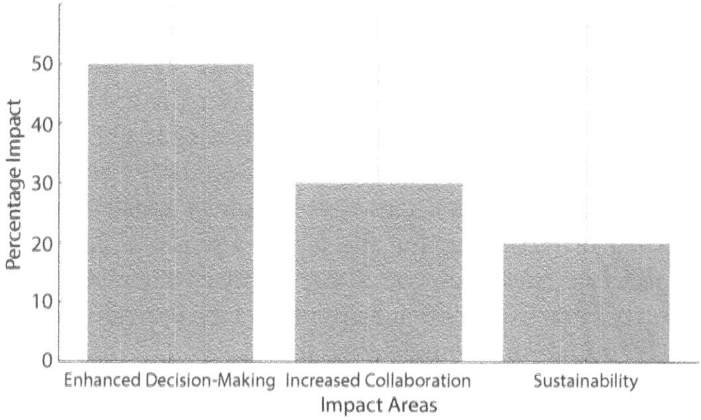

Figure 9.4 Impact of federated learning on agricultural productivity.

data remains secure. Such a cooperative attitude may result in improving and introducing new approaches to farming based on the general attained increase in yield.

9.5.5.3 *Sustainability*

We see that FL's real-world application of capitalising on yield data leads to more reliable yield forecasts, which can influence sustainable farming. Farmers can regulate the usage of inputs such as fertilizers and water and hence cut down on wastage, fully protecting the environment in ways that enhance food production [31].

Figure 9.4 illustrates the potential impacts of FL on agricultural productivity, showcasing benefits such as enhanced decision-making, collaboration, and sustainability.

9.6 Challenges and Limitations of Federated Learning in Crop Yield Predictions

Undoubtedly, FL can boost crop yield prediction by enabling data collection and analysis centrally while allowing distributed data collection. Nevertheless, some issues and limits constrain it from being widely commercialized in the agricultural domain.

9.6.1 Data Privacy Concerns

FL is proposed, as first, to improve data privacy by ensuring that data remains on local farms for each farm, but even this may not preclude reverse engineering of model updates. Adversaries might exploit model updates to infer sensitive information about local data. Consider a farmer who is unable to improve the accuracy of his model updates, but malicious actors succeed in reconstructing the farmer's sensitive data from these updated model updates. However, this potential vulnerability requires a strong security infrastructure that helps protect the individual data sources within the FL framework from external traffic.

9.6.2 Heterogeneity of Data

FL implementations are very challenging due to the variability of the agricultural data. Heterogeneous datasets [15] arise from different farms and different environmental conditions, farming practices, and crops. This reality makes the model aggregation process more complex, as local models trained to fit different data distributions may not generalize well in the end. An example scenario is that a model trained from data of a dry region will not be accurate in predicting yields for a region of high rainfall. Model biases stemming from the discrepancies between datasets jeopardize the final predictive performance and reliability of the aggregated model.

9.6.3 Communication Constraints

The success of FL relies on the ability to exchange information between farms and the central server. Unfortunately, rural agricultural areas tend to be connectivity challenged due to poor internet coverage and infrastructure barriers [13]. The heavy communication overhead can cause delays in training and updates of the model, degrading the responsiveness of the system. In addition, the modelling updates may become resource burdensome when the data sizes are large. To enable a successful deployment of FL in agriculture, it is therefore important to have a strong communication infrastructure.

9.6.4 Model Convergence Issues

Model convergence also presents a challenge for FL. Local models trained on heterogeneous data may converge at different rates, which may cause speed degeneration of overall training [14]. If some farms have substantially more data, or better-quality data, than other farms, their model update can bias the performance of the global model, resulting in suboptimal convergence and performance. The step of aggregating local models can be an arduous and expensive process that necessitates sophisticated techniques and tuning to achieve both good aggregation and good local model performance.

9.7 Future Directions in Federated Learning for Agriculture

FL has recently been gaining traction with the agricultural sector, and several future directions to improve FL's efficacy and applicability to crop yield prediction and surrounding tasks naturally emerge. The focus of these advancements is on enhancing the FL framework through improvements, integrating emerging technologies, and addressing current challenges.

9.7.1 Advancements in FL Techniques

Improving the efficiency and scalability of FL algorithms is ongoing research. Thus, federated transfer learning (FTL) models such as the one proposed by enable models to make better use of any collective knowledge learned from distinct tasks and domains regardless of the amount of available local datasets. It is especially worthwhile in the realm of agriculture because some farms may have insufficient data for a trained model. FTL can be applied to these farms to leverage insights from other regions with larger datasets so that the predictive models become more robust.

Furthermore, asynchronous FL holds a lot of promise. Updates from all clients are then synchronised in traditional FL, where communication delays and diversity in local data quality lead to inefficiencies. This enables asynchronous FL: Clients may send updates at different times and, as a result, improve their response to the flow and tolerate variations in data quality. This flexibility can accommodate the rapid changes in crop yield model prediction as demands from real-time change.

9.7.2 Integration with IoT and Blockchain

FL can further be integrated with IoT devices, which can change agricultural practices even further. Real-time information based on data from environmental conditions in the field, soil health, and crop status can be continuously collected by IoT devices and used for training models. Moreover, this integration not only results in more data (quality and quantity) for the use of FL but also better dynamic and responsive yield prediction models. For example, if we move to use the IoT sensor data, then real-time data can help refine those models when there is a sudden change in the weather.

Additionally, data security and trust among participating farms can be added to FL by combining it with blockchain technology. It can write a ledger for the recording of model updates and data transactions to the blockchain, which is decentralized and immutable, to enhance transparency and accountability. This allows farmers to trust data privacy and ownership and thus incentivizes more individuals to participate in collaborative model training. FL and blockchain used together could create a secure environment where farmers share insights while at the same time keeping their proprietary data safe.

9.7.3 Policy and Regulatory Considerations

With the advancement of FL technology, it is important to imagine policy and regulatory frameworks for data sharing and privacy in agriculture. Policy around collaboration by governments and agricultural organizations must ensure the securing of farmers' traditional rights. It would also encompass laying down explicit guidelines on how to protect data privacy and who will oversee intellectual property rights. Collaborative efforts, most notably from policymakers, agricultural stakeholders, and technologists, can promote the adoption of FL, which is key to making food security and sustainable farming a success.

AI disclosure statement: The author acknowledges the use of AI-based tools for minor language editing or grammatical correction. However, all intellectual contributions, analyses, and conclusions presented in this chapter are entirely the author's own.

References

1. Lobell, D.B. and Burke, M.B., On the use of statistical models to predict crop yield responses to climate change. *Agric. For. Meteorol.*, 150, 11, 1443–1452, 2010, https://doi.org/10.1016/j.agrformet.2010.07.008.

2. Shang, J., Liu, J., Zhang, Y., Zheng, H., Zhou, W., A comparison of machine learning models for predicting crop yields. *Comput. Electron. Agric.*, 145, 86–96, 2017, https://doi.org/10.1016/j.compag.2017.12.022.
3. Kamilaris, A. and Prenafeta-Boldú, F.X., Deep learning in agriculture: A survey. *Comput. Electron. Agric.*, 147, 70–90, 2018, https://doi.org/10.1016/j.compag.2018.02.016.
4. Kamilaris, A. and Prenafeta-Boldú, F.X., Deep learning in agriculture: A review. *Sensors*, 18, 1, 1–24, 2018, https://doi.org/10.3390/s18010021.
5. Tripathi, P., Kumar, N., Rai, M., Khan, A., Applications of deep learning in agriculture, in: *Advances in Computational Intelligence and Robotics*, M.A. Khan, R. Khan, P. Praveen (Eds.), pp. 17–28, IGI Global, USA, 2022, https://doi.org/10.4018/978-1-6684-5141-0.ch002.
6. Tripathi, P., Kumar, N., Rai, M., Shukla, P.K., Verma, K.N., Applications of machine learning in agriculture, in: *Advances in Electronic Government, Digital Divide, and Regional Development*, M.A. Khan, B. Gupta, A.R. Verma, P. Praveen, C.J. Peoples (Eds.), pp. 99–118, IGI Global, USA, 2023, https://doi.org/10.4018/978-1-6684-6418-2.ch006.
7. Rai, M., Tyagi, N., Tripathi, P., Kumar, N., Kumari, P., Advancement in agricultural techniques with the introduction of artificial intelligence and image processing, in: *Advances in Electronic Government, Digital Divide, and Regional Development*, M.A. Khan, B. Gupta, A.R. Verma, P. Praveen, C.J. Peoples (Eds.), pp. 47–68, IGI Global, USA, 2023, https://doi.org/10.4018/978-1-6684-6418-2.ch004.
8. McMahan, H.B., Moore, E., *et al.*, Communication-efficient learning of deep networks from decentralized data. *Artif. Intell. Stat.*, 54, 1273–1282, 2017. http://proceedings.mlr.press/v54/mcmahan17a/mcmahan17a.pdf.
9. Bonawitz, K., Eichner, H., Grieskamp, W., Huba, D., Ingerman, A., Ivanov, V., van Overveldt, T., Towards federated learning at scale: System design. *arXiv preprint arXiv:1902.01046,* https://arxiv.org/abs/1902.01046, 2019.
10. Godfray, H.C., Beddington, J.R., Crute, I.R., Haddad, L., Lawrence, D., Muir, J.F., Toulmin, C., Food security: The challenge of feeding 9 billion people. *Science*, 327, 5967, 812–818, 2010, https://www.science.org/doi/10.1126/science.1185383.
11. Challinor, A.J., Watson, J., Lobell, D.B., Howden, S.M., Smith, D.R., Chhetri, N., A meta-analysis of crop yield under climate change and adaptation. *Nat. Clim. Change*, 4, 4, 287–291, 2014, https://doi.org/10.1038/nclimate2153.
12. Lobell, D.B. and Asseng, S., Comparing estimates of climate change impacts from process-based and statistical crop models. *Environ. Res. Lett.*, 12, 1, 015001, 2017, https://doi.org/10.1088/1748-9326/aa518a.
13. Sheller, M.J., Reina, G.A., Edwards, B., Martin, J., Bakas, S., Multi-institutional deep learning model for predicting glioma subtypes from MRI: A proof of concept study. *Med. Image Anal.*, 65, 101–109, 2020, https://doi.org/10.1016/j.media.2020.101769.

14. Li, T., Sahu, A.K., Zaheer, M., Sanjabi, M., Talwalkar, A., Smith, V., Federated optimization in heterogeneous networks. *Proc. Mach. Learn. Syst.*, 2, 429–450, 2020, https://proceedings.mlsys.org/paper/2020/file/3b17750ecb61322c08e92143d35eafdf-Paper.pdf.
15. Kairouz, P., McMahan, H.B., Brendan, A., Advances and open problems in federated learning. *Found. Trends Mach. Learn.*, 14, 1–2, 1–210, 2021, http://dx.doi.org/10.1561/2200000083.
16. Zhang, C. and Wang, Y., A review on deep learning in agriculture: Current achievements and future challenges. *Comput. Electron. Agric.*, 153, 134–142, 2018, https://doi.org/10.1016/j.compag.2018.08.004.
17. Zhang, J., Guo, X., Zhang, Y., Xu, L., A novel approach to predicting crop yield based on deep learning. *Comput. Electron. Agric.*, 153, 232–239, 2018, https://doi.org/10.1016/j.compag.2018.08.004.
18. Jones, J.W., Hoogenboom, G., Porter, C.H., Boote, K.J., Batchelor, W.D., Hunt, L.A., Ritchie, J.T., The DSSAT cropping system model. *Eur. J. Agron.*, 18, 3-4, 235–265, 2003, https://doi.org/10.1016/S1161-0301(02)00107-7.
19. Hastie, T., Tibshirani, R., Friedman, J., *The Elements of Statistical Learning: Data Mining, Inference, and Prediction*, Springer Science & Business Media, Germany, 2009, https://doi.org/10.1007/978-0-387-84858-7.
20. Asseng, S., Ewert, F., Martre, P., Rötter, R.P., Lobell, D.B., Cammarano, D., Zhu, Y., Rising temperatures reduce global wheat production. *Nat. Clim. Change*, 3, 9, 827–832, 2013, https://doi.org/10.1038/nclimate1911.
21. Shahhosseini, M., Martinez-Feria, R.A., Hu, G., Archontoulis, S.V., Maize yield and nitrate loss prediction with machine learning algorithms. *Environ. Res. Lett.*, 16, 2, 024013, 2021, https://doi.org/10.1088/1748-9326/abd883.
22. Drusch, M., Del Bello, U., Carlier, S., Colin, O., Fernandez, V., Gascon, F., Rast, M., Sentinel-2: ESA's optical high-resolution mission for GMES operational services. *Remote Sens. Environ.*, 120, 25–36, 2012, https://doi.org/10.1016/j.rse.2011.11.026.
23. Jones, H.G., Serraj, R., Loveys, B.R., Xiong, L., Wheaton, A., Price, A.H., Thermal infrared imaging of crop canopies for the remote diagnosis and quantification of plant responses to water stress in the field. *Funct. Plant Biol.*, 36, 10–11, 978–989, 2017, https://doi.org/10.1071/FP09023.
24. Jeong, J.H., Resop, J.P., Mueller, N.D., Fleisher, D.H., Yun, K., Butler, E.E., Kim, S.H., Random forests for global and regional crop yield predictions. *PLoS One*, 11, 6, e0156571, 2016, https://doi.org/10.1371/journal.pone.0156571.
25. van Klompenburg, T., Kassahun, A., Catal, C., Crop yield prediction using machine learning: A systematic literature review. *Comput. Electron. Agric.*, 177, 105709, 2020, https://doi.org/10.1016/j.compag.2020.105709.
26. Shukla, S. and Jharkharia, S., Machine learning-based prediction of sugarcane yield: A comparison of algorithms. *Comput. Electron. Agric.*, 183, 106079, 2021, https://doi.org/10.1016/j.compag.2021.106079.

27. Shukla, S. and Jharkharia, S., Predictive modeling of agricultural yield: A machine learning approach. *Agric. Syst.*, 187, 102991, 2021, https://doi.org/10.1016/j.agsy.2020.102991.
28. Breiman, L., Random forests. *Mach. Learn.*, 45, 1, 5–32, 2001, https://doi.org/10.1023/A:1010933404324.
29. Kairouz, P., McMahan, B., *et al.*, Advances and open problems in federated learning. *arXiv preprint arXiv:1912.04977*, https://arxiv.org/abs/1912.04977, 2019.
30. Hard, A., Rao, K., Mathews, R., Brown, A., Federated learning for health care: A patient-centered model, in: *Proceedings of the 34th Conference on Neural Information Processing Systems (NeurIPS)*, 2018, https://arxiv.org/abs/1812.00735.
31. Mastrorillo, M., *et al.*, The role of data in achieving sustainable agriculture: A global perspective. *Sustain. Agric. Rev.*, 19, 239–266, 2016, https://doi.org/10.1007/978-3-319-33259-5_3.

10

Federated Learning in Smart Agriculture: Applications, Challenges, and Solutions

Abhishek Tyagi[1], Shekhar Tyagi[2]* and Guru Dayal Kumar[3]

[1]Department of Computer Science and Engineering, KIET Group of Institutions, Ghaziabad, Uttar Pradesh, India
[2]Department of Computer Science and Engineering, IIMT College of Engineering, Greater Noida, Uttar Pradesh, India
[3]School of Humanities and Social Sciences, Indian Institute of Technology Indore, Madhya Pradesh, India

Abstract

Federated learning (FL) offers a promising approach to enhancing smart agriculture by leveraging decentralized data to improve model accuracy and efficiency. This paper explores the integration of FL into various agricultural applications, including precision farming, pest and disease detection, crop yield prediction, soil health monitoring, and climate impact analysis. FL enables the aggregation of data from multiple farms without centralized data storage, preserving privacy while optimizing resource allocation and decisions. The paper addresses key challenges associated with implementing FL in agriculture, such as data heterogeneity, privacy and security concerns, infrastructure limitations, and communication overhead. Strategies to overcome these challenges are discussed, including advanced model algorithms, privacy-preserving techniques, and infrastructure improvements. Future directions emphasize the need for enhanced model robustness, interdisciplinary collaboration, and real-world validation to expand FL applications and ensure practical deployment. The insights provided aim to guide researchers and practitioners in leveraging FL to advance agricultural productivity and sustainability.

Keywords: Federated learning, smart agriculture, precision farming, data privacy, crop yield prediction, IoT in agriculture, machine learning

**Corresponding author*: shekhartyagirocks007@gmail.com

Padmesh Tripathi, Bhanumati Panda, Shanthi Makka, Reeta Mishra, S. Balamurugan and Sheng-Lung Peng (eds.) Federated Learning for Smart Agriculture and Food Quality Enhancement, (247–270) © 2026 Scrivener Publishing LLC

10.1 Introduction

10.1.1 Overview of Smart Agriculture

Smart agriculture [1–3], also called precision farming, is a new way of farming that uses advanced technology to improve how we grow food. It combines tools like the Internet-of-Things (IoT) devices, artificial intelligence (AI), machine learning, big data, drones, and satellite images with traditional farming methods. These technologies help farmers keep a close watch on their crops, soil, and animals, allowing them to manage their farms more accurately. Smart agriculture is very important because it helps solve big problems like the growing demand for food, climate change, and limited resources. By using technology to optimize the use of water, fertilizers, and pesticides, farmers can reduce waste, cut costs, and lessen their impact on the environment. This makes farming more efficient and sustainable.

10.1.2 Current State of Technology in Agriculture

The current technology in agriculture is rapidly changing and improving. IoT devices, like soil moisture sensors [4–6], weather stations, and GPS-equipped machinery, gather real-time data from the fields. This data is then analyzed by AI and machine learning systems [7–9] to provide useful insights. For example, predictive analytics can forecast weather patterns, pest outbreaks, and crop yields, helping farmers make better decisions. Drones and satellite images offer detailed maps that help monitor crop health and apply inputs like water and fertilizer precisely where needed. Blockchain technology [10–12] is also being explored to ensure traceability and transparency in the food supply chain. However, not all farms are adopting these technologies at the same rate. Factors like infrastructure, financial resources, and technical know-how affect how quickly different regions and farm sizes can implement these tools. Despite these challenges, the move toward digital farming is clear. There is growing investment and innovation aimed at making these technologies easier to use and more available to farmers everywhere.

10.1.3 Introduction to Federated Learning (FL)

Federated learning (FL) is a new approach in machine learning that helps protect privacy and data security. It allows models to be trained on decentralized devices or servers that hold local data, without needing to gather all the data in one place. In traditional machine learning, data from different sources is collected and centralized for model training. This centralization can cause privacy problems, especially in sensitive areas like healthcare, finance, and agriculture, where keeping data confidential is very important. FL addresses this by letting each device or node train the model on its own data locally. Instead of sharing the actual data, the devices only send the learned parameters (like weights and gradients) to a central server. The central server combines these parameters to update the global model and then sends this updated model back to the devices for more training. This process repeats until the model is fully trained [13–15].

10.1.4 Key Benefits of FL in General

FL offers several key benefits, especially for smart agriculture. Firstly, FL enhances data privacy and security by keeping raw data on the devices or farms where it is generated. This decentralized approach reduces the risk of data breaches and ensures that sensitive information stays confidential. Secondly, FL allows multiple stakeholders—such as farmers, researchers, and agribusinesses—to collaborate without sharing their proprietary data. They can all contribute to and benefit from a shared model, using diverse data sources to make the model more accurate and robust. Thirdly, FL helps develop more personalized and context-specific models. Because the training data remains on local devices, the models can better capture local conditions and variations, providing more relevant and effective recommendations for individual farms. Additionally, FL reduces the need for extensive data transfer and storage infrastructure. This is especially useful in rural and remote areas with limited connectivity and computational resources. By spreading the computational load across multiple devices, FL also makes efficient use of existing hardware and energy resources.

10.1.5 Contribution of this Article

This article explores how FL intersects with smart agriculture, offering a thorough review of its potential applications, challenges, and solutions. We aim to add to the existing literature by systematically identifying and discussing the various ways FL can enhance agricultural practices, including precision farming and pest and disease detection. Additionally, we address the specific challenges of implementing FL in agriculture, such as data heterogeneity, privacy concerns, and infrastructure limitations. We propose practical solutions and future research directions to overcome these challenges, providing a roadmap for researchers, practitioners, and policymakers to fully leverage FL in smart agriculture.

Through this detailed analysis, our article highlights the transformative impact of FL on the future of farming, promoting innovation and sustainability in agricultural practices worldwide.

10.2 Related Work

FL has gained significant attention in various fields, including agriculture, due to its ability to enhance data privacy and collaboration while avoiding the challenges of centralized data storage. Integrating FL in agriculture has shown promise in transforming how agricultural data is used and managed. The foundational work in [16] introduced FL, which allows models to be trained locally on devices and only share model updates, rather than raw data, with a central server. This approach addresses privacy concerns and is particularly relevant in agriculture, where protecting sensitive farm data is crucial.

In agriculture, FL has been applied to several problems, such as crop yield prediction and disease detection. For example, [17] proposed a FL-based approach for crop yield prediction, demonstrating how federated models can enhance prediction accuracy by integrating data from diverse sources while maintaining data privacy. Similarly, Mehta *et al.* [18] focused on disease detection using FL, specifically targeting mango leaf diseases. Their work utilized convolutional neural networks within a FL framework to improve disease detection accuracy while addressing data privacy and decentralization challenges. Durrant *et al.* [19] explored the role of FL in the agri-food sector, examining how cross-silo FL facilitates data sharing among agricultural stakeholders. Their study highlighted the potential of FL to improve data accessibility and collaboration while addressing privacy concerns. On the practical side, Bonawitz *et al.* [20] discussed secure aggregation techniques for privacy-preserving machine learning, which are critical for implementing FL systems. Their work introduced cryptographic methods to ensure data confidentiality during FL processes. Regarding technical advancements, Ooi *et al.* [21] reviewed hierarchical FL approaches and their applications in sensor networks relevant to agricultural monitoring systems. Their article explored challenges and opportunities in hierarchical FL, highlighting its applicability to agricultural sensor data.

Li *et al.* [22] provided a comprehensive review of deep learning methods for non-destructive evaluation of agricultural products, discussing various techniques and their effectiveness in assessing agricultural products, which complements the predictive capabilities of FL models. In the context of edge computing and FL, Tyagi and Tyagi [23] conducted an in-depth exploration of how these advanced technologies can enhance real-time data processing and connectivity in challenging terrain settings using drones. Their work, featured in the book Web 3.0, delves into the application of deep reinforcement learning to optimize the deployment and operation of drones. This research highlights the synergistic potential of combining edge computing with deep reinforcement learning, emphasizing its critical role in ensuring efficient communication and connectivity for drones operating in resource-constrained and rugged environments.

Moreover, Kumar *et al.* [24] conducted a comprehensive investigation into the use of deep learning techniques for predicting rainfall and flood impacts, utilizing long short-term memory (LSTM) neural networks. Their study underscores the significant predictive capabilities of these advanced neural architectures, particularly in forecasting applications. The research demonstrates the potential of LSTM networks to accurately predict environmental impacts, offering valuable insights for proactive flood management and agricultural planning. Tyagi and Tyagi [25] offered foundational insights into optimizing communication channels, which are crucial for enhancing the efficiency of FL systems. This work provides a deeper understanding of how to address bandwidth and latency issues, which are pivotal for effective cloud spectrum connectivity through channelization. By tackling these challenges, Tyagi and Tyagi's research contributes significantly to the improvement of FL applications, ensuring robust and efficient data exchange across decentralized networks. Kumar *et al.* [26] presented a thorough analysis of rainfall and flood vulnerability in Bihar, India, utilizing predictive machine learning techniques. Their work applies sophisticated machine learning models to assess and forecast the impacts of rainfall and flood events. By leveraging extensive datasets and advanced analytical methods, the study provides critical insights into flood risk management and disaster preparedness. Integrating FL into such frameworks could further enhance predictive capabilities by enabling the collaboration of multiple decentralized data sources without compromising data privacy. By using FL, diverse datasets from various regions can be combined to build more robust and accurate predictive models, ultimately improving disaster response and resource allocation. This integration underscores the potential of FL to augment traditional machine learning approaches in addressing complex environmental challenges.

Tyagi and Tyagi [27] presented a comprehensive study on the integration of edge computing with FL, titled Edge Computing: Empowering

Real-Time Data Processing and Connectivity. Published on Zenodo, this work explores how edge computing can significantly enhance the efficiency and responsiveness of FL systems, particularly in real-time applications. By processing data locally at the edge of the network, this approach reduces latency, conserves bandwidth, and ensures faster decision-making, which is crucial for applications requiring immediate responses. The study highlights the synergies between edge computing and FL, demonstrating how they can collectively improve data processing and connectivity in resource-constrained environments. This integration is particularly beneficial in scenarios such as smart agriculture, where real-time data from numerous sensors need to be processed swiftly to make timely decisions.

Overall, FL presents a promising approach to leveraging decentralized agricultural data while addressing privacy and collaboration challenges. Ongoing research and developments continue to advance the integration of FL in smart agriculture, leading to more effective and sustainable farming practices.

10.3 Federated Learning: Pioneering Precision Agriculture Applications

FL offers several promising applications in the field of agriculture, especially in enhancing data privacy and enabling collaborative learning across distributed farms without centralizing data. Here are some of the prominent applications:

Precision Farming: Precision farming [1] utilizes cutting-edge technologies to enhance agricultural practices at a granular level. FL offers a transformative approach to optimizing precision farming by enabling the integration and analysis of data from diverse sources, including sensors and IoT devices deployed across multiple farms. This decentralized model of data aggregation facilitates in-depth and nuanced insights into agricultural operations while maintaining data privacy and security. In practical terms, FL allows for the collection of extensive datasets encompassing various parameters such as soil moisture, nutrient concentrations, crop growth stages, and meteorological conditions. These datasets are generated from distributed sensor networks across different agricultural sites. FL frameworks are then employed to train machine learning models collaboratively without the need to centralize the data at a single location. This approach ensures that sensitive farm-specific information remains confidential while still contributing to the collective learning process.

The trained federated models can deliver highly localized recommendations by analyzing aggregated insights from diverse farm environments [14]. For example, the models can optimize irrigation schedules based on real-time soil moisture levels and weather forecasts, suggest precise fertilizer applications tailored to nutrient deficiencies detected in the soil, and implement targeted pest control strategies by identifying pest patterns across different regions. This level of optimization enhances not only crop management but also resource allocation efficiency, contributing to increased farm productivity and sustainability. By leveraging FL, precision farming becomes more adaptive and responsive to varying field conditions, ultimately leading to more informed decision-making and improved agricultural outcomes.

Pest and Disease Detection: Early detection and management of pests and diseases are paramount in minimizing crop losses and ensuring agricultural productivity. FL presents an effective approach to pest and disease detection by facilitating the integration and analysis of data from multiple farms, thus enhancing the precision and timeliness of detection. Typically, data pertinent to plant health, environmental conditions, and pest activity are collected through an array of sensors and manual inspections. This data encompasses a range of variables, including leaf coloration, growth anomalies, weather patterns, and pest population dynamics.

FL frameworks enable the development of sophisticated machine learning models that leverage this distributed data to identify subtle patterns and predict potential pest and disease outbreaks. By aggregating data from diverse geographic regions, federated models can detect early indicators of pest infestations or disease occurrences through advanced anomaly detection and pattern recognition algorithms. For instance, deviations in leaf spectral signatures or unusual patterns of pest activity, which may be indicative of emerging threats, can be discerned with high accuracy.

Through FL, these models can be trained collaboratively across various farms without the need for centralizing sensitive data. The insights gained from such models are then disseminated back to the participating farms, providing actionable intelligence in the form of early warning alerts. Farmers can thus implement timely interventions, such as targeted pesticide applications or the deployment of biological control agents, based on the predictive insights generated by the federated models. This proactive approach significantly mitigates the impact of pests and diseases, enhances crop protection strategies, and contributes to overall agricultural resilience.

Crop Yield Prediction: Accurate prediction of crop yield is crucial for effective agricultural management and planning. FL offers a significant advancement in yield forecasting by integrating and analyzing data from diverse geographical locations. This comprehensive dataset typically includes historical yield records, weather conditions, soil characteristics, and agricultural practices. By employing FL techniques, predictive models can be developed that synthesize this aggregated data to account for a broad spectrum of variables and localized factors. FL enables the training of models across a federated network of farms, allowing for the incorporation of extensive and varied data sources without the necessity of centralizing sensitive information. These models can employ advanced algorithms to forecast crop yields with high precision by leveraging historical trends and current conditions.

For instance, federated models can analyze patterns in historical yield data in conjunction with real-time environmental inputs to provide reliable forecasts. Such predictions empower farmers to make data-driven decisions regarding planting schedules, resource allocation, and harvesting timelines. Knowledge of anticipated yields facilitates optimized use of resources such as labor, machinery, and storage capacities. Furthermore, accurate yield forecasts inform market strategies and financial planning, thereby reducing uncertainties and enhancing economic viability for farmers.

Soil Health Monitoring: Soil health is fundamental to the sustainability and productivity of agricultural systems, and FL can significantly enhance soil health monitoring by enabling the integration and analysis of data across a decentralized network of farms. This comprehensive dataset generally includes critical soil metrics such as nutrient concentrations, pH levels, organic matter content, and microbial activity. FL supports the development of sophisticated machine learning models by aggregating soil health data from diverse agricultural environments. These federated models leverage distributed learning to synthesize information from multiple sources, allowing for a more nuanced understanding of soil conditions and interactions. The models are trained on this federated dataset to identify complex patterns and correlations among soil health indicators, thereby providing granular insights and actionable recommendations for soil management.

For instance, federated models can utilize advanced algorithms to detect deficiencies or toxicities in soil nutrients by analyzing aggregated data from various farms. This capability enables the models to recommend precise soil amendments or suggest crop rotation strategies tailored to specific nutrient profiles and soil conditions. Furthermore, by examining data across a wide range of farm environments, FL models can identify emerging trends and potential soil health issues that might not be evident from isolated datasets.

The integration of data through FL allows for the detection of broader soil health patterns and facilitates early intervention and adaptive management strategies. This comprehensive approach ensures that farmers receive accurate, context-specific recommendations, enhancing soil fertility management and supporting sustainable farming practices. Consequently, FL applications in soil health monitoring contribute to increased agricultural productivity and environmental sustainability.

Climate Impact Analysis: Climate change presents formidable challenges to agriculture, influencing crop production through altered weather patterns and necessitating adaptive strategies. FL offers a robust framework for analyzing the impact of climatic variations on agricultural outcomes by synthesizing climate and crop performance data across multiple decentralized sources. FL enables the development of advanced predictive models that aggregate extensive datasets, including meteorological data, soil conditions, and crop performance metrics, from a network of farms. These models utilize distributed learning algorithms to analyze how varying climatic factors—such as temperature fluctuations, rainfall variability, and extreme weather events—affect crop growth and yield. By leveraging federated models, researchers can discern complex interactions between climatic conditions and agricultural productivity. For instance, the models can identify how different crops respond to specific climatic stresses, including prolonged heatwaves or unseasonal rainfall. Such analyses facilitate a deeper understanding of crop resilience and vulnerability to climate-induced changes.

The insights generated from federated models enable the formulation of targeted adaptive measures. These may include recommendations for selecting crop varieties with enhanced climate resilience, optimizing planting schedules to align with shifting climatic patterns, or adjusting irrigation strategies to accommodate changes in precipitation. By applying these recommendations, farmers can proactively address the adverse effects of climate change, thereby maintaining stable crop production and ensuring agricultural sustainability in the face of environmental uncertainties.

10.3.1 Advanced Federated Learning Applications in Agriculture

10.3.1.1 Water Management

Application: Efficient water management is crucial for sustainable agriculture. FL can optimize irrigation practices by integrating data from various sources, such as soil moisture sensors, weather forecasts, and crop water requirements.
Details: Federated models can analyze aggregated data to predict optimal irrigation schedules and amounts. By leveraging data from diverse farms, these models can account for regional variations in water needs and soil characteristics, leading to more precise and efficient water usage. This reduces water waste and ensures crops receive the appropriate amount of moisture for optimal growth.

10.3.1.2 Yield Forecasting for Supply Chain Optimization

Application: Accurate yield forecasting is essential for supply chain management in agriculture. FL can improve yield predictions by combining data on weather, soil conditions, and historical yields from multiple sources.
Details: Federated models can analyze data from various farms and regions to provide more accurate forecasts of crop yields. These forecasts can help optimize supply chain operations, such as inventory management, transportation logistics, and market pricing strategies. By improving forecast accuracy, farmers and supply chain managers can better align production with market demand.

10.3.1.3 Precision Livestock Farming

Application: FL can enhance livestock management by integrating data from wearable sensors and monitoring systems used on animals.
Details: By analyzing data on animal health, behavior, and environmental conditions, federated models can provide insights into optimal feeding strategies, health monitoring, and breeding practices. For example, models can detect early signs of illness or stress in livestock, enabling timely intervention and improving overall herd health.

10.3.1.4 Optimizing Fertilizer Use

Application: Effective fertilizer application is key to maximizing crop yield and minimizing environmental impact. FL can optimize fertilizer recommendations by aggregating soil nutrient data and crop response information.
Details: Federated models can analyze data from various farms to determine the most effective fertilizer types and application rates for different soil conditions and crop types. This approach helps to reduce over-fertilization, enhance soil health, and increase crop yield while minimizing environmental pollution.

10.3.1.5 Integrated Pest Management (IPM)

Application: FL can support integrated pest management strategies by combining data from pest traps, weather conditions, and crop health observations.
Details: Federated models can analyze this data to predict pest population dynamics and identify effective control measures. By integrating data from multiple sources, FL can provide more accurate forecasts of pest outbreaks and recommend targeted interventions, such as biological controls or selective pesticide applications.

10.3.1.6 Farm Equipment Optimization

Application: FL can improve the efficiency and maintenance of farm equipment by analyzing usage data and performance metrics.
Details: By aggregating data from various equipment sources, federated models can predict maintenance needs, optimize equipment usage, and identify potential failures before they occur. This proactive approach enhances equipment reliability and reduces downtime, leading to more efficient farm operations.

10.3.1.7 Climate Adaptive Crop Breeding

Application: FL can support crop breeding programs by analyzing data on crop performance under varying climatic conditions.
Details: Federated models can integrate data on crop genetics, environmental factors, and performance metrics to identify traits associated with climate resilience. This information can guide the development of new crop varieties that are better adapted to changing climate conditions, improving long-term agricultural sustainability.

10.3.1.8 Automated Field Mapping

Application: FL can enhance field mapping and land management by combining data from various sensors and remote sensing technologies.
Details: Federated models can analyze data from drones, satellites, and ground sensors to create detailed maps of field conditions, including soil properties, crop health, and weed distribution. These maps can inform precision agriculture practices and help farmers make data-driven decisions about field management.

10.3.1.9 Predictive Maintenance for Agricultural Machinery

Application: FL can enhance predictive maintenance systems for agricultural machinery by integrating data from sensors embedded in different machines.
Details: Federated models can analyze real-time data on machinery performance, usage patterns, and operational conditions to predict potential failures or maintenance needs. By leveraging data from diverse sources, FL can provide early warnings and optimize maintenance schedules, reducing downtime and extending the lifespan of machinery.

10.3.1.10 Optimization of Greenhouse Environments

Application: FL can optimize greenhouse environments by aggregating data from various sensors that monitor factors such as temperature, humidity, light levels, and carbon dioxide concentrations.
Details: Federated models can analyze this data to fine-tune environmental controls, such as ventilation, heating, and lighting systems. This optimization ensures ideal growing conditions for plants, improving yields and resource efficiency in greenhouse operations.

10.3.1.11 Remote Sensing and Image Analysis

Application: FL can improve remote sensing and image analysis by combining data from various satellite and drone imagery sources.
Details: Federated models can analyze high-resolution images to detect crop health issues, estimate biomass, and monitor growth stages. By integrating data from different sources, FL enhances the accuracy of image analysis and provides actionable insights for precision agriculture.

10.3.1.12 Crop Breeding and Genetic Research

Application: FL can support crop breeding and genetic research by aggregating genetic and phenotypic data from different research institutions and farms.
Details: Federated models can identify desirable genetic traits and their interactions with environmental factors, accelerating the development of new crop varieties with enhanced traits such as disease resistance or drought tolerance.

10.3.1.13 Resource Allocation and Farm Management

Application: FL can enhance farm management by optimizing resource allocation, such as labor, inputs, and equipment.
Details: Federated models can analyze data from various farms to recommend optimal resource allocation strategies. This includes scheduling labor, allocating equipment, and managing inputs to maximize efficiency and productivity.

10.3.1.14 Consumer Demand Forecasting

Application: FL can assist in forecasting consumer demand for agricultural products by analyzing sales data, market trends, and consumer preferences.
Details: Federated models can provide insights into market trends and predict future demand for different crops. This information helps farmers align production with market needs, optimizing supply and reducing food waste. These additional applications illustrate the broad and evolving potential of FL in addressing complex challenges and driving innovation across various sectors of agriculture.

10.4 Implementing Federated Learning in Smart Agriculture: Challenges and Solutions

FL offers transformative potential for smart agriculture by enabling collaborative model training across decentralized data sources. However, its implementation in agricultural settings presents several challenges. This section delves into key obstacles related to data heterogeneity, privacy and security concerns, infrastructure constraints, and communication overhead while also proposing solutions and best practices for overcoming these challenges.

10.4.1 Data Heterogeneity

10.4.1.1 Differences in Data Types and Formats

In agricultural settings, data collected from various sources, such as soil sensors, weather stations, and satellite imagery, can vary significantly in type and format. For instance, sensor data might include numerical readings of soil moisture and temperature, whereas satellite images provide visual data of crop health. Additionally, data collected from different farms or regions may have varying formats and units of measurement.

10.4.1.2 Impact on FL Model Performance

Data heterogeneity can significantly impact the performance of FL models. Variations in data types and formats can lead to inconsistencies in the training process, potentially causing model inaccuracies or biases. For example, if one farm's soil moisture data is recorded in different units compared to another, then the aggregated model may struggle to interpret and integrate these inputs correctly. Furthermore, variations in data quality and granularity can result in skewed or less reliable model predictions.

10.4.1.3 Solutions and Best Practices

To address data heterogeneity, several strategies can be employed:

Data Preprocessing: Standardize data formats and units before federated model training. Implement preprocessing steps that convert diverse data types into a unified format.
Feature Engineering: Develop robust feature extraction and transformation techniques that can handle heterogeneous data types and enhance model integration.
Adaptive Models: Use adaptive FL models that can adjust to varying data characteristics and improve performance despite data inconsistencies.
Data Normalization: Normalize data to reduce variability and ensure consistent input for the federated models.

10.4.2 Privacy and Security Concerns

10.4.2.1 Importance of Data Privacy in Agriculture

Agricultural data often contains sensitive information, such as specific crop yields, soil conditions, and farm practices. Ensuring data privacy is crucial to protect farmers' proprietary information and maintain trust in FL systems.

10.4.2.2 Security Challenges in FL

FL faces several security challenges, which include the following:

Data Leakage: Even though data remains decentralized, FL processes could still expose sensitive information through model updates.
Adversarial Attacks: Malicious actors may attempt to manipulate model training or extract sensitive data through adversarial attacks, potentially compromising the integrity and confidentiality of the data.

10.4.2.3 Techniques to Enhance Privacy and Security

To address privacy and security concerns, the following techniques can be employed:

Differential Privacy: Implement differential privacy mechanisms to ensure that individual data points cannot be re-identified from model updates.
Secure Multi-Party Computation (SMPC): Use SMPC protocols to perform computations on encrypted data, safeguarding data confidentiality during FL processes.
Homomorphic Encryption: Apply homomorphic encryption to allow computations on encrypted data without needing to decrypt it, protecting data privacy.
Federated Averaging with Privacy Preserving Methods: Utilize federated averaging algorithms combined with privacy-preserving methods to mitigate risks of data exposure during model aggregation.

10.4.3 Infrastructure and Resource Constraints

10.4.3.1 Limitations of Existing Infrastructures in Rural Areas

Many agricultural regions, especially rural areas, face limitations in infrastructure such as internet connectivity and computing resources. These constraints can hinder the effective implementation of FL systems.

10.4.3.2 Computational and Storage Requirements for FL

FL requires significant computational power and storage resources for model training and updates. Rural farms may lack the necessary infrastructure to support these demands, leading to potential inefficiencies in model performance and scalability.

10.4.3.3 Strategies to Overcome these Challenges

Several strategies can be implemented to overcome infrastructure and resource constraints:

Edge Computing: Utilize edge computing to perform data processing and model training locally on farm devices, reducing the need for constant connectivity and central processing.
Hybrid FL Models: Develop hybrid FL models that combine local computations with periodic synchronization to balance resource usage and efficiency.
Lightweight Models: Design lightweight and efficient models that require less computational power and storage, making them suitable for deployment in resource-constrained environments.
Infrastructure Investment: Invest in improving rural infrastructure, including internet connectivity and computational resources, to better support FL applications.

10.4.4 Communication Overhead

10.4.4.1 Network Requirements for FL

FL relies on network connectivity to facilitate communication between the central server and distributed clients (e.g., farms). The performance of FL systems can be significantly affected by network constraints.

10.4.4.2 Challenges Due to Limited Connectivity

Limited or intermittent connectivity in rural areas can lead to high communication overheads and delays in model updates. Frequent synchronization between the central server and distributed clients can be challenging in environments with unreliable or low-bandwidth connections.

10.4.4.3 Solutions to Reduce Communication Overhead

To mitigate communication overhead, consider the following solutions:

Communication Compression: Apply techniques to compress model updates and reduce the amount of data transmitted between clients and the central server.
Asynchronous Updates: Implement asynchronous FL approaches that allow clients to update the model independently and periodically, reducing the dependency on continuous connectivity.
Efficient Aggregation: Optimize model aggregation algorithms to minimize the amount of data exchanged and improve the efficiency of communication processes.
Local Model Training: Increase the frequency of local model training to reduce the need for frequent global communication and updates.

Implementing FL in smart agriculture presents several challenges, including data heterogeneity, privacy and security concerns, infrastructure and resource constraints, and communication overhead. Addressing these challenges through standardized data practices, privacy-enhancing techniques, infrastructure improvements, and efficient communication strategies is crucial for the successful deployment and scalability of FL systems in agricultural contexts. By overcoming these obstacles, FL can realize its potential to revolutionize agricultural practices, improve productivity, and support sustainable farming practices.

Case Study: Federated Learning for Weather Forecasting in Agriculture

10.4.5 Problem Statement

Accurate weather forecasting is vital for effective agricultural management, influencing critical decisions such as planting schedules, irrigation, and harvesting. With the growing complexity of weather patterns and the need for localized forecasts, FL emerges as a promising approach to enhance weather prediction accuracy while addressing privacy concerns. This case study explores a conceptual framework for applying FL to weather forecasting in agriculture, aiming to integrate data from diverse sources while preserving data privacy and delivering actionable insights for farmers.

10.4.6 Objectives

The primary goals of this case study are as follows:

To Enhance Forecast Accuracy: Improve the precision of weather forecasts by leveraging diverse weather data from various agricultural sites.
To Ensure Data Privacy: Maintain confidentiality of individual farm data throughout the model training and aggregation processes.
To Deliver Localized Insights: Provide actionable, location-specific weather forecasts to support better decision-making in agriculture.

10.4.7 Conceptual Framework

10.4.7.1 Data Collection

In this conceptual framework, weather data would be collected from a network of farms equipped with IoT sensors. The data would include the following:

Temperature: Daily maximum, minimum, and average temperatures.
Humidity: Relative humidity levels.
Precipitation: Amount and frequency of rainfall.
Additional Metrics: Wind speed, soil moisture, and other relevant climatic variables.

Each farm would retain its data locally, which would prevent the need for data centralization and reduce privacy risks.

Federated Learning Framework
The proposed FL framework would involve the following steps:

Initial Model Training: A baseline weather forecasting model would be developed and trained on a central server using aggregated data from various sources. This model would serve as the starting point for FL.
Local Model Training: Farms would independently train the model on their local data, adapting it to specific regional weather patterns while keeping the raw data on-site.
Secure Aggregation: Periodically, model updates from each farm would be securely aggregated at the central server using methods like Federated Averaging. This process combines updates from local models into a global model without exposing individual farm data.

Model Redistribution: The enhanced global model would be redistributed to the farms for further local training and refinement, facilitating continuous improvement.

10.4.7.2 Privacy and Security Measures

To ensure data privacy and security, the framework would employ the following:

Secure Multi-Party Computation: Techniques to protect data during aggregation and prevent unauthorized access.
Differential Privacy: Methods to provide guarantees about the privacy of individual data points while aggregating model parameters.

10.4.8 Anticipated Outcomes

Enhanced Forecast Accuracy: By integrating diverse data sources, the FL model is expected to improve the accuracy of weather forecasts. This integration helps capture local weather variations more effectively than models trained on limited datasets.
Localized Forecasting: The FL model would provide farmers with localized weather forecasts, enabling better planning for irrigation, planting, and harvesting based on specific regional conditions.
Scalability and Adaptability: The federated approach is scalable, allowing new farms to join the network and contribute data without significant reconfiguration of the existing model. This adaptability supports continuous enhancement of forecasting capabilities.
Privacy Preservation: FL's decentralized approach ensures that data remains local and secure, addressing privacy concerns and making it easier for farms to participate in the forecasting network.

10.4.9 Challenges and Considerations

Data Heterogeneity: Variability in data quality and formats across different farms could affect model performance. Standardization techniques would need to be implemented to manage these differences.
Infrastructure Limitations: Computational and connectivity constraints in some rural areas may impact the efficiency of local model training and updates. Strategies to optimize resource usage and communication efficiency would be necessary.

Communication Overhead: Aggregating and redistributing model updates involves communication overhead. Effective strategies to manage and reduce this overhead would be crucial for maintaining system performance.

This conceptual case study demonstrates the potential of FL to improve weather forecasting in agriculture by integrating diverse datasets while preserving privacy. By providing localized and accurate forecasts, FL can enhance agricultural decision-making and productivity. As technology and methodologies continue to advance, the implementation of FL in weather forecasting holds promise for supporting more effective and sustainable agricultural practices.

10.5 Conclusion

The integration of FL into smart agriculture represents a transformative approach to modern farming practices, offering substantial improvements in data management, predictive accuracy, and operational efficiency. This article has explored various facets of FL applications in agriculture, focusing on precision farming, pest and disease detection, crop yield prediction, soil health monitoring, and climate impact analysis. Each of these applications benefits from FL's unique ability to process decentralized data while preserving privacy and enhancing model robustness through collaborative learning. In precision farming, FL enables the optimization of resource allocation and farm productivity by aggregating data from diverse sensors and IoT devices. This decentralized approach allows for tailored recommendations without the need for data centralization, thereby maintaining the privacy of individual farms. Similarly, in pest and disease detection, FL's ability to analyze data from multiple sources helps in early identification of outbreaks, facilitating timely and effective interventions. The use of FL for crop yield prediction demonstrates its capability to synthesize data from various regions, leading to more accurate forecasts and improved decision-making for planting and harvesting. In soil health monitoring, FL integrates data on soil conditions from different farms, offering actionable insights for maintaining and enhancing soil fertility. Furthermore, FL's role in climate impact analysis provides a comprehensive understanding of how changing weather patterns affect agriculture, supporting adaptive measures to mitigate the impacts of climate change. Despite its potential, implementing FL in smart agriculture presents several challenges. Data heterogeneity, privacy and security concerns, infrastructure constraints, and communication overhead are critical issues that need to be addressed.

Strategies such as developing robust model architectures, employing privacy-preserving techniques, enhancing infrastructure, and optimizing communication protocols are essential for overcoming these challenges and fully realizing the benefits of FL in agriculture. The advancements in FL technologies and their application to agriculture hold the promise of more efficient, sustainable, and resilient farming practices. Future research and development should focus on refining FL models, improving interoperability among diverse data sources, and ensuring robust security measures to protect sensitive agricultural data. By addressing these challenges and leveraging the strengths of FL, the agricultural sector can achieve significant advancements in productivity and sustainability, ultimately contributing to global food security and environmental stewardship.

In summary, FL offers a powerful framework for enhancing various aspects of smart agriculture, and continued innovation in this field will be crucial for addressing the evolving challenges and opportunities in modern farming.

10.6 Future Directions

Future directions for FL in smart agriculture include enhancing model robustness and efficiency to better handle diverse and heterogeneous agricultural data. Advancements in privacy and security measures are essential to safeguard sensitive information while maintaining high performance. Expanding FL applications into areas such as crop breeding and integrating it with technologies like blockchain could further enhance agricultural practices. Interdisciplinary collaboration is crucial for addressing complex challenges, including climate change adaptation and resource management. Additionally, improving infrastructure to support FL in rural and resource-constrained environments, along with extensive field validation, will be vital for practical implementation and scalability. By focusing on these areas, FL has the potential to significantly advance agricultural productivity, sustainability, and resilience.

AI disclosure statement: The author acknowledges the use of AI-based tools for minor language editing or grammatical correction. However, all intellectual contributions, analyses, and conclusions presented in this chapter are entirely the author's own.

References

1. AlZubi, A.A. and Galyna, K., Artificial intelligence and internet of things for sustainable farming and smart agriculture. *IEEE Access*, 2023.
2. Cesco, S., Sambo, P., Borin, M., Basso, B., Orzes, G., Mazzetto, F., Smart agriculture and digital twins: Applications and challenges in a vision of sustainability. *Eur. J. Agron.*, *146*, 126809, 2023.
3. Koshariya, A.K., Kalaiyarasi, D., Jovith, A.A., Sivakami, T., Hasan, D.S., Boopathi, S., AI-enabled IoT and WSN-integrated smart agriculture system, in: *Artificial Intelligence Tools and Technologies for Smart Farming and Agriculture Practices*, pp. 200–218, IGI Global, 2023.
4. Mane, S., Das, N., Singh, G., Cosh, M., Dong, Y., Advancements in dielectric soil moisture sensor Calibration: A comprehensive review of methods and techniques. *Comput. Electron. Agric.*, *218*, 108686, 2024.
5. Schwamback, D., Persson, M., Berndtsson, R., Bertotto, L.E., Kobayashi, A.N.A., Wendland, E.C., Automated low-cost soil moisture sensors: Trade-off between cost and accuracy. *Sensors*, *23*, 5, 2451, 2023.
6. Wilson, T.B., Kochendorfer, J., Diamond, H.J., Meyers, T.P., Hall, M., French, B., Saylor, R.D., A field evaluation of the SoilVUE10 soil moisture sensor. *Vadose Zone J.*, *22*, 2, e20241, 2023.
7. Rai, M., Tyagi, N., Tripathi, P., Kumar, N., Kumari, P., Advancement in Agricultural Techniques with the Introduction of Artificial Intelligence and Image Processing, in: *Smart Village Infrastructure and Sustainable Rural Communities*, pp. 47–68, IGI Global, 2023, https://doi.org/10.4018/978-1-6684-6418-2.ch004.
8. Tripathi, P., Kumar, N., Rai, M., Khan, A., Applications of Deep Learning in Agriculture, in: *Artificial Intelligence Applications in Agriculture and Food Quality Improvement*, pp. 17–28, IGI Global, 2022, https://doi.org/10.4018/978-1-6684-5141-0.ch002.
9. Tripathi, P., Kumar, N., Rai, M., Shukla, P.K., Verma, K.N., Applications of Machine Learning in Agriculture, in: *Smart Village Infrastructure and Sustainable Rural Communities*, pp. 99–118, IGI Global, 2023, https://doi.org/10.4018/978-1-6684-6418-2.ch006.
10. Alam, S., Security concerns in smart agriculture and blockchain-based solution, in: *2022 OPJU International Technology Conference on Emerging Technologies for Sustainable Development (OTCON)*, IEEE, pp. 1–6, 2023, February.
11. Kumar, P.R., Meenakshi, S., Shalini, S., Devi, S.R., Boopathi, S., Soil Quality Prediction in Context Learning Approaches Using Deep Learning and Blockchain for Smart Agriculture, in: *Effective AI, Blockchain, and E-Governance Applications for Knowledge Discovery and Management*, pp. 1–26, IGI Global, 2023.

12. Rehman, K.U., Andleeb, S., Ashfaq, M., Akram, N., Akram, M.W., Blockchain-enabled smart agriculture: Enhancing data-driven decision making and ensuring food security. *J. Clean. Prod., 427,* 138900, 2023.
13. Imteaj, A., Thakker, U., Wang, S., Li, J., Amini, M.H., A survey on federated learning for resource-constrained IoT devices. *IEEE Internet Things J., 9,* 1, 1–24, 2021.
14. Le, D.D., Dao, M.S., Tran, A.K., Nguyen, T.B., Le-Thi, H.G., Federated Learning in Smart Agriculture: An Overview, in: *2023 15th International Conference on Knowledge and Systems Engineering (KSE),* IEEE, pp. 1–4, 2023, October.
15. Žalik, K.R. and Žalik, M., A review of federated learning in agriculture. *Sensors, 23,* 23, 9566, 2023.
16. McMahan, B., Moore, E., Ramage, D., Hampson, S., y Arcas, B.A., Communication-efficient learning of deep networks from decentralized data, in: *Artificial Intelligence and Statistics,* pp. 1273–1282, PMLR, 2017, April.
17. Manoj, T., Makkithaya, K., Narendra, V., A federated learning-based crop yield prediction for agricultural production risk management. *2022 IEEE Delhi Section Conference (DELCON),* IEEE, pp. 1–7, 2022.
18. Mehta, S., Kukreja, V., Vats, S., Advancing agricultural practices: Federated Learning-based CNN for Mango Leaf Disease Detection, in: *2023 3rd International Conference on Intelligent Technologies (CONIT),* pp. 1–6, IEEE, 2023.
19. Durrant, A., Markovic, M., Matthews, D., May, D., Enright, J., Leontidis, G., The role of cross-silo federated learning in facilitating data sharing in the agri-food sector. *Comput. Electron. Agric., 193,* 106648, 2022.
20. Bonawitz, K., Ivanov, V., Kreuter, B., Marcedone, A., McMahan, H.B., Patel, S., Seth, K., Practical secure aggregation for privacy-preserving machine learning, in: *Proceedings of the 2017 ACM SIGSAC Conference on Computer and Communications Security,* pp. 1175–1191, 2017, October.
21. Ooi, M.P.L., Sohail, S., Huang, V.G., Hudson, N., Baughman, M., Rana, O., *et al.*, Measurement and applications: Exploring the challenges and opportunities of hierarchical federated learning in sensor applications. *IEEE Instrum. Meas. Mag., 26,* 9, 21–31, 2023.
22. Li, Z., Wang, D., Zhu, T., Tao, Y., Ni, C., Review of deep learning-based methods for non-destructive evaluation of agricultural products. *Biosyst. Eng., 245,* 56–83, 2024.
23. Tyagi, S. and Tyagi, A., Deep reinforcement learning-based framework for tactical drone deployment in rigorous terrains: From modeling to real-world implementation, in: *Web 3.0,* pp. 39–53, CRC Press, 2024, https://doi.org/10.1201/9781003461418.
24. Kumar, G.D., Tyagi, S., Pradhan, K.C., Deep learning forecasting: An LSTM neural architecture-based approach to rainfall and flood impact

predictions in Bihar. *Procedia Comput. Sci.*, *235*, 1455–1466, 2024. https://doi.org/10.1016/j.procs.2024.04.137.

25. Tyagi, S. and Tyagi, H., A Comparative Analysis of Proposed Schemes for Increasing Channel Utilization In 3G Networks. http://dx.doi.org/10.13140/RG.2.2.35188.36480, 2015.
26. Kumar, G.D., Tyagi, S., Pradhan, K.C., Predictive ML analysis: Rainfall & flood vulnerability in Bihar, India, in: *Artificial Intelligence and Information Technologies*, pp. 447–453, CRC Press, 2024, https://doi.org/10.1201/9781032700502-71.
27. Tyagi, S. and Tyagi, A., Edge Computing: Empowering Real-Time Data Processing & Connectivity, 2024, https://doi.org/10.5281/zenodo.12738009.

11

Federated Learning and Its Impact on Decision-Making in Smart Agriculture

Divita Jain[1]*, Nikita Bhati[2] and Nisha Bhardwaj[3]

[1]Department of Nutrition and Dietetics, SAHS, Sharda University, Greater Noida, India
[2]Department of Management, SBSS, Sharda University, Greater Noida, India
[3]Department of Psychology, SSHSS, Sharda University, Greater Noida, India

Abstract

As global food demand rises with population growth, enhancing agricultural productivity while meeting diverse sustainability needs becomes increasingly complex. Federated learning (FL) offers a transformative approach by enabling decentralized data analysis without centralizing it, which supports precision farming, improves pest and disease detection, enhances soil health monitoring, and provides localized weather forecasts. By integrating data from multiple sources, FL optimizes crop management practices and supports better decision-making through collaborative model development. FL also revolutionizes food quality assurance by aggregating data from sensors, cameras, and tests to detect quality deviations and ensure consistent standards across facilities. It enhances traceability and transparency in food supply chains and aids in reducing contamination risks by developing predictive models based on comprehensive data sources. Incorporating artificial intelligence (AI) and the Internet of Things (IoT) further enhances these advancements. IoT provides real-time monitoring of environmental conditions and supply chain logistics, whereas AI automates quality control, predicts equipment failures, and optimizes crop management through predictive analytics. AI also forecasts market demand and streamlines logistics, ensuring timely delivery and quality of food products. Federated learning systems (FLS) address the psychological aspects of decision-making by reducing cognitive load through data-driven recommendations and improving risk perception by providing valuable insights. FLS fosters social learning and collective intelligence,

**Corresponding author*: divitajain99@gmail.com

Padmesh Tripathi, Bhanumati Panda, Shanthi Makka, Reeta Mishra, S. Balamurugan and Sheng-Lung Peng (eds.) Federated Learning for Smart Agriculture and Food Quality Enhancement, (271–292) © 2026 Scrivener Publishing LLC

encouraging collaboration and knowledge sharing among farmers. Together, FL, AI, and IoT form a robust framework for improving food quality, safety, and agricultural efficiency, fostering a more sustainable and resilient food system.

Keywords: Federated learning, artificial intelligence, traceability, environmental conditions, social learning

11.1 Introduction to Federated Learning in Agriculture

Worldwide food consumption is rising together with the growth of the world's population. Increasing food production capacity in response to rising demand is a difficult endeavor, made more difficult by the need to take into account the various needs of distinct multicultural communities. As a result, efforts have been made to identify strategies for improving food production, such as making the most use of land and water. Sustainability in agriculture can be achieved by effectively using data in decision-making [1]. This highlights the need for a global food system that is resource-efficient and addresses the diverse and precise demands of sustainability. Efforts have been made recently to improve the efficiency and quality of agricultural produce by utilizing technology to support smart farming. For instance, we can sense, gather, and mine data in a variety of settings and circumstances by utilizing technologies like artificial intelligence (AI) and the Internet of Things (IoT) [2]. Then, by undergoing a more thorough examination, this data can help with situation awareness and decision-making in real time, leading to the production of high-quality products and improved decision-making. Throughout the entire process of production, storage, processing, distribution, retailing, and consumption, food is moved from farmer to consumer as part of the food supply chain [3]. The industry is actively embracing a wide range of emerging technologies and advanced networking designs, such as edge computing, blockchain technology, and software-defined networking, because traditional systems and network architecture are unable to match these demands. Although these technologies have benefits, they also contain a number of security vulnerabilities that may be exploited to cause significant harm [4].

11.2 Applications of Federated Learning in Smart Agriculture

11.2.1 Crop Management and Precision Farming

Federated learning (FL) enhances precision farming by enabling the development of robust models that integrate data from multiple sources, such as satellite imagery, field sensors, and farm machinery. These models can optimize crop management practices, including planting schedules, fertilization, and irrigation, by learning from diverse data without requiring centralized data storage [5]. This approach allows farmers to make data-driven decisions tailored to their specific conditions, ultimately improving crop yields and resource efficiency.

11.2.2 Pest and Disease Detection

In pest and disease management, FL facilitates the creation of predictive models using data from various sources, including field observations, remote sensing, and historical records. By training these models collaboratively across different farms and regions, FL can enhance the accuracy of pest and disease detection systems. This collaborative approach helps in early identification and forecasting of outbreaks, allowing for timely and targeted interventions to minimize crop damage and reduce the use of chemical treatments [6].

11.2.3 Soil Health Monitoring

FL supports soil health monitoring by aggregating data from soil sensors, satellite imagery, and agronomic studies without centralizing the data. This collective learning enables the development of models that can assess soil quality, track nutrient levels, and monitor soil moisture [7]. By analyzing soil health across diverse agricultural settings, FL helps in understanding soil variability and improving soil management practices, leading to better crop performance and sustainable farming practices.

11.2.4 Weather Prediction and Climate Adaptation

FL enhances weather prediction and climate adaptation by integrating weather data from multiple sources, including local weather stations, satellite observations, and climate models. Collaborative training of predictive models enables more accurate and localized weather forecasts, which are crucial for planning agricultural activities. Additionally, FL helps in assessing climate impacts on crop growth and developing adaptive strategies to mitigate the effects of climate change, ensuring resilience and sustainability in agriculture [8].

11.3 Enhancing Food Quality through Federated Learning

11.3.1 Quality Assurance and Control in Food Production

FL revolutionizes quality assurance and control in food production by enabling the development of advanced models that analyze data from various production stages. By aggregating data from multiple sources such as sensors, production line cameras, and laboratory tests, FL helps create more accurate models for detecting quality issues and deviations [9]. This collaborative approach ensures consistent monitoring and improvement of food quality standards across different production facilities without sharing sensitive data, leading to enhanced product consistency and compliance with quality standards.

11.3.2 Traceability and Transparency in Food Supply Chains

In food supply chains, FL improves traceability and transparency by allowing the integration of data from various stakeholders, including suppliers, processors, and retailers. By collaboratively analyzing data on product origin, processing conditions, and distribution routes, FL models can track the movement of food products more effectively. This enhanced visibility helps in verifying the authenticity of products, identifying potential sources of contamination, and ensuring adherence to regulatory requirements [10]. As a result, consumers gain confidence in the safety and quality of their food.

11.3.3 Improving Food Safety and Reducing Contamination Risks

FL plays a crucial role in improving food safety by facilitating the development of predictive models that analyze data related to contamination risks. By pooling data from food safety inspections, environmental monitoring, and historical contamination records, FL models can identify patterns and

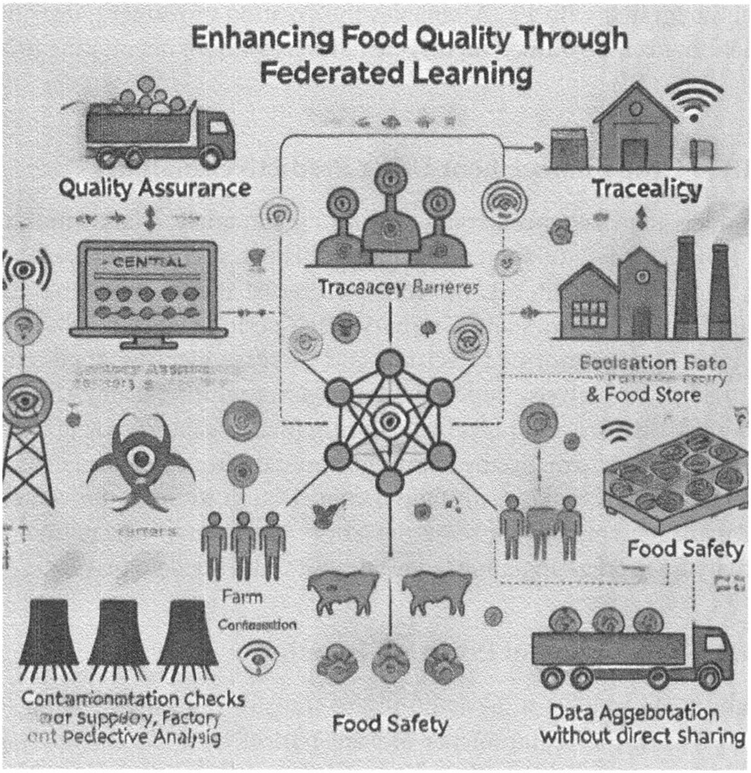

Figure 11.1 Enhancing food quality through federated learning.

predict potential contamination events. This proactive approach enables timely interventions and preventive measures, reducing the risk of food-borne illnesses and enhancing overall food safety [11]. Collaborative learning also aids in refining safety protocols and optimizing response strategies across different food production and processing environments as shown in Figure 11.1.

11.4 Using AI to Make Decisions in Smart Agriculture

The integration of AI is important in the transformation of traditional agriculture methods into smart, data-oriented process. By utilizing AI models, framers and other agriculture stakeholders may make informed decision that led to enhance efficiency, reduce cost, and improve crop production.

When integrated with FL, AI enables the creation of resilient, distributed models that can be trained on diverse datasets without negotiating privacy [12].

11.4.1 Crop Management Using Predictive Analytics

AI models that have been trained using FL can analyze large amount of data from a variety of sources, including historical crop data, soil condition, and weather patterns, in order to forecast the most optimal planting and irrigation times and also harvesting periods. This insight enables farmers to make informed decision and increase yield by conducting a comprehensive analysis. Additionally, image recognition using AI can easily identify diseases and pests by analyzing images captured by drones. FL enables their performance by learning from data from various farms [13]. These AI systems make a substantial contribution to the improvement of crop management, early disease detection, and overall agricultural productivity by aggregating knowledge without reducing data privacy.

11.4.2 Techniques of Precision Farming

Advanced AI models in agriculture are transforming the farming techniques to suit the requirements of each plot of land. By the analysis of localized parameters like soil, moisture levels, weather patterns, health, and nutrient, these models provide important suggestions to maximized farming methods. The approach applied in this context aims to reduce wastage by providing the targeted and applied only where and when required, for instance, fertilizers, water, and pesticides, thereby maximizing both overall production and sustainability. Tractors and other machinery can be controlled by AI, which can adjust their functions to match the unique requirement of the crop variety and the environment [14].

11.4.3 Supply Chain Optimization with the Help of AI

AI can forecast market demand for a variety of crops by analyzing consumer trends, economic indicators, and climate conditions. This information helps farmers in making informed decisions related to plant a crop and production quality [15]. This helps in reducing overproduction or underproduction, aligning supply efficiently with demand. Additionally, AI is important for optimizing distribution and logistics, which ensures fresh and on time delivery of agriculture products from farmers to sellers. With the help of effective and efficient supply chain, AI ensures that consumers get good quality of food and also provide it on time.

11.5 Improving Food Quality with IoT, AI, and Blockchain

Application of IoT, AI, and blockchain in agriculture helps in enhancing the quality and safety of food products. These technologies facilitate automated decision-making, real-time monitoring, and predictive maintenance, ensuring that a good food quality is maintained from field to plate.

11.5.1 Real-Time Monitoring Based on IoT

The integration of IoT sensors in the field and storage facilities enables the regular monitoring of key environmental parameters, including humidity, light exposure, and temperature. The availability of real- time data is essential for providing ideal conditions that support healthy crop development and proper storage of food, helping to maintain the quality and freshness of agriculture goods [16]. By ensuring suitable growing conditions and storage after harvesting, these sensors effectively reduce the risk of rotting or spoilage and enhancing shelf life.

In addition to support environment monitoring, these devices offer significant supply chain visibility by tracking the transportation of agricultural goods from farmers to markets. These devices provide real-time information on the location and state of inventory and enable users to track factors like potential delays and temperature.

11.5.2 AI-Driven System for Quality Control

AI models connected with IoT technology, like sensors and cameras, play a significant role in the automated evaluation of food products, as it helps in real-time detection of contamination or signs of spoilage [8]. By continuously examining data on variables like texture, temperature, and color, this AI system can promptly detect potential quality issue, therefore ensuring the that only superior quality items pass through supply chain. The application of this automated process greatly diminishes human error and accelerates the quality control, therefore providing customers with safer and fresher food.

Additionally, AI-driven system utilizes IoT data to predict when storage tools, like freezers and temperature control system, are likely to fail. By detecting the trends in temperature fluctuation, AI can predict failures before they occur [17]. Adopting this predictive maintenance approach allows timely repairs, preventing equipment failures that may cause degradation of stored food items. As a result, the preservation of food quality, waste reduction, and operational efficiency is improved, which ensures a stable supply of good-quality food from farmers to end consumers.

11.5.3 Implementation of Blockchain in Agriculture and Food Quality Management

Blockchain enriches the AI functioning by creating safe and immutable records of every data transaction. Blockchain plays an important role in improving transparency, data security, and trust in the system of FL for agriculture and management of food quality. In case of FL, where data remains decentralized yet model updates are exchanged among participants, blockchain guarantees the security of these methods by creating unchangeable records of every transaction [18]. This approach also guarantees the preservation of data transparency, therefore reducing potential hazards like manipulation during the training of models and data breaches.

In context to supply chain transparency, blockchain allows all parties in the food chain, such as farmers, distributors, retailers, and processors to keep traceable record of each transaction. Combining this with FL provides verified, real-time data on the condition and quality of food products, therefore enhancing food safety, minimizing fraud, and allowing the rapid detection of issues such as spoiling or contamination. Moreover, blockchain technology foster cooperation and trust among all parties involved, like food producers and small-scale farmers, who may otherwise be reluctant to exchange statistical data. By enabling transparent and secured exchange of model updates, blockchain promotes broader involvement and enhance the capacity of FL.

11.6 Federated Learning Enhances the Detection of Food Adulterants

Food adulteration, the practice of adding inferior or harmful substances to food products, is a significant concern for public health and food safety. The detection of food adulterants is crucial for ensuring that food products are safe and meet regulatory standards. Traditional methods of detection, while effective, often face challenges related to data privacy, high costs, and centralized data management [19]. FL offers a promising alternative by addressing these issues through a decentralized approach to machine learning. Here is how FL can enhance the detection of food adulterants.

11.6.1 Maintaining Data Privacy

One of the most significant advantages of FL is its ability to maintain data privacy. In conventional centralized models, data from multiple sources is collected and analyzed in a single location, which can raise privacy concerns, especially when dealing with sensitive or proprietary data from food safety laboratories. FL mitigates these concerns by keeping data decentralized. Each institution or organization involved in the detection process trains its machine learning model locally on its own data and only shares aggregated model updates (such as gradients or weights) with a central server [20]. This approach ensures that raw data remains within the institution, safeguarding sensitive information from potential breaches.

11.6.2 Reducing Data Transfer and Storage Costs

Centralized data processing often involves significant costs related to data transfer and storage. Institutions must send large volumes of data to a central server, which can be both expensive and logistically challenging. FL reduces these costs by minimizing the amount of data that needs to be transferred. Instead of transferring entire datasets, institutions only send model updates to the central server. This reduces the bandwidth required and decreases storage needs, making it a more cost-effective and efficient solution for large-scale data analysis [21].

11.6.3 Leveraging Collaborative Data for Enhanced Model Accuracy

FL allows multiple institutions to collaborate in training a global model without sharing their data. This collaborative approach harnesses diverse datasets from different sources, which enhances the model's ability to detect a wide range of adulterants. Each institution contributes valuable data, which helps in building a more robust and accurate model. By aggregating updates from various institutions, the global model benefits from a broader spectrum of data, leading to improved detection accuracy and a better understanding of different types of adulterants [22].

11.6.4 Scalability and Flexibility

FL is inherently scalable and flexible, making it well-suited for the dynamic nature of food safety monitoring. As new institutions join the federated network, they can contribute their data to further refine and improve the global model. This scalability is crucial in the food safety sector, where numerous entities may be involved in testing and monitoring different food products [23]. Additionally, FL can be adapted to various types of data and detection methods, providing the flexibility needed to address diverse adulteration scenarios and evolving food safety challenges.

11.6.5 Real-Time Updates and Monitoring

FL enables real-time model updates and monitoring. Institutions continuously train their local models as new data becomes available, and these updates are periodically aggregated to enhance the global model. This capability supports timely detection and response to emerging adulteration patterns. For instance, if a new type of adulterant is identified by one institution, then its local model updates can be quickly integrated into the global model, improving the collective ability to detect and address new threats in real time [24].

11.6.6 Handling Data Heterogeneity

Food safety data can vary widely in quality and format across different institutions. FL is designed to handle this heterogeneity effectively. Each institution trains its model on its own data, which may differ in quality and representation. The central server aggregates these diverse models into a unified global model, managing variations in data quality and ensuring that the global model benefits from the rich, diverse datasets [25]. This approach leads to improved detection performance across different types of food products and adulterants.

11.7 Federated Learning Enhances Food Inspection and Grading

Food inspection and grading are critical processes in ensuring food safety, quality, and compliance with regulatory standards. These processes involve evaluating food products to detect contamination, adulteration, and to assess quality attributes such as freshness, ripeness, and texture. Traditionally, food inspection and grading have relied on centralized systems and manual methods, which can be resource-intensive and prone to inconsistencies [26]. FL offers a promising alternative by addressing several challenges associated with traditional methods. Here is how FL can enhance food inspection and grading:

11.7.1 Improving Data Privacy and Security

FL is particularly advantageous in preserving data privacy and security. In traditional systems, data from multiple sources such as food inspection agencies, processing plants, and grading facilities is often aggregated and analyzed centrally. This centralization raises concerns about data security and privacy, especially when dealing with sensitive information. FL addresses these concerns by enabling institutions to train machine learning models locally on their own data. The central server only aggregates model updates, rather than raw data, ensuring that sensitive information remains secure and private [27].

11.7.2 Reducing Costs and Operational Burdens

Centralized food inspection and grading systems often involve high costs related to data transfer, storage, and infrastructure. Institutions must send large amounts of data to a central location for analysis, which can be both costly and logistically challenging. FL mitigates these costs by minimizing the amount of data that needs to be transferred. Instead of sending entire datasets, institutions send only model updates to the central server. This approach reduces bandwidth requirements and storage needs, making food inspection and grading more cost-effective and efficient [28].

11.7.3 Enhancing Model Accuracy through Collaboration

FL allows multiple institutions to collaborate in training a global model without sharing their data. This collaborative approach enables the aggregation of diverse datasets from various sources, which enhances the model's ability to accurately inspect and grade food products. For example, different institutions may have expertise in evaluating different types of food products or attributes [29]. By pooling their knowledge through federated model training, institutions can develop a more robust and comprehensive model that improves the accuracy of food inspection and grading.

11.7.4 Facilitating Scalability and Adaptability

The decentralized nature of FL makes it highly scalable and adaptable to different contexts. As new institutions or inspection sites join the federated network, they can contribute their data to improve the global model. This scalability is particularly valuable in food inspection and grading, where the diversity of food products and inspection scenarios can be vast. FL's adaptability allows it to handle various types of food inspections, from detecting contaminants in raw ingredients to assessing the quality of processed foods [30].

11.7.5 Supporting Real-Time Updates and Monitoring

FL supports real-time model updates and monitoring, which is crucial for timely food inspection and grading. Institutions continuously train their local models as new data becomes available, and these updates are periodically aggregated to refine the global model [31]. This real-time capability enables rapid detection of emerging issues, such as new contaminants or quality defects, and ensures that inspection and grading processes remain up-to-date with the latest information.

11.7.6 Managing Data Heterogeneity

Food inspection and grading data can vary significantly across different institutions in terms of quality, format, and type. FL is designed to handle this heterogeneity effectively. Each institution trains its model using its own data, which may differ in representation and quality. The central server aggregates these diverse models into a unified global model, accommodating variations in data and ensuring that the global model benefits from a wide range of information. This leads to more accurate and reliable inspection and grading outcomes [32].

11.7.7 Enabling Collaborative Improvement

FL fosters a collaborative environment where institutions can work together to enhance food inspection and grading systems. By sharing model improvements and insights rather than raw data, institutions can jointly develop better detection and grading models. This collaborative approach encourages knowledge sharing and innovation, leading to continuous improvements in food safety and quality assessment.

11.8 The Impact of Federated Learning Systems on Farmer Decision-Making: A Psychological Perspective

The advancement in technology has modified how decision is taken through numerous industries, including agriculture. Among the technologies and innovations that are currently revolutionizing agricultural decision-making are federated learning system (FLS). By allowing decentralized date analysis, FIS presence privacy while utilization collective intelligence. Numerous farmers may gain profit from pooled insights without having to share their own private information [33].

The psychological aspect of farmer on decision-making, such as information processing, technology trust, autonomy, perception of risk, maintaining, and social learning, could be significantly affected by the current system.

11.8.1 Cognitive Load and Decision Complexity

Farmers deal the numerous complex variables of intricate factors in agriculture, including crop cycles, price of the market, soil conditions, and weather patterns. Maintaining so much data can be overwhelmed to decision-making capacity, particularly for small-scale farmers. FLS produces data-driven, real-time recommendations that clarify various aspects, thereby reducing the level of the complexity [34]. Farmer's cognitive load is lessened by FLS, which integrates historical and environmental data and implements predictive analytics. This allows farmers free to concentrate on useful information rather than trying to understand complex data on their own.

Through FLS streamlines decision-making, it brings a new level of complex by requiring comprehension and interpretation of the various outputs produced by machine learning algorithms. Farmers who are not proficient in technology could find it challenging to use these insights initially, which could increase cognitive burden rather than lessen it. The appropriate education is required to facilitate this shift. Farmers are required to be able to properly analyze the data and have faith in the reliability and accuracy of the system. According to psychological studies on cognitive load, decision-making enhances both precision and effectiveness when mental effort is lessened. By realizing that some of their mental burden can be reduce by FLS, it can increase the quality of agricultural decision with sufficient funding and support.

11.8.2 Risk Perception and Behavioral Change

Risk sensitivity is central to agricultural decision-making, as farmers regularly evaluate the risks related to various practices, which includes crop selection, pesticides use, or incorporating new technologies. FLS uses mass data to recognize risks and deliver insights to farmers so that they make well-informed decisions. However, farmers perceive these risks differently that some people find it questionable and doubtful on the other hand some of them may feel confident with the data-driven recommendations [35]. According to prospect theory, people are more likely to emphasize the possible losses over gains, making them care for the negative outcomes over the positive. To drive behavioral change, FLS must provide the risk data that addresses the concerns and building confidence among insecure farmers and promoting broader adoption of data-driven methods.

11.8.3 Social Learning and Collective Intelligence

FLS depends on collective intelligence, where various kinds of farmers provide data that is pooled and examined in order to generate insights that are beneficial to all the parties engaged [4]. This is consistent with the theory of social learning, which states that people learn new things by seeing the action and the outcomes of others. FLS fosters collaboration by allowing farmers to exchange knowledge and insights. FLS encourages teamwork and enhancing decision-making through providing accessibility to a broader range of data [36]. The social dynamics of FLS motivates farmers to contribute and provide accurate data, promoting a cooperative and community-oriented style of farming. The quality of insights increases with the number of farmers participating, supporting the process of social learning and leading to better decision-making and outcomes.

11.9 Challenges in Implementing Federated Learning

The major challenge in implementing FL in both food quality management and agriculture is the presence of data heterogeneity. Developing AI models that are universally effective is a difficult task because of the significant variation in technological framework, environmental conditions, and quality control standards across farms and food supply chains. Additionally, the infrastructure needed to operate local data processing and models of AI is lacking in many rural farms and food storage stores. Implementation is further complicated by unreliable supply of power, connectivity issues, and lack of technology advancements. FL, which is meant to secure local data, is prone to breaches and model inversion crimes, which may compromise the integrity of sensitive agriculture and data of food quality. As a result, security and privacy are very important. Another issue is communication overhead, particularly in regions which are geographically dispersed where the regular communication among local devices and central servers is needed for model updates. This is especially difficult in food supply chains, where maintaining product quality is dependent upon the monitoring and tracking of real-time data. Additionally, the logistical challenges related to scaling FLS throughout a multitude of food products and farms are important, as small-scale food processors and farmers may be unwilling due to competition fears. Finally, the integration of FL with other growing technologies, like blockchain, IoT, and AI-driven automations is expensive as well as complex, especially in the food quality sector, where accurate control and monitoring are crucial for ensuring safety.

11.10 Limitations of Federated Learning

There are various limitations associated with the application of FL in smart agriculture and food quality management. Data heterogeneity is one of the major limitations, as the format, quality, and environment conditions of data from various food processing facilities and farms are highly variable. Consequently, it is difficult to create AI models that are globally effective. A significant number of farms, especially those located in rural regions, have a shortfall in the necessary computational power and network infrastructure to actively participate in FL. Furthermore, security and privacy concerns do exist, as FL is susceptible to threats like leakage of data and model inversion, despite the potential of blockchain to reduce some risks. The high energy consumption required for both blockchain and FL may not be sustainable. There are regulatory and legal obstacles, as there are no clear universal standards that regulate the use of these technology. This highlight concerns regarding liability, data ownership, and cross border data sharing. The adoption of these technologies—IoT, AI, and blockchain—within agriculture and food quality system is also complex and expensive, which presents a challenge for smaller players to participate. These limitations underscore the need for further research, technological advancements and supportive policies to address the alleviated theme and enhance the efficacy of FL in the management of food quality management and agriculture.

11.11 Future Directions for Research

In order to overcome these challenges, future improvements must prioritize the standardization of data formats and collection protocols within a variety of agriculture and food production environments, thus enhancing the scalability and adaptability of AI models. Designing lightweight AI models and implementing edge computing can efficiently tackle the limited computational resources in smaller and rural food facilities. Integrating stronger privacy protocols, like differential privacy and secure multi-party computing, can improve the security and reliability of data in FLS. The adoption of model compression techniques, the extension of 5G networks, and local broadband programs are important for addressing communication overhead. By establishing transparent governance structure and giving incentives for involvement in FL networks, collaborations may be encouraged in both food supply chains and agriculture. Governments and organizations have the potential to contribute significantly to the assistance of smaller food producers and farms by giving timely access to essential technology. Finally, the collaboration of FL with AI, IoT, and blockchain technologies will offer a more uniform environment for both food quality management and agriculture productivity.

11.12 Conclusion

FL offers transformative potential for both smart agriculture and food quality enhancement. Key insights include its ability to facilitate collaborative, privacy-preserving data analysis across diverse agricultural and food production systems. This approach improves precision farming by enabling data-driven crop management, enhances pest and disease detection through collective learning, and supports soil health monitoring by integrating data from various sources. In food quality, FL strengthens quality assurance, enhances traceability and transparency in supply chains, and improves food safety by predicting contamination risks. Overall, FL empowers stakeholders with more accurate, localized, and actionable insights while maintaining data privacy and security. The future of smart agriculture and food quality is set to benefit greatly from FL. As this technology evolves, it promises to drive more personalized and efficient agricultural practices, optimize resource use, and enhance sustainability. For food quality, FL will likely lead to more robust quality control systems, greater transparency in supply chains, and improved safety measures.

This advancement will foster stronger collaboration among stakeholders, leading to more resilient and adaptive agricultural systems and a safer, more transparent food supply. As the technology matures, its integration into these sectors will be crucial for addressing global challenges related to food security and sustainability. AI-powered decision-making in smart agriculture, facilitated by FL allows precise farming by offering tailored suggestions and automating equipment modifications based on real-time information. This improves crop productivity maximise resource utilization and reduce waste. Additionally, the combination of IoT and AI transforms food quality by enabling time to time monitoring, preventive repairs, and automated inspections. To maintain food quality and reducing spoilage, these technologies ensure that crops are grown, stored, and distributed under the ideal conditions. Collectively, they contribute to a sustainable agricultural method and enhance the efficiency and transparency of food supply chain. FL offers a transformative approach to detecting food adulterants by addressing key challenges associated with traditional centralized methods. By preserving data privacy, reducing costs, enhancing model accuracy through collaboration, and providing scalability and flexibility, FL represents a significant advancement in food safety technology. As FL continues to evolve, its integration into food safety systems has the potential to improve the detection and monitoring of food adulterants, ultimately contributing to safer and more reliable food supply chains. FL offers significant advancements in food inspection and grading by addressing key challenges associated with traditional methods. By preserving data privacy, reducing costs, enhancing model accuracy, and facilitating scalability and adaptability, FL provides a powerful tool for improving the efficiency and effectiveness of food inspection and grading processes. As FL technology continues to evolve, its integration into food safety systems promises to enhance the accuracy and reliability of food inspections and grading, contributing to safer and higher-quality food products. FLS offers the potential to completely transform agriculture by offering farmers access to data-driven, socially relevant knowledge. However, the level of success of their technologies interlock with farmer's psychological processes is just as important to their success as their ability to innovate. Farmer's interactions with FL models are influenced by number of variables, such as cognitive load, emotional responses, and behavioural inclinations. Partners in the agriculture technology sector can create FLS that enhance farmer's decision-making processes and optimise the productivity of farms by using a psychological approach. As a result, this could lead to increased acceptance rates and successful outcomes, ensuring that AI and human knowledge collaborate for the advancement of agriculture in the long run.

Acknowledgments

The authors are thankful to Sharda University, Greater Noida, Uttar Pradesh, India, for intense curiosity, leadership, and provision of research facilities.

AI disclosure statement: The author acknowledges the use of AI-based tools for minor language editing or grammatical correction. However, all intellectual contributions, analyses, and conclusions presented in this chapter are entirely the author's own.

References

1. Ahmad, A., Liew, A.X., Venturini, F., Kalogeras, A., Candiani, A., Di Benedetto, G., Martos, V., AI can empower agriculture for global food security: challenges and prospects in developing nations. *Front. Artif. Intell.*, 7, 1328530, 2024, https://doi.org/10.3389/frai.2024.1328530.
2. Asharindavida, F., Nibouche, O., Uhomoibhi, J., Liu, J., Vincent, J., Wang, H., Miniature spectrometer data analytics for food fraud. *J. Consum. Prot. Food Saf.*, 18, 4, 415–431, 2023, https://doi.org/10.1007/s00003-023-01439-8.
3. Bhosale, Y.H., Zanwar, S.R., Ali, S.S., Vaidya, N.S., Auti, R.A., Patil, D.H., Multi-plant and multi-crop leaf disease detection and classification using deep neural networks, machine learning, image processing with precision agriculture-A review. *2023 International Conference on Computer Communication and Informatics (ICCCI)*, IEEE, pp. 1–7, Jan. 2023, 10.1109/ICCCI56745.2023.10128246.
4. Boobalan, P., Ramu, S.P., Pham, Q.V., Dev, K., Pandya, S., Maddikunta, P.K.R., Gadekallu, T.R., Huynh-The, T., Fusion of federated learning and industrial Internet of Things: A survey. *Comput. Netw.*, 212, 109048, 2022, https://doi.org/10.1016/j.comnet.2022.109048.
5. He, Y., Bai, X., Xiao, Q., Liu, F., Zhou, L., Zhang, C., Detection of adulteration in food based on nondestructive analysis techniques: A review. *Crit. Rev. Food Sci. Nutr.*, 61, 14, 2351–2371, 2021, https://doi.org/10.1080/10408398.2020.1777526.
6. Kulkarni, P., Karwande, A., Kolhe, T., Kamble, S., Joshi, A., Wyawahare, M., *iv preprint arXiv:2106.10698*, 2021, https://doi.org/10.48550/arXiv.2106.10698.
7. Gavai, A., Bouzembrak, Y., Mu, W., Martin, F., Kaliyaperumal, R., Van Soest, J., Choudhury, A., Heringa, J., Dekker, A., Marvin, H.J., Applying federated learning to combat food fraud in food supply chains. *npj Sci. Food*, 7, 1, 46, 2023, https://doi.org/10.1038/s41538-023-00220-3.
8. Muhammad, B.L. and Kusharki, M.B., Federated Learning for Collaborative Crop Disease Monitoring in Wheat Production, in: *2023 2nd International*

Conference on Multidisciplinary Engineering and Applied Science (ICMEAS), vol. 1, IEEE, pp. 1–5, Nov. 2023, 10.1109/ICMEAS58693.2023.10379272.

9. Farmar-Bowers, Q. and Lane, R., Understanding farmers' strategic decision-making processes and the implications for biodiversity conservation policy. *J. Environ. Manage.*, 90, 2, 1135–1144, 2009, https://doi.org/10.1016/j.jenvman.2008.05.002.
10. Friha, O., *et al.*, FELIDS: Federated learning-based intrusion detection system for agricultural Internet of Things. *J. Parallel Distrib. Comput.*, 165, 17–31, 2022, https://doi.org/10.1016/j.jpdc.2022.03.003.
11. Kumar, P., Gupta, G.P., Tripathi, R., PEFL: Deep privacy-encoding-based federated learning framework for smart agriculture. *IEEE Micro*, 42, 1, 33–40, 2021. 10.1109/MM.2021.3112476.
12. Le, D.D., *et al.*, Federated Learning in Smart Agriculture: An Overview, in: *2023 15th International Conference on Knowledge and Systems Engineering (KSE)*, IEEE, pp. 1–4, Oct. 2023, 10.1109/KSE59128.2023.10299497.
13. Leal, D., de Queiroz, D.T., Rodrigues, R.A., A behavior analysis of Brazilian farmers based on sunk cost effects and prospect theory. *Rev. Produção e Desenvolvimento*, 6, 2020, https://doi.org/10.32358/rpd.2020.v6.473.
14. Mallinger, K. and Baeza-Yates, R., Responsible AI in Farming: A Multi-Criteria Framework for Sustainable Technology Design. *Appl. Sci.*, 14, 1, 437, 2024, https://doi.org/10.3390/app14010437.
15. McCown, R.L., Changing systems for supporting farmers' decisions: problems, paradigms, and prospects. *Agric. Syst.*, 74, 1, 179–220, 2002, https://doi.org/10.1016/S0308-521X(02)00026-4.
16. Mehta, S., Kukreja, V., Vats, S., Empowering Farmers with AI: Federated Learning of CNNs for Wheat Diseases Multi-Classification, in: *2023 4th International Conference for Emerging Technology (INCET)*, IEEE, pp. 1–6, May 2023, 10.1109/INCET57972.2023.10170091.
17. Nguyen, D.C., Pham, Q.V., Pathirana, P.N., Ding, M., Seneviratne, A., Lin, Z., Dobre, O., Hwang, W.J., Federated learning for smart healthcare: A survey. *ACM Comput. Surv. (CSUR)*, 55, 3, 1–37, 2022, https://doi.org/10.1145/3501296.
18. Puri, V., Solanki, V.K., Aponte, G.J.R., Blockchain and Federated Learning Based Integrated Approach for Agricultural Internet of Things, in: *The International Conference on Intelligent Systems & Networks*, Springer, pp. 240–246, Mar. 2023, https://doi.org/10.1007/978-981-99-4725-6_30.
19. Neto, H.A., Tavares, W.L., Ribeiro, D.C., Alves, R.C., Fonseca, L.M., Campos, S.V., On the utilization of deep and ensemble learning to detect milk adulteration. *Biodata Min.*, 12, 1–13, 2019, https://doi.org/10.1186/s13040-019-0200-5.
20. Parmar, I., Gehlot, P., Modi, V., Patel, D., Recent developments in food adulteration analytical techniques. *Int. J. Pharm. Sci. Res.*, 13, 5, 2001–2012, 2022, http://dx.doi.org/10.13040/IJPSR.0975-8232.13(5).2001-12.

21. Suryavanshi, A., Kukreja, V., Srivastava, P., Mehta, S., Joshi, K., Smart Farming: Integrated Federated Learning CNNs for Brinjal Leaf Disease Detection, in: *2024 IEEE International Conference on Interdisciplinary Approaches in Technology and Management for Social Innovation (IATMSI)*, vol. 2, IEEE, pp. 1–6, Mar. 2024, 10.1109/IATMSI60426.2024.10503064.
22. Žalik, K.R. and Žalik, M., A review of federated learning in agriculture. *Sensors*, 23, 23, 9566, 2023, https://doi.org/10.3390/s23239566.
23. Mohamed, E.S., Belal, A.A., Abd-Elmabod, S.K., El-Shirbeny, M.A., Gad, A., Zahran, M.B., Smart farming for improving agricultural management. *Egypt. J. Remote Sens. Space Sci.*, 24, 3, 971–981, 2021, https://doi.org/10.1016/j.ejrs.2021.08.007.
24. Pournader, M., Ghaderi, H., Hassanzadegan, A., Fahimnia, B., Artificial intelligence applications in supply chain management. *Int. J. Prod. Econ.*, 241, 108250, 2021, https://doi.org/10.1016/j.ijpe.2021.108250.
25. Rahaman, M., Lin, C.Y., Pappachan, P., Gupta, B.B., Hsu, C.H., Privacy-Centric AI and IoT Solutions for Smart Rural Farm Monitoring and Control. *Sensors*, 24, 13, 4157, 2024, https://doi.org/10.3390/s24134157.
26. Akintuyi, O.B., Adaptive AI in precision agriculture: a review: investigating the use of self-learning algorithms in optimizing farm operations based on real-time data. *Res. J. Multi. Stud.*, 7, 02, 016–030, 2024, https://doi.org/10.53022/oarjms.2024.7.2.0023.
27. Rane, J., Mallick, S.K., Kaya, O., Rane, N.L., Scalable and adaptive deep learning algorithms for large-scale machine learning systems. *Future Res. Opportunities Artif. Intell. Ind. 4.0*, 5, 2–40, 2024.
28. Bhat, S.A. and Huang, N.F., Big data and AI revolution in precision agriculture: Survey and challenges. *IEEE Access*, 9, 110209–110222, 2021, 10.1109/ACCESS.2021.3102227.
29. Bhagoji, A.N., Chakraborty, S., Mittal, P., Calo, S., Analyzing federated learning through an adversarial lens, in: *International Conference on Machine Learning*, PMLR, pp. 634–643, May 2019.
30. Pirmani, A., Oldenhof, M., Peeters, L.M., De Brouwer, E., Moreau, Y., Accessible Ecosystem for Clinical Research (Federated Learning for Everyone): Development and Usability Study. *JMIR Formative Res.*, 8, e55496, 2024, doi: 10.2196/55496.
31. Zhang, T., Gao, L., He, C., Zhang, M., Krishnamachari, B., Avestimehr, A.S., Federated learning for the internet of things: Applications, challenges, and opportunities. *IEEE Internet Things Mag.*, 5, 1, 24–29, 2022. 10.1109/IOTM.004.2100182.
32. Tahir, M. and Ali, M., II, On the performance of federated learning algorithms for IoT. *IoT*, 3, 2, 273–284, 2022, https://doi.org/10.3390/iot3020016.
33. Rose, D.C., Keating, C., Morris, C., Understanding how to influence farmers' decision-making behaviour: a social science literature review, unpublished, 2018.

34. Fenoglio, D., Josifovski, D., Gobbetti, A., Formo, M., Gjoreski, H., Gjoreski, M., Langheinrich, M., Federated Learning for Privacy-aware Cognitive Workload Estimation, in: *Proceedings of the 22nd International Conference on Mobile and Ubiquitous Multimedia*, pp. 25–36, Dec. 2023, https://doi.org/10.1145/3626705.3627783.
35. Choudhary, S.K., Kar, A.K., Dwivedi, Y.K., How does Federated Learning Impact Decision-Making in Firms: A Systematic Literature Review. *Commun. Assoc. Inf. Syst.*, 54, 1, 22, 2024. 10.17705/1CAIS.05419.
36. Yellepeddi, S.M., Katari, P., Alluri, V.R.R., Bonam, V.S.M., Venkata, A.K.P., Federated Learning for Collaborative Threat Intelligence Sharing: A Practical Approach. *Distrib. Learn. Broad Appl. Sci. Res.*, 5, 146–167, 2019.

12

A Federated Differential Privacy Model with Pyramid Residual Network for Predicting Crop Yields

Reddy Saisindhutheja[1]*, Shanthi Makka[2], Reeta Mishra[3] and Padmesh Tripathi[4]

[1]*Department of Computer Science and Engineering, Vardhaman College of Engineering, Hyderabad, Telangana, India*
[2]*Department of Computer Science and Engineering (Data Science), Vardhaman College of Engineering, Hyderabad, India*
[3]*School of Computer Science and Engineering, IILM University, Greater Noida, India*
[4]*Department of Computer Science and Engineering, Delhi Technical Campus, Greater Noida, India*

Abstract
Agriculture remains a cornerstone of human civilization, playing a crucial role in influencing various sectors of the trade, public welfare, and biodiversity. It extends beyond providing nourishment, fostering economic advancement, encouraging long-term progress, and safeguarding traditions and societal values. Precise agricultural forecasting is crucial for enabling strategic decisions, enhancing efficiency, minimizing uncertainties, and promoting sustainable communities and ecosystems to address the needs of an expanding global population. Diverse agricultural forecasts, including climate predictions, yield assessments, pest and disease control, soil quality evaluations, crop condition surveillance, market and pricing analytics, and harvest optimization, support farmers and stakeholders in dealing with every aspect of farming, from cultivation to commercial planning. This research utilizes a federated differential privacy method to train Pyramid Residual Networks (Pyramid ResNets) within a decentralized architecture for estimating crop yields, with performance benchmarked against other machine

**Corresponding author*: thejasindhu@gmail.com

Padmesh Tripathi, Bhanumati Panda, Shanthi Makka, Reeta Mishra, S. Balamurugan and Sheng-Lung Peng (eds.) Federated Learning for Smart Agriculture and Food Quality Enhancement, (293–318) © 2026 Scrivener Publishing LLC

learning approaches. This research revealed that, with Pyramid ResNet and the Nadam optimizer, the proposed model, federated differential privacy with pyramid residual networks (FDPRN) outperformed centralized models in crop yield prediction, achieving 95.08% accuracy and 93.49% precision, with false positive and false negative rates of 4.012% and 3.861%, respectively. This approach shows strong potential for implementation in federated scenarios.

***Keywords*:** Prediction of crop yield, machine learning, nadam optimizer, pyramid residual networks, federated learning, differential privacy model

12.1 Introduction

Agriculture forms the foundation of India's economy, providing a source of income for more than half of the population and establishing itself as a vital economic domain [1, 2]. The agricultural sector contributes 17%–20% of the country's GDP, a notable yet reduced share compared to earlier decades [3]. Approximately 40%–45% of the workforce is engaged in agriculture and related industries, with millions employed in farming and allied activities [4]. India ranks second in the world for the production of crops such as rice, wheat, and sugarcane [5] and leads in pulse production. It also excels in livestock, being the largest producer of milk and poultry [6]. Renowned as a leader in organic farming and advancements in digital agriculture, India's rich agricultural heritage, diverse climate, cultural importance, and dominance in spice trade have earned it the title "Land of Spices" [7]. With substantial contributions to global agricultural production and exports, India consistently ranks among the top three agricultural nations worldwide.

Farming plays a vital role in the modern world, greatly influencing economic development, food availability, dietary health, cultural traditions, international commerce, market balance, community well-being, rural advancement, and innovation in technology [8]. Also, It encompasses diverse activities, from resource management to international trade, to address the increasing demands of a rapidly evolving world. Therefore, precise agricultural forecasting is crucial for advancing sustainable practices, managing risks, adapting to climate change, facilitating global trade, supporting communities, and enabling informed decision-making [9].

Agricultural estimates is a set of reports dealing with issues concerning farming and crop production, which aids in resource management, projected growth, and decision-making. Figure 12.1 displays some of the major types of agricultural assessments.

(i) Crop Yield Prediction: Crop yield prediction estimates the total agricultural output for a given area or worldwide.

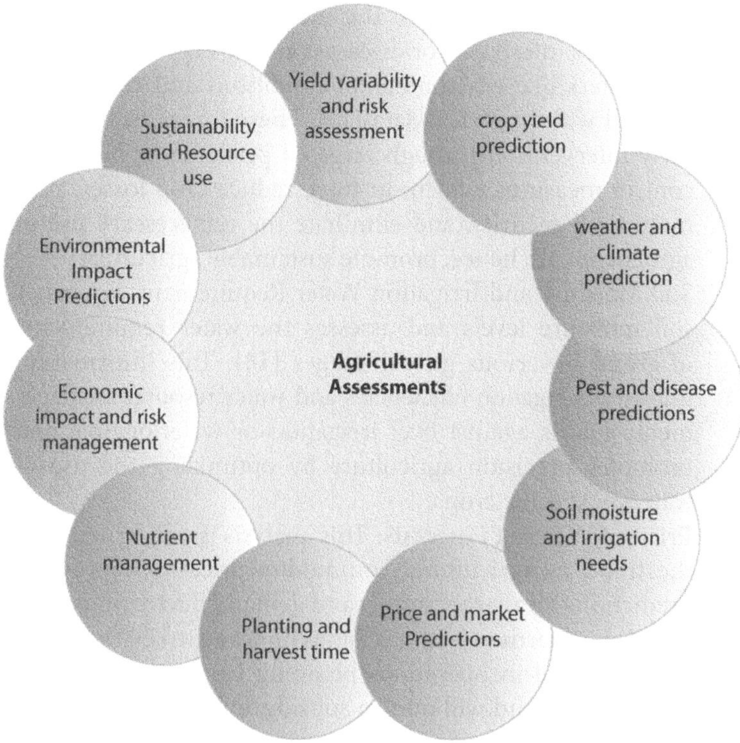

Figure 12.1 Agricultural assessments.

It considers influences such as soil, weather, farming systems, and pest or disease effects on crops [10]. Predicting yields accurately provides useful information for policymakers and farmers alike to help use resources efficiently, plan food security, and in trade policy decisions. Moreover, they recognize areas with low-yielding fields, requiring special attention.

(ii) Weather and Climate Forecasts: Weather and climate forecasts are short-term predictions about rainfall and temperature and long-term projections such as trends in drought and seasonal shift. These provide potential impacts, which are very important in agricultural planning [11]. It manages the risks of extreme weather events like floods, storms, and drought. It guides farmers in adjusting planting schedules and selection of crop and informing strategies for coping with new climatic conditions.

(iii) Disease and Pest Forecasts: The likelihood of severity of pests or diseases infestation or diseases on crops based on critical parameters like environmental conditions and crop cycles coupled with historical data [12]. These forecasts allow the early intervention through targeted pesticide or biological control measures, which, in turn, reduce crop losses, promote food security, and eliminate the unnecessary use of agrochemicals; hence, promote sustainable agriculture.

(iv) Soil Moisture and Irrigation Water Requirement: Estimates soil moisture levels and assesses the water requirements of crops at various growth stages [13]. This information improves irrigation efficiencies and water resource management, guards against over-irrigation or waterlogging, and promotes precision agriculture by optimizing the use of water on specific crops.

(v) Price and Market Forecasts: This analysis is concerned with the trend in agricultural commodity prices and provides predictions for market supply and demand fluctuations [14]. It will assist farmers with decisions on when to sell their produce, will aid in operations planning for agri-value chain stakeholders, and will inform subsidy, imports, and exports policy decisions.

(vi) Planting and Harvest Timing: Determines the favorable planting seasons in relation to the environmental conditions, characteristics of the crop, and market demand [15]. Such information enables the farmer to enhance crop production and quality, reduce losses that occur due to poor planting or harvesting periods, and thus swim with the market trends by synchronizing harvests with peak periods.

(vii) Nutrient Management: It detects nutrient deficiencies in the soil and provides an appropriate fertilizer strategy to rectify nutritional deficiencies of crops [16]. This approach promotes balanced fertilizer application, prevents overapplication, increases soil health, enhances sustainability in agriculture over longer periods, and optimizes growth and yield.

(viii) Economic Impact and Risk Management: Measures the economic viability of agricultural undertakings by projecting farm income and loss and possible economic risk [17]. Such an appraisal enables farmers and investors to take wise financial decisions and helps insurance companies produce specially designed strategies to mitigate risk, while agricultural policies are informed for better economic resilience.

(ix) Environmental Impact Predictions: It helps to foresee the risks of soil erosion, water contamination, greenhouse gas emissions, and other such impacts associated with agriculture practice [18]. Such findings act as the impetus for taking environmentally friendly farming practices, preventing agricultural pollution that harms natural ecosystems, and helping regulations and policies regarding land use sustainably.

(x) Sustainability and Resource Use: Determines the ecological footprint and resource use of agriculture [19]. It measures the environmental implications of agriculture and evaluates the way in which resources are used effectively, such as land, water, and energy use. This standard encourages the change toward sustainable farming systems, promotes an appropriate balance between agricultural productivity and conservation of the environment while being part of the global sustainability initiative, such as attainment of sustainable development goals (SDGs).

(xi) Yield Variability and Risk Analysis: Examines the heterogeneity of risks and rewards in agriculture [20]. This analysis analyzes variations in crop yields across areas or seasons and addresses potential risks to agricultural productivity. It directs policymakers to high-risk areas that call for risk mitigation efforts, informs insurance models, and inspires resiliency through the adaptation of farming practices toward minimal variability.

Together, these categories provide a holistic framework in addressing the challenges and opportunities that confront modern agriculture. It gives particular importance to sustainability, risk management, and improved productivity. Each of these categories can provide a basis and innovation in the agricultural domain.

Predicting crop production is one among the most crucial agricultural forecasts due to its massive and wide-reaching influence on a large scale of cultivation, food security, and financial stability. Perhaps for this reason, crop yield predictions have been one of the most essential aspects of agricultural forecasts to date [21]. In addition to negatively affecting crop yields, global warming can lead to food poverty or famine. Over here, the latter becomes a key challenge in regions prone to floods or other weather-related disasters, which is the impact of climate on agriculture. Factors such as temperature, humidity, ultraviolet radiation, and rainfall significantly influence crop yields. While climate variables tend to stay consistent in a specific location, the environmental needs of different crops and their growth stages can vary. This suggests that each crop's ability to withstand weather conditions differs [22].

Various approaches have been proposed in multiple studies to tackle the challenges posed by weather variability on crops. Numerous researchers have shown how climate irregularities impact crops by examining the connection between weather conditions and agricultural output. Some studies have even identified the specific thresholds of rainfall, temperature, and water required for plant development. In light of the global climate crisis, this paper emphasizes the importance of gaining a deeper understanding of local agricultural histories to ensure the creation of equitable and secure rural futures [23].

In [24], the authors provide a suggestion that several aspects, like global warming, biofuels, and monetary concerns, which may influence farming costs in both positive and negative ways. In [25], the researchers utilized climate projections from the coupled model intercomparison project phase 5 (CMIP5), a group of simulations of climate model, known as global climate models (GCMs), under two greenhouse gas emission scenarios, RCP_4.5 and RCP_8.5, to assess changes in the occurrence, intensity, and geographic distribution of ten unfavorable weather indices across 25 locations in the UK.

Numerous studies have focused on predicting agricultural yields. In [26], the researchers designed a trained algorithm to identify patterns in the data and used machine learning (ML) methods, including gradient boosting, random forest (RF), decision tree, and support vector machine (SVM) for crop predictions. The predictions targeted crops such as wheat, rice, maize, and potatoes.

The researchers in [27] assessed the production of agriculture using ML methods such as XGBoost and RF combined with deep learning techniques [49] like convolution neural network and long short-term memory (LSTM). For a clearer comparison between the errors of the proposed method and other techniques, the outcomes are evaluated in the concluding section using the root mean square error (RMSE) measure.

This study in [28] trains Pyramid Residual Network (Pyramid ResNet) within a decentralized crop production forecasting framework utilizing a federated differential privacy (FDP) method, and the performance is compared against other ML models. Experimental results show that FDP with Pyramid ResNet using the Nadam optimizer yields the best performance when compared to ML models, making it highly applicable for yield forecasting in a federated setting.

The highlights of the proposed study are as follows:

(i) Model Design: The application of Pyramid ResNets for efficient feature extraction and improved model performance on heterogeneous data.
(ii) Privacy Assurance: Integration of a differential privacy (DP) module to ensure robust privacy guarantees during local model updates.
(iii) Optimization Strategy: Implementation of the Nadam optimizer to accelerate model convergence and stability.
(iv) Secure Communication: Secure aggregation of local model updates to prevent unauthorized access and ensure secure global model integration.

12.2 Crop Yield Prediction Using Federated Learning

ML and federated learning (FL), both branches of artificial intelligence, focus on building models from data, but they approach data handling and model training in distinct ways.

12.2.1 Machine Learning

Centralized ML can be deployed in situations where the need for data and privacy is minimal [48, 50, 51]. Such scenarios involve centralized systems for disease detection and large-scale crop yield predictions based on past data.

Figure 12.2 represents an ML system. In this context, data forms the core element, gathered from different devices or sources that generate information and stored in a unified location on an organization's server or computing system. Various techniques, tools, and frameworks are then utilized to apply ML to this aggregated data, resulting in the creation of an ML model. This model can subsequently be used to analyze new input data and produce the required results [52, 53].

ML models are trained over historical data [29]. In the context of crop yield prediction, there may not be good adaptation of historical data toward future changes; as far as crop varieties are considered, farming practices, soil conditions, pest, and climate changes can be very broad and unpredictable. Agriculture data use raises privacy and security concerns. Fear of manipulation and vulnerability may make farmers and other agricultural enterprises unwilling to give out data. To overcome this challenge,

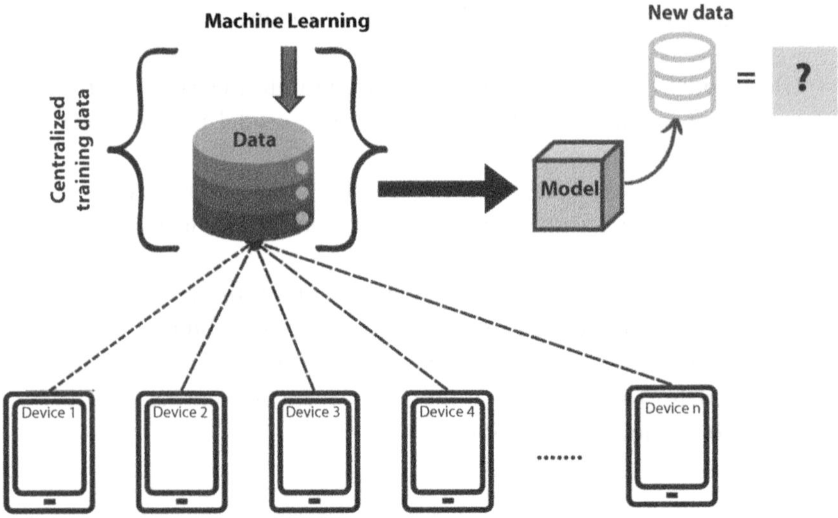

Figure 12.2 Machine learning model.

it becomes necessary to obtain data from a variety of areas without disrupting its original location.

Because all records are stored centrally, ML carries at a larger risk [42–44]. It performs best where centralizing data can be done without raising significant privacy issues. ML has many positive advantages over FL, but it has several limitations as well, especially where scalability, privacy, and data distribution are concerned.

12.2.2 Federated Learning

Collaborative model training without relying on centralized data storage is achieved through a decentralized framework known as FL [30], which is depicted in Figure 12.3. In FL, more devices or nodes cooperatively train a global model with ensured data privacy by retaining their local data.

It initiates from the central server or aggregator's side, which is globally initialized\ and then distributed to the participating devices. Each of the participant devices performs local training with its private data, which leads to locally updated model parameters. The central server receives these updates and combines them, often using methods like weighted averaging, to adjust the global model. The refined global model is then distributed back to the devices for the next phase of local training. This process involves repeated cycles of local training and global integration, continuing

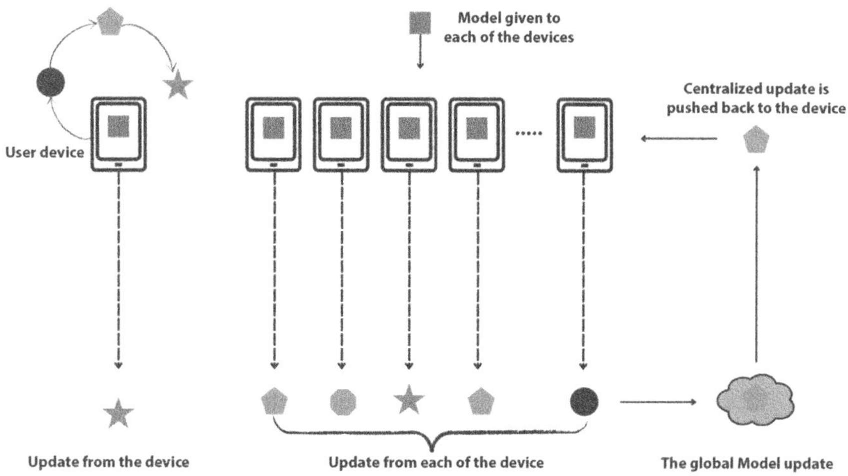

Figure 12.3 Federated learning approach.

until the model achieves the required performance level, allowing for a balance between personalized learning and a generalized, high-performing global model.

FL is ideal for use cases prioritizing data privacy and security or where information is distributed across multiple locations. For instance, it enables analysis of medical images from various hospitals without compromising patient confidentiality, optimizes traffic flow by processing data from dispersed city sensors, and supports collaborative pest management systems with minimal data sharing. In addition, crop yield forecasting integrates data from numerous sensitive sources [31, 45], such as the following:

(i) Economic Data: It entails information such as farm income, investment patterns, market strategies, and sales figures. This data is vital for linking economic conditions to crop productivity and market trends and for developing economic models and policy recommendations. However, its sensitivity arises from the risk of financial fraud or exploitation due to unauthorized access, making regulatory compliance essential for protecting this data.

(ii) Farmer and Farm Details: This category comprises personal identifiers (e.g., names and contact details), farm-specific geographic data (e.g., GPS coordinates), farm size, ownership, and tenure arrangements. Such data is critical for spatial analysis and understanding access to resources.

Nevertheless, breaches can lead to privacy violations, identity theft, or misuse of GPS data, whereas ownership details may be exploited in disputes, necessitating robust security measures.

(iii) Agricultural Practices: Information about farming techniques encompasses past harvest records, irrigation strategies, crop rotation schedules, pest management practices, and fertilizer application details. This data plays a vital role in enhancing agricultural methods, boosting efficiency, and promoting sustainable practices. However, it is sensitive because it can reveal proprietary practices, risking exploitation by competitors or legal challenges from non-compliance with privacy or sustainability regulations.

(iv) Environmental Data: It encompasses soil composition, fertility, moisture content, health indicators (e.g., pH and nutrient levels), and climatic conditions. This data is essential for precision agriculture, predicting yields, and mitigating climate risks. Its sensitivity stems from potential misuse in disputes over land valuation, exploitation by commercial interests, and reputational risks for regions with negative environmental trends.

(v) Geospatial Data: Includes land use patterns, topography, terrain features, and agricultural land mapping. It provides insights into land productivity, supports land-use planning, and aids in optimizing resource allocation. However, it is highly sensitive as it may expose strategic information, risk misuse for commercial purposes, or lead to disputes, necessitating adherence to regulatory protections.

12.3 Methodologies of the Proposed Work

12.3.1 Federated Differential Privacy Algorithm

FDP integrates DP with FL to provide a privacy-focused method for training ML models on distributed data [32]. The subsequent steps outline the FDP algorithm.

(i) **Initialization:** After the global model is configured, it is transmitted to the designated clients by the central server.

The privacy parameters, represented as (ϵ, ϑ), represent the level of privacy protection and the likelihood of failure.

(ii) **Local Training for Selected Participants:** The group of clients C_g is identified by the central server to engage in local training. Upon finishing the local training, each client generates an update for their individual model.

(iii) **Aggregation and Transfer of Noisy Updates:** Every client L adds noise with a Gaussian distribution to its local model update before sending it to the central server, with the aim of achieving DP. Equation (12.1) corresponds to noisy addition of the local models. The nomenclature of the below equation is as follows: $\tilde{\Delta}\propto_L$ represents the local updates, and $\mathbb{N}(0, \beta^2)$ stands for the Gaussian noise regarding (ϵ, ϑ). After obtaining the noisy update $\tilde{\Delta}\propto_L$, it sends update to the server.

$$\tilde{\Delta}\propto_L = \Delta \propto_L + \mathbb{N}(0, \beta^2) \tag{12.1}$$

(iv) **Privacy-Preserving Aggregation:** The central server combines the noisy updates from the clients to refine the global model. Equation (12.2) illustrates how the server aggregates these perturbed updates.

$$\propto_{g+1} = \propto_g + \frac{1}{|C_g|} \sum_{L \epsilon C_g} \tilde{\Delta}\propto_L \tag{12.2}$$

(v) **Update of the Global Model:** The central server updates the global model using the parameters of $\propto_{g+1}$. This procedure is iterated as long as the model converges.

12.3.2 Pyramid Residual Networks

It is an enhancement of the traditional ResNets [33]. To optimize the learning and performance of the DNN, particularly in the fields of image recognition and classification, mathematically, one must initialize a global model, perform local training with residual links, add some noise for DP, and iterate the process until convergence. The following are the steps involved in Pyramid ResNets.

(i) **Input Layer:** The network starts with a convolutional layer that transforms the input image PI_0 into a set of feature maps F_0 with CH_0 channels as expressed in Equation (12.3).

$$F_0 = Cnv(PI_0; Wgt_0) \tag{12.3}$$

In Equation (12.3), $Cnv(PI_0; Wgt_0)$ depicts the convolution operation carried out with weights Wgt_0.

(ii) **Blocks Form Residual Architecture:** Pyramid ResNets are constituted by such blocks, with each having several convolutional layers, with residual connections. The incoming nature of each block to the other expands the feature maps and makes it pyramidally structured. The definition of the nth block with residual function RFc is given by Equation (12.4).

$$RFc_n(IP_{n-1}, \{Wgt_{n,j}\}) = \rho(BTHN(Cnv(IP_{n-1}, Wgt_{n,1}))) + \rho(BTHN(Cnv(IP_n; Wgt_{n,2}))) \tag{12.4}$$

Here, IP_{n-1} and IP_n are input and output, respectively; $Wgt_{n,1}$ and $Wgt_{n,2}$ depict the convolutional layer weights in the nth block; the batch normalization can be denoted as $BTHN$ activation function with ρ. Equation (12.5) depicts the output from n^th block.

$$IP_n = RFc_n(IP_{n-1}, \{Wgt_{n,j}\}) + IP_{n-1} \tag{12.5}$$

(iii) **Dimensionality Reduction:** Dimensional reduction is realized through down sampling without the change in channels: As an example, emission of the feature maps IP_n down sampled to IP_n' as defined in Equation (12.6).

$$IP_n' = Dsample(IP_n) \tag{12.6}$$

(iv) **Global Mean Pooling:** GMP minimizes the parameter count and helps prevent overfitting. The pooled aggregated feature PAF is expressed in Equation (12.7).

$$PAF = GMP(IP_L) \tag{12.7}$$

(v) **The Complete Bipartite Layer:** The aggregated feature is passed into a complete bipartite layer for the classification task. In the case of multi-class classification, a SoftMax activation function, denoted as *Smax*(), is frequently applied in the output layer to produce a probability distribution over the output classes, as illustrated in Equation (12.8).

$$OP = Smax(Wgt_{fc}.PAF + b_{fc}) \tag{12.8}$$

Here, *OP* is the output class, and Wgt_{fc} and b_{fc} denote the weights and bias of the complete bipartite layer, respectively.

12.3.3 Nadam Optimizer

To increase the effectiveness of deep neural network training, the Nadam optimizer merges the benefits of Adaptive Moment Estimation (Adam) and Nesterov Accelerated Gradient (NAG) [34]. Like Adam, Nadam adjusts the learning rate based on estimates of the first- and second-order mean gradients, which allows for stable training, especially in cases while gradients over parameters fluctuate significantly. However, Nadam also incorporates momentum, using not just the current gradient but also the accumulated gradients from previous iterations, which allows for smoother and faster convergence.

Additionally, the use of Nesterov's momentum gives the optimizer a look-ahead feature, improving the accuracy of parameter updates. This combination of adaptive learning rates and momentum makes Nadam highly effective in high-dimensional, large-scale datasets typically found in deep learning challenges including natural language processing and image classification. By accelerating the optimization process and reducing oscillations, Nadam is especially well suited for complex, deep network architectures, helping them converge more quickly and reliably.

The primary goal is to alter the existing gradient to provide faster convergence, as shown in Equation (12.9).

$$\rho_n = \rho_{n-1} - \frac{¥}{\sqrt{\hat{\varrho}_t} + \epsilon}\left(\varsigma_1 \hat{\omega}_t + \frac{(1-\varsigma_1)h_t}{1-\varsigma_1^t}\right) \tag{12.9}$$

Here, ¥ denotes the learning rate; $\hat{\varrho}_t = \frac{\varsigma_t}{1-\varsigma_2^t}$ and $\hat{\omega}_t = \frac{\eta_t}{1-\varsigma_1^t}$ denotes change in the bias improvement of initial and secondary momentum, respectively; ς_1 and ς_2 represent the rates of decay; and ϵ is a constant.

12.3.4 Architecture of the Proposed Work

This work uses three datasets: (i) crop yield [35], (ii) crop recommendation [36], and (iii) water quality [37]. The federated model processes each dataset with Pyramid ResNets for image classification. To ensure data privacy and prevent unauthorized access, a DP module is applied. The Nadam optimizer accelerates convergence, and the results are securely aggregated for a global model update. After five iterations, the yield predictions achieved 96% accuracy.

The following outlines the sequential procedure of the proposed methodology. The proposed framework follows a systematic methodology where local datasets are processed using Pyramid ResNets, which is depicted in Figure 12.4. The trained model updates are then subjected to DP mechanisms

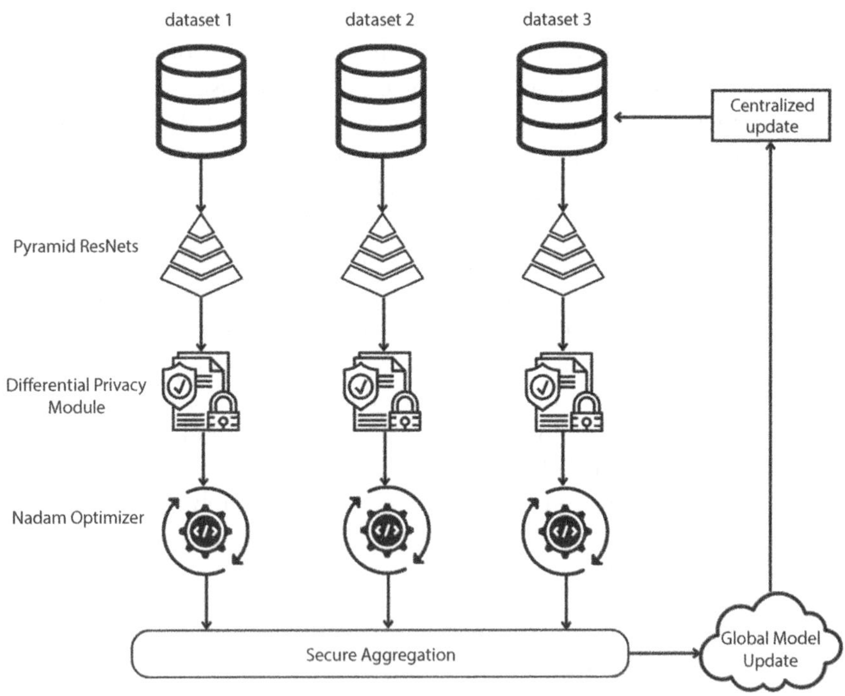

Figure 12.4 Architecture of the proposed work.

and optimized using the Nadam optimizer. Finally, a secure aggregation protocol consolidates the local updates into a global model, which is iteratively refined through multiple training rounds.

(i) **Data Distribution:** Each dataset represents the local data stored on individual devices or nodes. These datasets are not shared directly with a central server to protect user privacy.
(ii) **Local Model Training (Pyramid ResNets):** Deep learning architecture is employed for feature extraction and categorization. Each node trains its local model (Pyramid ResNet) on its dataset. This ensures that computation happens locally, leveraging user data without transmitting it elsewhere.
(iii) **Differential Privacy Module:** DP ensures that any information extracted during training does not reveal specific details about individual data points. It adds noise to the local model updates, ensuring privacy-preserving learning.
(iv) **Nadam Optimizer:** An optimization algorithm used to train the local models efficiently. Nadam combines the benefits of adaptive moment estimation and Nesterov momentum for faster and more stable convergence.
(v) **Secure Aggregation:** After training locally, the models generate updates (gradients or parameters). These updates are encrypted and securely aggregated to protect the privacy of individual models. Secure aggregation combines model updates from all nodes into a single global update without revealing individual contributions.
(vi) **Global Model Update:** The aggregated updates are transmitted to the central server, which integrates them into a unified model. This global model reflects the collective insights gained from the entire local datasets.
(vii) **Centralized Update and Iteration:** The global model is sent to the individual nodes for the next training cycle, repeating until convergence or the desired performance is achieved.

Key advantages of the methodology include: (i) **Privacy-Preserving:** Data remains decentralized, and DP ensures no sensitive information is leaked. (ii) **Efficiency:** Pyramid ResNets and Nadam Optimizer ensure fast and effective local training. (iii) **Security:** Secure aggregation guarantees encrypted transmission and secure contributions from all nodes.

12.4 Execution and Outcomes

The proposed approach is evaluated against various ML models namely RF, naive Bayes (NA), artificial neural network (ANN), and SVM and implemented using python. Performance metrics include accuracy, precision, false positive rate (FPR), and false negative rate (FNR).

12.4.1 Dataset Exploration

(i) **Crop Yield:** This dataset provides comprehensive information on crop production statistics for India by state and district and covers the four main crop seasons from 1997 to 2023: kharif, spring, summer, and fall. Data on the yield and yearly production of different crops cultivated around the country are also included in the collection. Crop year, season, state, area production, annual rainfall, fertiliser, pesticide, and yield are among the dataset's properties. It offers insight into how various ecological factors (such rainfall, fertilizer use, and pesticide intake) relate to crop yield across a range of states and crop types.

(ii) **Crop Recommendation:** The creation of this dataset relied on improving India's current rainfall, temperature, and fertilizer statistics. Based on the different features in the dataset, it enables the user to cultivate more appropriate crops on a certain farm. N, P, and K stand for nitrogen, phosphorus, and potassium concentration in soil, as well as temperature, humidity, soil pH, and rainfall.

(iii) **Water Quality:** Station code, location, state, temperature, dissolved oxygen (D.O.), pH, conductivity, biochemical oxygen demand (B.O.D.), nitrate, and fecal coliform are among the characteristics that are included in this collection, which offers integrated historical water quality data from certain sites in India.

12.4.2 Evaluation of Proposed Work with Other State-of-the-Art Models

Table 12.1 displays the measures used to compare the suggested model's performance to other conventional models. In ML model evaluation, positive metrics such as accuracy and precision provide insights into the model's effectiveness in correctly identifying relevant instances.

Table 12.1 Performance metrics.

Metrics	Overview
Accuracy	The proportion of correct predictions out of the total predictions made.
Precision	The ratio of true positive outcomes to the total number of cases predicted as positive.
FPR	The proportion of true negative examples incorrectly classified as positive among all true negatives.
FNR	The proportion of true positive examples incorrectly classified as negative among all true positives.

Accuracy measures the overall ratio of true values out of all predictions made, reflecting the general routine of the model. Conversely, precision focuses on the relevance of positive predictions, calculating the proportion of true positives out of all examples classified as positive, which is particularly important in scenarios where false positives are costly [46, 54, 55].

The proposed FDPRN model demonstrated superior performance compared to conventional models in both accuracy and precision. It achieved an accuracy of 95.09%, significantly higher than RF [38] with 92.4%, ANN [39] with 91.2%, NA [40] with 89.03%, and SVM [41] with 86.27% as depicted in Figure 12.5. This highlights the robustness of the FDPRN approach in effectively handling the dataset and providing reliable predictions.

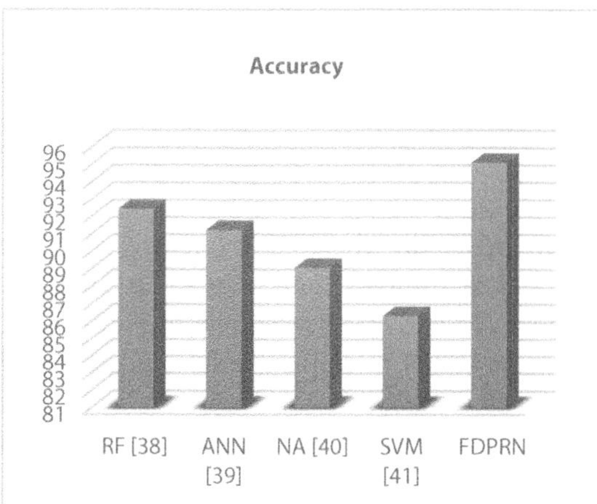

Figure 12.5 Accuracy.

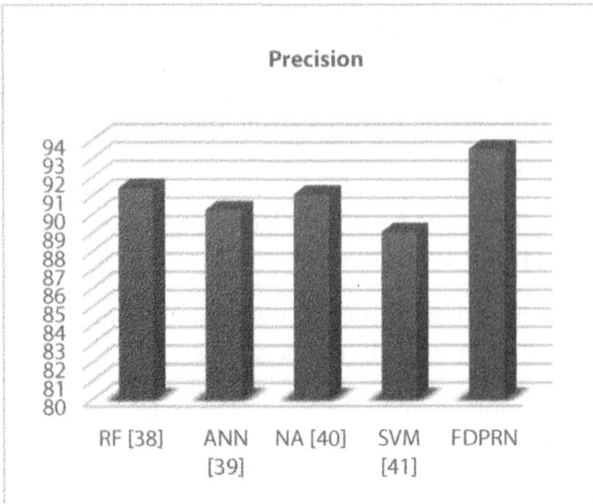

Figure 12.6 Precision.

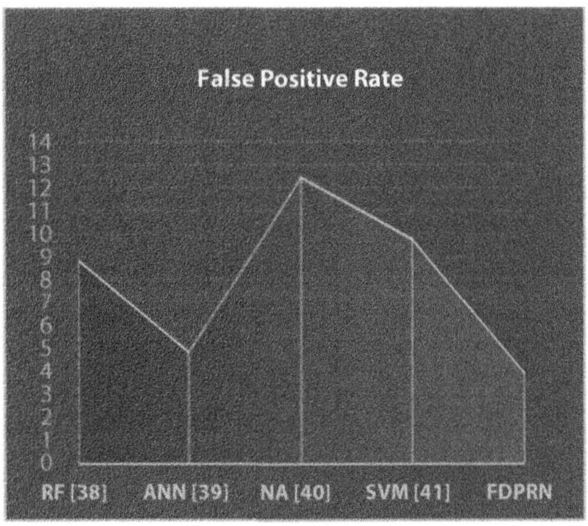

Figure 12.7 FPR.

With respect to precision, the FDPRN model also excelled, scoring 93.49%, which is a clear improvement over RF at 91.4%, ANN at 90.2%, NA at 91.06%, and SVM at 89.02% as shown in Figure 12.6. These results underline the model's ability to minimize false positives while maintaining high-quality predictions. Such consistent performance across metrics showcases the FDPRN model's potential as a more accurate and precise alternative to traditional ML techniques.

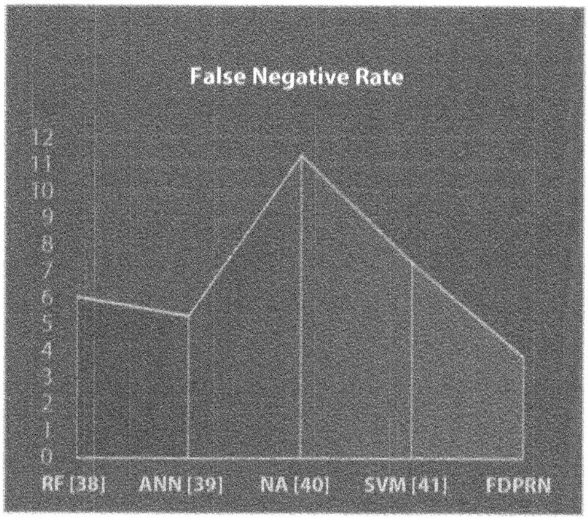

Figure 12.8 FNR.

In contrast, negative metrics like FPR and FNR are used to estimate errors. FPR measures the ratio of real negatives erroneously classified as positives, highlighting the model's tendency to falsely identify negatives as positives, which can lead to unnecessary actions or misallocations. FNR, conversely, indicates the ratio of actual positives lost by the model, pointing out its failure to detect true positives, which can have serious consequences in tasks like disease detection or fraud prevention. These metrics provide a

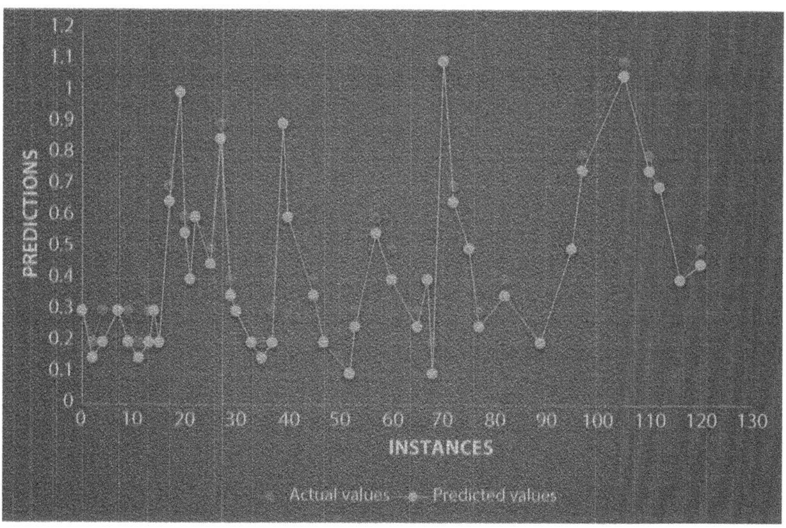

Figure 12.9 Difference between predicted and actual data points.

well-rounded perspective on a model's performance, highlighting both its advantages and disadvantages [47, 50, 51].

The FDPRN yielded the most feasible results in the negative metric terms, having 4.012% of FPR, whereas the other models have 8.91% for RF, 4.92% for ANN, 12.51% for NA, and 9.81% for SVM. Again, the proposed model gave the minimum FNR of 3.861%, whereas the remaining models had value of 6.091% for RF, 5.36% for ANN, 11.361% for NA, and 7.382% for SVM. Figures 12.7 and 12.8 exhibit the results for the metrics of FPR and FNR, respectively.

The gap between real and predicted outcomes can be observed in Figure 12.9. The root mean squared error % (RMSE %) was computed on the basis of these findings with a mean of 0.432331 and an RMSE of 0.007451; the RMSE% is 1.7231%. RMSE expressed as a percentage of the mean of the actual values. It is calculated using the following formula represented in Equation (12.10).

$$R\% = \frac{R}{\bar{m}} \times 100 \tag{12.10}$$

where R is the RMSE represented in Equation (12.11), and $\bar{m}$ is average of the actual values represented in Equation (12.12).

$$R = \sqrt{\frac{1}{y}\sum_{i=1}^{y}(m_i - \hat{m}_i)^2} \tag{12.11}$$

$$\bar{m} = \frac{1}{y}\sum_{i=1}^{y} m_i \tag{12.12}$$

where y denotes the total count regarding data points, and m_i and $\hat{m}_i$ refer to the actual and predicted standards for i^{th} data point, respectively.

12.5 Conclusions and Future Scope

In conclusion, this research demonstrates the effectiveness of a FDP approach combined with Pyramid ResNets and the Nadam optimizer for accurate crop yield prediction. By leveraging a decentralized architecture, the proposed method outperforms traditional centralized models, achieving high accuracy and precision of 95.08% and 93.49% while maintaining low FPR and FNR of 4.012% and 3.861%. The results highlight the potential of this approach in improving agricultural forecasting, supporting sustainable farming practices, and addressing the challenges of a growing global population. This methodology holds promise for future implementation in FL scenarios, offering a privacy-preserving and efficient solution for agricultural stakeholders. Furthermore, the RMSE% is calculated to evaluate the model's performance, yielding an exceptionally low value of 1.7231%.

Future studies will aim to integrate real-time sensor data and multimodal sources, such as satellite imagery, to improve prediction accuracy while also deploying models on edge devices to enhance scalability in resource-constrained environments. To boost data security, advanced privacy techniques like homomorphic encryption will be explored. The model will also be adapted for use across different crops and regions to increase its flexibility. Moreover, explainable AI techniques will be integrated to enhance the interpretability of predictions, and the framework will be compared with advanced algorithms like transformers and convolutional LSTMs. Efforts will also include automated hyperparameter optimization and energy-efficient training to improve model performance and sustainability. Expanding the model to support global, multi-institutional collaborations, ensuring robustness against noisy or imbalanced data, and evaluating its economic impact on farming practices will further enhance its practical applications. We also employ the cross-validation technique to evaluate the model's accuracy.

AI disclosure statement: The author acknowledges the use of AI-based tools for minor language editing or grammatical correction. However, all intellectual contributions, analyses, and conclusions presented in this chapter are entirely the author's own.

References

1. Javed, M.A., Saleem, A., Hafeez, A., Ali, B., Microbial Traits in Plant Stress Tolerance and Importance for Agriculture, in: *Advances in Plant Microbiome Research for Climate-Resilient Agriculture*, pp. 179–198, Apple Academic Press, Florida, U.S.A, 2024.
2. Riseh, R.S., Vazvani, M.G., Hassanisaadi, M., Thakur, V.K., Agricultural wastes: a practical and potential source for the isolation and preparation of cellulose and application in agriculture and different industries, in: *Industrial*

Crops and Products, vol. 208, p. 117904, Elsevier, Amsterdam, Netherlands, 2024.
3. Goldar, B. and Das, P.C., Share of manufacturing in India's GDP: Stagnant or increasing?, in: *Structural Change and Economic Dynamics*, vol. 68, pp. 75–85, Elsevier, Amsterdam, Netherlands, 2024.
4. Indwar, T., Empowering Rural India: The Role of Agriculture and Government Programs. *Prismatic Horiz. J. Social Sci. Humanit.*, 2, 1, 48–56, 2024.
5. Singh, K., Mishra, S.K., Gupta, N., Singh, K., Residue Retention Effects on Productivity of Sugarcane (Plant)–Sugarcane (Ratoon)–Wheat and Rice–Wheat Cropping Systems in North-West India, in: *Sugar Tech*, pp. 1–11, Springer, New Delhi, India, 2024.
6. Behera, R., *et al.*, Challenges and Future Prospects of Livestock and Poultry Farming in Indian Coastal Ecosystem-An Overview. *J. Indian Soc. Coastal Agric. Res.*, 41, 1, 53–63, 2023.
7. Ravindran, P.N., Spices: Definition, Classification, History, and Role in Indian Life, in: *Handbook of Spices in India: 75 Years of Research and Development*, pp. 1–102, Springer Nature Singapore, Singapore, 2023.
8. Pathak, H., Impact, adaptation, and mitigation of climate change in Indian agriculture, in: *Environmental Monitoring and Assessment*, vol. 195, pp. 52–68, Springer, Netherlands, 2023.
9. Sabu, K.M. and Kumar, TK.M., Predictive analytics in Agriculture: Forecasting prices of Arecanuts in Kerala, in: *Procedia Computer Science*, vol. 171, pp. 699–708, Elsevier, Amsterdam, Netherlands, 2020.
10. Kolipaka, V.R.R. and Namburu, A., An automatic crop yield prediction framework designed with two-stage classifiers: a meta-heuristic approach, in: *Multimedia Tools and Applications*, vol. 83, pp. 28969–28992, Springer, Dordrecht, Netherlands, 2024.
11. Wimmer, S., Stetter, C., Schmitt, J., Finger, R., Farm-level responses to weather trends: A structural model, in: *American Journal of Agricultural Economics*, vol. 106, pp. 1241–1273, Wiley Online Library, New Jersey, USA, 2024.
12. Ganeshkumar, C., David, A., Sankar, J.G., Saginala, M., Application of drone Technology in Agriculture: A predictive forecasting of Pest and disease incidence, in: *Applying drone technologies and robotics for agricultural sustainability*, pp. 50–81, IGI Global, Hershey, Pennsylvania, USA, 2023.
13. Chandrappa, V.Y., Ray, B., Ashwatha, N., Shrestha, P., Spatiotemporal modeling to predict soil moisture for sustainable smart irrigation, in: *Internet of Things*, vol. 21, p. 100671, Elsevier, Amsterdam, Netherlands, 2023.
14. Wang, Y., Agricultural products price prediction based on improved RBF neural network model, in: *Applied artificial intelligence*, vol. 37, p. 2204600, Taylor & Francis, Abingdon, UK, 2023.
15. Elbasi, E., Zaki, C., Topcu, A. E., Abdelbaki, W., Zreikat, A. I., Cina, E., Saker, L., Crop prediction model using machine learning algorithms. *Appl. Sci.*, 13, 16, 9288, 2023.

16. Ennaji, O., Vergütz, L., El Allali, A., Machine learning in nutrient management: A review, in: *Artificial Intelligence in Agriculture*, vol. 9, pp. 1–11, Elsevier, Amsterdam, Netherlands, 2023.
17. Benami, E., Jin, Z., Carter, M.R., Ghosh, A., Hijmans, R. J., Hobbs, A., Lobell, D.B., Uniting remote sensing, crop modelling and economics for agricultural risk management. *Nat. Rev. Earth Environ.*, 2, 2, 140–159, 2021.
18. Khang, A., Rath, K.C., Panda, S., Sree, P. K., Panda, S.K., Revolutionizing agriculture: exploring advanced technologies for plant protection in the agriculture sector, in: *Handbook of Research on AI-Equipped IoT Applications in High-Tech Agriculture*, pp. 1–22, IGI Global, Hershey, Pennsylvania, USA, 2023.
19. Sachithra, V. and Subhashini, L.D.C.S., How artificial intelligence uses to achieve the agriculture sustainability: Systematic review, in: *Artificial Intelligence in Agriculture*, vol. 8, pp. 46–59, Elsevier, Amsterdam, Netherlands, 2023.
20. Hu, T., Xuesong Z., Gil B., Yanlan L., Yuyu Z., Jay M., Yang L., Kaiguang Z., Crop yield prediction *via* explainable AI and interpretable machine learning: Dangers of black box models for evaluating climate change impacts on crop yield, in: *Agricultural and Forest Meteorology*, vol. 336, p. 109458, Elsevier, Amsterdam, Netherlands, 2023.
21. Kuradusenge, M., Hitimana, E., Hanyurwimfura, D., Rukundo, P., Mtonga, K., Mukasine, A., Uwitonze, C., Ngabonziza, J., Uwamahoro, A., Crop yield prediction using machine learning models: Case of Irish potato and maize. *Agriculture*, 13, 1, 225, 2023.
22. Beillouin, D., Schauberger, B., Bastos, A., Ciais, P., Makowski, D., Impact of extreme weather conditions on European crop production in 2018: Random forest - Yield anomalies. *Philos. Trans. R. Soc. B Biol. Sci.*, 375, 20190510, 2020.
23. Matthan, T., Beyond bad weather: Climates of uncertainty in rural India, in: *Climate Change and Critical Agrarian Studies*, pp. 164–185, Routledge, Taylor & Francis, Abingdon, UK, 2023.
24. Nguyen, H., Randall, M., Lewis, A., Factors Affecting Crop Prices in the Context of Climate Change—A Review, in: *Agriculture*, vol. 14, p. 135, 2024.
25. Harkness, C., Semenov, M.A., Areal, F., Senapati, N., Trnka, M., Balek, J., Bishop, J., Adverse weather conditions for UK wheat production under climate change. *Agric. For. Meteorol.*, 282, 107862, 2020.
26. Pant, J., Pant, R.P., Singh, M.K., Singh, D.P., Pant, H., Analysis of agricultural crop yield prediction using statistical techniques of machine learning, in: *Materials Today: Proceedings*, vol. 46, pp. 10922–10926, Elsevier, Amsterdam, Netherlands, 2021.
27. Sharma, P., Dadheech, P., Aneja, N., Aneja, S., Predicting agriculture yields based on machine learning using regression and deep learning. *IEEE Access*, 11, 111255–111264, 2023.

28. Wu, Y., Qin, Y., Wang, Z., Ma, X., Cao, Z., Densely pyramidal residual network for UAV-based railway images dehazing, in: *Neurocomputing*, vol. 371, pp. 124–136, Elsevier, Amsterdam, Netherlands, 2020.
29. Benos, L., Tagarakis, A.C., Dolias, G., Berruto, R., Kateris, D., Bochtis, D., Machine learning in agriculture: A comprehensive updated review. *Sensors*, 21, 11, 3758, 2021.
30. Friha, O., *et al.*, FELIDS: Federated learning-based intrusion detection system for agricultural Internet of Things, in: *Journal of Parallel and Distributed Computing*, vol. 165, pp. 17–31, Elsevier, Amsterdam, Netherlands, 2022.
31. Muruganantham, P., *et al.*, A systematic literature review on crop yield prediction with deep learning and remote sensing. *Remote Sens.*, 14, 9, 1990, 2022.
32. El Ouadrhiri, A. and Abdelhadi, A., Differential privacy for deep and federated learning: A survey. *IEEE Access*, 10, 22359–22380, 2022.
33. Yao, G., Wang, C., Wu, Y., Wang, Y., Pyramid fully residual network for single image de-raining. *Neurocomputing*, 456, 168–178, 2021.
34. Sharma, J., Soni, S., Paliwal, P., Saboor, S., Chaurasiya,P.K., Sharifpur, M., Khalilpoor, N., Afzal, A., A novel long term solar photovoltaic power forecasting approach using LSTM with Nadam optimizer: A case study of India. *Energy Sci. Eng.*, 10, 8, 2909–2929, 2022.
35. https://www.kaggle.com/datasets/akshatgupta7/crop-yield-in-indian-states-dataset, Accessed on 02-03-2024.
36. https://www.kaggle.com/datasets/atharvaingle/crop-recommendation-dataset, Accessed on 02-03-2024.
37. https://www.kaggle.com/datasets/anbarivan/indian-water-quality-data, Accessed on 02-03-2024.
38. Prasad, N.R., Patel, N.R., Danodia, A., Crop yield prediction in cotton for regional level using random forest approach. *Spatial Inf. Res.*, 29, 195–206, 2021.
39. Nosratabadi, S., Szell, K., Beszedes, B., Imre, F., Ardabili, S., Mosavi, A., Comparative analysis of ANN-ICA and ANN-GWO for crop yield prediction. *2020 RIVF International Conference on Computing and Communication Technologies (RIVF)*, IEEE, pp. 1–5, 2020.
40. Chandana, C. and Parthasarathy, G., Efficient machine learning regression algorithm using naïve Bayes classifier for crop yield prediction and optimal utilization of fertilizer. *Int. J. Performability Eng.*, 18, 1, 47, 2022.
41. Gupta, S., Geetha, A., Sankaran, K.S., Zamani, A.S., Ritonga, M., Raj, R., Ray, S., Mohammed, H.S., Machine Learning-and Feature Selection-Enabled Framework for Accurate Crop Yield Prediction. *J. Food Qual.*, 1, 6293985, 2022.
42. Tripathi, P., Kumar, N., Rai, M., Shukla, P.K., Verma, K.N., Applications of Machine Learning in Agriculture, in: *Smart Village Infrastructure and Sustainable Rural Communities*, vol. edited, M.A. Khan and (Eds.), pp. 99–118, IGI Global, Hershey, Pennsylvania, USA, 2023.

43. Rai, M., Tyagi, N., Tripathi, P., Kumar, N., Kumari, P., Advancement in Agricultural Techniques with the Introduction of Artificial Intelligence and Image Processing, in: *Smart Village Infrastructure and Sustainable Rural Communities*, M.A. Khan (Ed.), pp. 47–68, IGI Global, Hershey, Pennsylvania, USA, 2023.
44. Tripathi, P., Kumar, N., Rai, M., Khan, A., Applications of Deep Learning in Agriculture, in: *Artificial Intelligence Applications in Agriculture and Food Quality Improvement*, M.A. Khan, R. Khan, P. Praveen (Eds.), pp 17–28, IGI Global, USA, 2022
45. Li, A., Markovic, M., Edwards, P., Leontidis, G., Model pruning enables localized and efficient federated learning for yield forecasting and data sharing. *Expert Syst. Appl.*, 242, 122847, 2024.
46. SaiSindhuTheja, R. and Shyam, G.K., An efficient metaheuristic algorithm-based feature selection and recurrent neural network for DoS attack detection in cloud computing environment. *Appl. Soft Comput.*, 100, 106997, 2021.
47. Saisindhutheja, R., Shyam, G.K., Makka, S., A Vulnerability based Attack Detection and Mitigation in Cloud SaaS Framework. *J. Eng. Sci. Technol. Rev.*, 15, 5, 158–169, 2022.
48. Makka, S., Lingam, S., Reddy, S., Arora, G., The evaluation of abstractive text summarization using deep learning. *AIP Conference Proceedings*, vol. 2938, AIP Publishing, 2023.
49. Makka, S., Arora, G., Reddy, S.S.T., Lingam, S., Use of machine learning models for analyzing the accuracy of predicting the cancerous diseases. *International Conference on Innovations in Data Analytics*, vol. 1442, Springer Nature Singapore, Singapore, pp. 169–180, 2022.
50. Makka, S., Popuri, R.B., Sandiri, R., Velpuru, M.S., Arora, G., Intelligent Disease Prediction and Personalized Health Management System. *Nanotechnol. Perceptions*, 20, S6, 907–920, 2024.
51. Makka, S., Reddy, S.B., Jaidka, P., Kumar, R.R., Begum, A., Empowering Health Prognosis: Quine Mccluskey Binary Classifier Algorithm for Predicting Heart Disease. *Nanotechnol. Perceptions*, 2024, 893–906, 2024.
52. Sunitha, L., Bal Raju, M., Makka, S., Ramasahayam, S., Outlier Detection from Mixed Attribute Space Using Hybrid Model, in: *International Conference on Hybrid Intelligent Systems*, Springer Nature Switzerland, Cham, pp. 940–947, 2022.
53. Sunitha, L., Makka, S., Madhu, S., Bheemeswara Sastry, J., Study on influence of outliers on the performance of various classification algorithms, in: *Innovations in Electronics and Communication Engineering: Proceedings of the 9th ICIECE*, Springer Singapore, Singapore, pp. 437–445, 2022.
54. SaiSindhuTheja, R,. and Shyam, G.K., A machine learning based attack detection and mitigation using a secure SaaS framework. *J. King Saud Univ. Comput. Inf. Sci.*, 34, 7, 4047–4061, 2022.

55. Saisindhutheja, R. and Shyam, G.K., A Hybridized Machine Learning Model for Optimal Feature Selection and Attack Detection in Cloud SaaS Framework. *IoT and Analytics for Sensor Networks: Proceedings of ICWSNUCA 2021*, Springer Singapore, pp. 403–413, 2022.

13

A Review on Detection of Adulteration in Food Using Federated Learning

Jagamohan Meher[1]* and Rajanandini Meher[2]

[1]*Krishi Vigyan Kendra Kalahandi, Odisha University of Agriculture and Technology, Bhawanipatna, Kalahandi, Odisha, India*
[2]*Department of Biotechnology, Kalasalingam Academy of Research and Education, Krishnankoil, Srivilliputhur, Tamil Nadu, India*

Abstract
Food adulteration poses serious risks to public health and consumer trust, necessitating the use of effective detection tools. Traditional centralized systems for identifying adulteration frequently face issues relating to data privacy, scalability, and the requirement to manage a variety of data sources. Federated learning (FL), a decentralized machine learning paradigm, provides a viable solution by allowing for collaborative model training across dispersed data sources while ensuring data privacy. This article gives a thorough examination of FL's use in food adulteration detection. We begin by defining FL's essential principles, emphasizing its advantages over traditional centralized approaches, particularly in terms of user privacy and compliance with data protection rules. The topic covers several approaches and algorithms utilized in FL frameworks, with an emphasis on their applicability and adaption for food adulteration detection. We investigate the many forms of data and features used in adulteration research, such as chemical compositions, spectral data, and imaging methods. The paper highlights how FL may use numerous data types from many sources, including food manufacturers, quality control laboratories, and regulatory bodies, to create strong and generalizable models. The key technological issues of adopting FL for adulteration detection are addressed, including data heterogeneity, communication efficiency, and model convergence. Finally, future research prospects are highlighted, with a focus on integrating emerging technologies such as blockchain and Internet-of-Things devices to strengthen FL's capabilities in food safety and public health. This review

**Corresponding author*: jmeher.nitrr@gmail.com

Padmesh Tripathi, Bhanumati Panda, Shanthi Makka, Reeta Mishra, S. Balamurugan and Sheng-Lung Peng (eds.) Federated Learning for Smart Agriculture and Food Quality Enhancement, (319–346) © 2026 Scrivener Publishing LLC

highlights FL's transformative potential for constructing advanced, accurate, and reliable adulteration detection systems.

***Keywords*:** Federated learning, food adulteration, privacy preservation, data integration, detection accuracy

13.1 Introduction

Food adulteration (FA) is the immoral practice of modifying food items by adding, deleting, or substituting ingredients. This approach frequently jeopardizes the quality, safety, or nutritional value of the food. FA is driven by economic incentives, with the primary goal of increasing profit margins by lowering production costs or improving the product's look to attract customers. Unfortunately, FA endangers consumers' health and safety. It can cause a variety of short- and long-term health problems, including food poisoning, allergic reactions, and long-term diseases including cancer as a result of toxic chemical exposure. These health dangers emphasize the vital need for strict food safety rules and monitoring systems to detect and prevent adulteration. Adulteration can be divided into three categories: intentional, incidental, and economic. Intentional adulteration is the intentional addition of harmful components or dilution with inferior materials, such as adding water to milk or using counterfeit coloring in sweets. Poor handling, processing, storage, or transportation techniques cause unintended adulteration, resulting in pesticide, heavy metal, and pathogen contamination [1]. Economic adulteration is the practice of exchanging high-quality materials for lower-cost equivalents to cut production costs, such as using synthetic dyes instead of natural colors or blending expensive oils with less expensive ones. The health consequences of consuming contaminated foods can range from moderate gastrointestinal distress to serious illnesses such as cancer, organ damage, and neurological issues [2]. Regulatory authorities around the world, including the Food and Drug Administration (FDA) in the United States, Food Safety and Standards Authority of India (FSSAI) in India, and European Food Safety Authority (EFSA) in Europe, have set rules and laws to combat FA [2]. However, the ever-changing nature of adulteration tactics, as well as the complexity of food supply systems, continues to offer hurdles to regulatory attempts.

Traditional methods of detecting FA use a variety of chemical, physical, and microscopic investigations. Chemical tests, physical inspection, and microscopic analysis are among the most used approaches. Chemical tests utilize reagents to detect specific adulterants, such as iodine solution for detecting starch in milk. Physical examination is visually checking food items for evidence of adulteration, such as color changes or strange textures. Microscopic analysis uses microscopes to assess food samples for foreign chemicals or contaminants that are invisible to the naked eye [3]. While these old approaches are useful in detecting certain types of adulteration, they have some limitations. Many traditional tests are time-consuming and labor-intensive, necessitating experienced workers, which raises the overall cost and complexity of the detection process. Furthermore, some adulterants may be present in low amounts, making them difficult to identify with standard procedures. Furthermore, previous approaches are often intended to detect certain adulterants and may be inefficient against novel or emerging adulteration tactics [4]. These limitations underscore the need for more effective, efficient, and adaptive ways of detecting FA, hence maintaining food safety and integrity.

Federated learning (FL) is a novel technique of machine learning (ML) that allows several entities to train models together without sharing their raw data. This decentralized technique allows data to remain on local devices or inside organizational bounds, increasing privacy and security while harnessing the collective intelligence of varied datasets [5]. FL has numerous significant advantages in the detection of FA. FL protects sensitive and private information about food production operations by keeping data local and only sharing model changes [6]. Combining data from numerous sources allows for the development of more robust and accurate detection models that can identify a broader range of adulterants and adulteration strategies. FL can scale across multiple entities and geographies, allowing for wider collaboration and increasing the overall effectiveness of adulteration detection systems. Furthermore, this strategy enables models to continuously learn and adapt to new data, hence improving their ability to detect emergent adulteration methods and patterns. The use of FL in FA detection systems offers a significant leap in food safety [7]. By exploiting the capabilities of this technology, stakeholders in the food sector can collaborate to develop more resilient and effective procedures for detecting and combating FA.

13.2 Fundamentals of FL

FL is a game changer in the field of ML since it introduces a decentralized way to train models across several data sources. Unlike traditional centralized ML, which aggregates data into a single repository for model training, FL allows different devices or organizations to train a model together without sharing raw data [8]. This decentralized technique requires each participant, such as a mobile device, hospital, or corporation, to train a local model with its data. Only model updates, such as weights and gradients, are sent to a central server, which then aggregates them to develop the global model. This approach is repeated recursively, resulting in a continuously improved model while preserving data privacy.

FL's main principles include decentralized data processing, model aggregation, iterative training, and design-based privacy. These concepts distinguish FL from traditional techniques, providing increased privacy and security by keeping raw data locally and lowering the likelihood of data breaches. FL can potentially use several data sources, resulting in more robust and generalizable models. It is also bandwidth-efficient, scalable, and complies with strict data privacy requirements such as General Data Protection Regulation (GDPR), Health Insurance Portability and Accountability Act (HIPAA), and Central Consumer Protection Authority (CCPA) [9].

The architecture of FL minimizes data exposure by default, which addresses several privacy and security concerns. Model training takes place on local devices, which reduces the possibility of data interception during transfer. Only model updates are shared with the central server, which protects individual privacy. Techniques such as differential privacy and secure aggregation techniques improve security by ensuring that personal information is protected [5]. FL, which adheres to data localization principles, assists enterprises in meeting regulatory standards, making it a useful tool for constructing privacy-preserving models in areas where data sensitivity and compliance are critical.

13.2.1 Working Procedure of FL

Figures 13.1 and 13.2 depict the working procedure of FL using a decentralized approach in which numerous entities or devices collaborate to train a common model without directly transferring sensitive data. Initially, a central server creates a global model and distributes it to participating clients, which may include individual devices, businesses, or data centres. Each client trains the model individually with its local dataset, extracting key features and computing model updates based on its specific data properties. These updates, which are often gradients or model parameters, are then securely collected by the central server and used to develop the global model. This iterative technique repeats many rounds of local training, update aggregation, and model refinement until convergence requirements are reached.

FL protects data privacy and security by keeping raw data locally, hence reducing concerns about data sharing and confidentiality. This strategy not

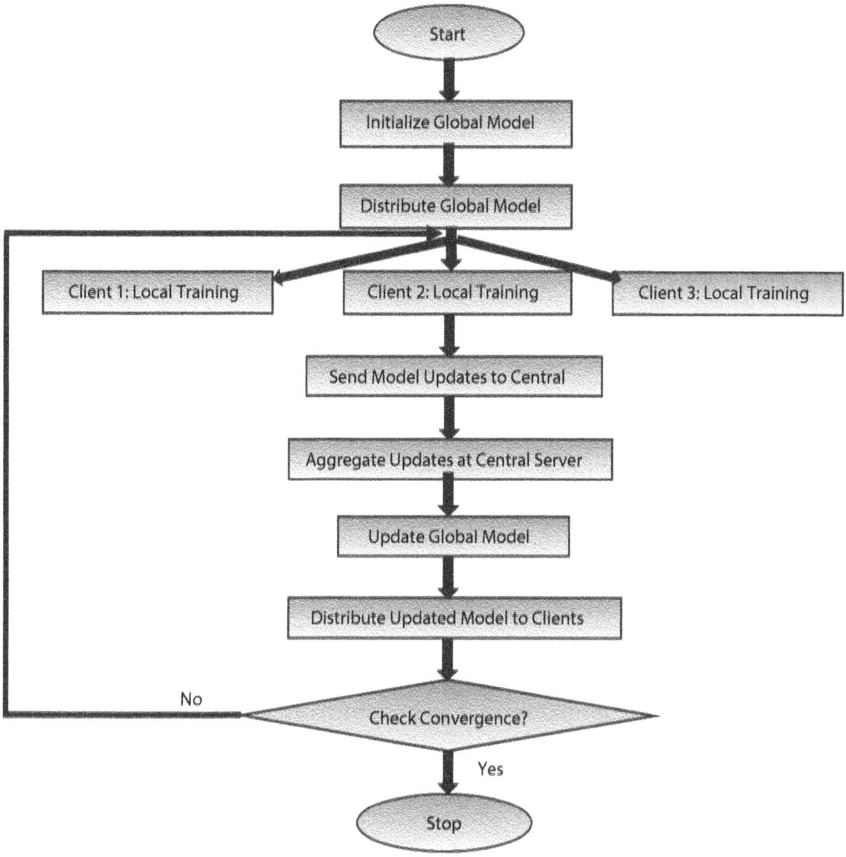

Figure 13.1 Working procedure of FL system.

only improves the resilience and accuracy of the global model by integrating varied datasets, but it also allows compliance with regulatory frameworks that govern data protection and privacy [5].

13.2.2 FL in Food Adulteration Detection

An FA detection system designed using FL uses decentralized data from multiple sources to improve food safety while maintaining data privacy. Clients in this system include food laboratories, producers, and regulatory organizations, all of whom have distinct datasets and expertise in various types of FA. Initially, a central server sets up a global detection model and distributes it to the clients. Each client trains the model locally with its dataset, which may consist of spectroscopic data, chromatographic results,

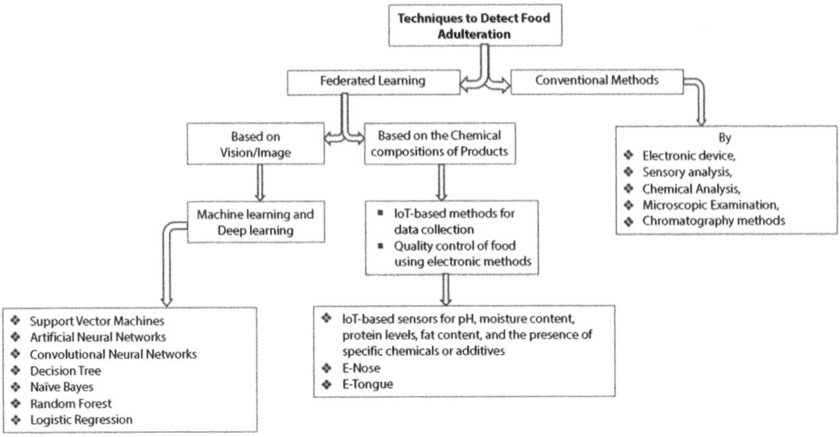

Figure 13.2 Working procedure of FL system.

mass spectrometry discoveries, chemical assays, sensor readings, or image data. Local models learn to recognize specific adulterants or pollutants in their domains. Instead of exchanging raw data, clients send encrypted model changes (such as weights or gradients) to the central server, which aggregates them to improve the overall model [6]. This collaborative method enables the global model to benefit from the different and broad datasets while protecting sensitive information. It ensures that the detection system stays versatile and accurate across a variety of food kinds and contamination scenarios, all while adhering to strict data privacy rules. FL thus appears as an effective instrument for protecting food quality and public health, allowing stakeholders to resist adulteration while adhering to data privacy requirements collectively [9].

13.3 Data Types and Features in FA Detection

FA detection with FL requires a variety of data kinds and features to construct reliable and robust models. FL trains a global model using decentralized data from a variety of sources, including laboratories, food producers, and quality control organizations. This method protects data privacy while improving the model's capacity to detect adulteration across a variety of food goods. FL allows for the usage of these various data kinds and characteristics without requiring data to be centralized. Each participating entity (e.g., lab and food producer) can train local models using their data [10]. The local models then send only the learned parameters (e.g., weights and gradients) to a central server. This server aggregates the parameters to update a global model, making use of the numerous and rich data sources while protecting data privacy.

FA detection models can be trained *via* FL on a large range of data, resulting in robust and generalizable models. This decentralized method keeps sensitive data local, lowering privacy concerns and complying with data protection rules while yet leveraging the collective knowledge and diverse datasets from several companies [11].

13.3.1 Data Types

13.3.1.1 Enhancing FA Detection with FL Using Chemical Assay Data

Enhancing chemical assay data for FA detection *via* FL enhances accuracy while maintaining data privacy. In this structure, several institutions, such as food laboratories, manufacturers, and regulatory authorities, conduct independent chemical testing on food samples to detect specific adulterants. These organizations use techniques such as high-performance liquid chromatography (HPLC), gas chromatography–mass spectrometry (GC-MS), and enzyme-linked immunosorbent assay to provide comprehensive chemical data. Each organization collects relevant features from their test data, such as concentration levels of specific substances and chromatographic properties, and utilizes these features to build local ML models that can recognize adulteration tendencies [10].

Chemical composition data gives specific information on food products' molecular makeup, enabling for accurate identification of contaminants. Researchers can improve adulterant detection by integrating chemical composition data with spectral data obtained using near-infrared (NIR) or Fourier transform infrared spectroscopy (FTIR) techniques. For example, in milk adulteration investigations, chemical analysis can reveal the presence of certain contaminants such as water or urea, and spectrum data can validate these findings and provide additional insights into other potential contaminants [11].

13.3.1.2 Enhancing FA Detection with FL Using Spectroscopic Data

FL enhances spectroscopic data for FA detection, dramatically improving accuracy while maintaining data privacy. In this approach, food laboratories, manufacturers, and regulatory authorities conduct independent spectroscopic investigations on food samples to detect specific adulterants. These entities generate detailed chemical data using techniques such as infrared (IR) spectroscopy, NIR spectroscopy, and ultraviolet-visible (UV-Vis). Each organization pulls key variables from its spectroscopic data, such as absorption peaks and spectral signatures, and then trains local ML algorithms to detect adulteration tendencies.

Spectral data analysis has developed as a key tool for identifying FA, relying on the interaction of light with food ingredients to identify toxins. Different substances have unique absorption and reflection properties, resulting in distinct spectral fingerprints. NIR, FTIR, and Raman spectroscopy have all proven useful in getting these spectral fingerprints. For further information on individual methodologies and their applications in FA detection, it summarizes numerous analytical techniques used in recent research investigations (Table 13.1).

Table 13.1 Spectral data used in FA detection.

Food	Adulterant	Techniques	References
Bovine meat	Detecting bovine meat adulteration with turkey meat	UV-Vis, NIR, and MIR spectroscopy coupled with chemometric techniques	[12]
Beef meat	Discrimination of beef and horsemeat	Raman spectroscopy	[13]
Honey	Sugar syrup	IR-based analysis, NMR, and Raman spectroscopy	[14]
Olive oils (extra virgin olive oil (EVOO) and virgin olive oil (VOO))	Vegetable oils (sunflower oil (SFO), palm olein oil (PLMOO), and coconut oil (CO))	NMR spectroscopy	[15]
High-quality olive oil	Lower-grade oils or illegal additives	NMR spectroscopy and chromatography	[16]
Extra virgin olive oil	Lower-quality olive oils	Trapped ion mobility spectrometry coupled to liquid chromatography and high-resolution mass spectrometry	[17]
Mustard and canola oil	Argemone oil	fluorescence spectroscopy coupled with chemometrics	[18]

13.3.1.2.1 Key Spectral Techniques

a. **NIR Spectroscopy:** It is widely used to detect adulteration in dairy, honey, and oils by detecting the absorption of NIR light, which offers information on molecular vibrations associated with certain chemical bonds [19]. For example, research on milk adulteration has used NIR in conjunction with ML methods such as support vector machines (SVMs) and partial least squares regression (PLSR) to accurately detect adulterants such as water, starch, and urea.
b. **FTIR Spectroscopy:** It is used to identify items such as honey, olive oil, and spices by analysing the IR spectra of absorption or emission [20]. FTIR research has successfully discovered and quantified honey adulterants such as high fructose corn syrup and inverted sugar by evaluating the distinct spectrum signatures of pure and contaminated samples.
c. **Raman Spectroscopy:** It is important for high-value items such as saffron and essential oils [21]. It uses inelastic scattering of light to provide comprehensive information about molecular vibrations. This technique has been successfully used to identify melamine in milk and adulteration in saffron, honey, and butter by distinguishing between pure and contaminated samples with high accuracy [22].
d. **Hyperspectral Imaging:** Hyperspectral imaging combines conventional imaging and spectroscopy to extract spatial and spectral data from a sample. This approach has been used to identify adulteration in fruits, vegetables, and grains. Researchers used hyperspectral imaging with convolutional neural networks (CNNs) to get precise spectral and spatial data, allowing for accurate adulterant identification and freshness assessment [23].

Additional approaches, such as UV-Vis, NIR, and Mid-Infrared (MIR) spectroscopy, which are frequently combined with chemometric methods, have been used to detect bovine meat adulteration with turkey meat. These procedures are perfect for detecting adulteration because they are non-destructive and require only a small sample amount [12].

13.3.1.2.2 Benefits of FL in Integrating Spectral Data

By combining spectral data from several sources, the global model may detect subtle and variable patterns of adulteration that would not be visible in a single dataset, resulting in more accurate detection of adulterants. The data diversity makes the global model more resilient and generalizable across different food products and adulterants, successfully handling variances in spectral data caused by different food types, processing conditions, and adulteration methods.

FL enables data privacy and adherence to data protection rules because raw spectral data never leaves the local devices. This is especially important in the food business, where private information must be protected. FL can expand to include additional participants over time, continuously improving the model as new data sources are added, which is critical for keeping up with the changing nature of FA strategies.

13.3.1.3 Enhancing FA Detection with FL Using Chromatographic Data

FL enhances the accuracy of chromatographic data for FA detection while maintaining data privacy. In this approach, food laboratories, manufacturers, and regulatory agencies perform separate chromatographic studies on food samples to detect specific adulterants. These organizations use techniques such as HPLC, GC, and GC-MS to create precise chemical profiles. Each organization pulls key variables from their chromatographic data, such as retention periods, peak regions, and component concentrations, and utilizes them to build local ML models capable of detecting adulteration tendencies.

13.3.1.4 Enhancing FA Detection with FL Using Sensor Data

FL enhances accuracy and protects data privacy when using sensor data to detect FA. In this system, several entities, including food laboratories, producers, and regulatory authorities, conduct sensor-based analysis of their food samples to detect specific adulterants. These beings use a variety of sensors, including electronic noses, electronic tongues, and biosensors, to generate detailed chemical profiles and other pertinent information. Each company pulls key elements from their sensor data, such as unique sensor responses, signal patterns, and concentration levels, and then trains local ML models to detect adulteration tendencies.

13.3.1.5 *Enhancing FA Detection with FL Using Imaging Data*

FL enhances accuracy and data privacy while processing imaging data for FA detection. In this approach, food laboratories, producers, and regulatory authorities undertake independent imaging-based tests on food samples to detect specific adulterants. To create detailed visual and spectral data, these entities use a variety of imaging techniques such as hyperspectral imaging, multispectral imaging, and X-ray imaging. Each organization pulls key elements from its imaging data, such as color histograms, texture patterns, and spectral fingerprints, and then trains local ML models to detect adulteration trends.

Imaging techniques are useful in detecting visual adulterations and assessing the quality of food goods. High-resolution cameras and powerful image processing algorithms may detect anomalies including foreign particles, discoloration, and texture variations. Hyperspectral imaging is a technique that combines conventional imaging with spectral data to provide precise information on the chemical and physical aspects of food products [24]. For example, hyperspectral imaging may detect surface contamination in fruits and vegetables, as well as non-visible adulterants in powdered spices. ML algorithms can also be trained on image datasets to automatically classify and detect contaminated food items. Imaging techniques are increasingly being used in FA detection due to their capacity to provide thorough and non-destructive analysis of food samples. These approaches, when combined with ML algorithms, significantly improve the accuracy and efficiency of adulterant detection [25].

FL makes it easier to create advanced image analysis models for detecting FA by combining imaging data from multiple sources. This strategy improves the models' ability to detect modest adulteration indicators *via* a variety of mechanisms [6]. FL guarantees that training data covers a broader range of food varieties, adulteration methods, and environmental variables by collecting image data from multiple sources. This diversity allows the model to learn from a larger collection of instances, boosting its capacity to generalize and detect minor adulteration cues in a variety of scenarios.

13.3.1.5.1 How Imaging Techniques Work

The food sample is prepared and placed into the imaging apparatus. Several imaging techniques, including hyperspectral imaging, NIR imaging, FTIR imaging, and Raman imaging, are used to acquire detailed images of the material at various wavelengths. To improve image quality, collected photos are pre-processed using techniques such as noise reduction, normalization, and correction. Images are analyzed to extract important elements such as spectral fingerprints, spatial patterns, and texture information. Spectral data from photos is frequently coupled with other forms of data, such as chemical compositions and data collected by Internet-of-Things (IoT) devices, to provide a more comprehensive analysis.

CNNs, SVMs, and Random Forests are trained on labelled datasets to identify patterns in adulterated and non-adulterated samples. The trained algorithms examine the fresh photos to identify and detect adulterants based on previously learned patterns. The ML model generates detection results, indicating if the sample is contaminated and determining the kind

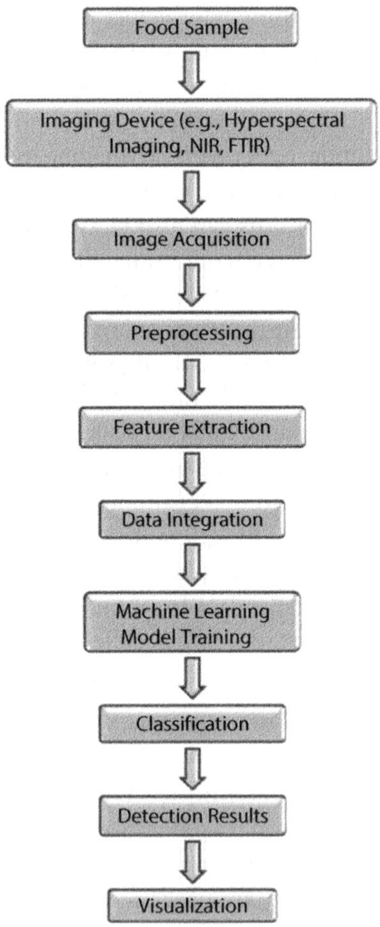

Figure 13.3 Stepwise working process of imaging techniques.

and extent of adulteration. The results are frequently visualized to aid in evaluating the findings and making sound judgments. Here is a simplified Figure 13.3 that demonstrates the process.

Imaging techniques have become critical in the detection of FA because they provide non-destructive, quick, and highly accurate results. These techniques, when paired with ML algorithms, improve detection by evaluating visual data to identify adulterants. Here are some important imaging techniques used in FA detection.

a. **Hyperspectral Imaging**
Hyperspectral imaging is often used to detect adulteration in fruits, vegetables, cereals, and other processed goods. This technology records images at several wavelengths of light, beyond the visible spectrum, delivering both spatial and spectral data. It can detect small variations in the composition and quality of food products. In a study on fruit freshness, hyperspectral imaging was integrated with CNNs to detect adulteration and evaluate fruit quality. The technique allowed for the identification of chemical and physical changes in fruits, resulting in great accuracy in identifying rotten or contaminated items [25].

b. **NIR Imaging**
It is widely used to identify adulteration in dairy goods, meat, grains, and oils. NIR imaging detects the sample's absorption of NIR light, which reveals information about its molecular composition. The approach is excellent at detecting adulterants that change the chemical composition of food products. Researchers detected milk adulteration using NIR imaging by evaluating spectral data and identifying changes in chemical composition caused by adulterants such as water and urea. The spectral data was classified using ML algorithms such as SVM, which resulted in high detection accuracy [10].

c. **FTIR Imaging**
It can detect adulteration in honey, oils, spices, and other food products. FTIR imaging captures a sample's IR spectra, revealing detailed molecular fingerprints that can be used to detect adulterants. To detect honey adulteration, FTIR imaging was integrated with ML algorithms to determine the presence of high fructose corn syrup and other sugar-based contaminants. The approach accurately classified

pure and contaminated honey samples by evaluating their unique spectral patterns [25].

d. **Raman Imaging**
It is applied to high-value items such as saffron, spices, and essential oils. This approach detects inelastic scattering of light, which provides information on molecular vibrations that can be used to identify chemical substances. A study on saffron adulteration used Raman imaging and ML algorithms to distinguish between pure saffron and adulterated samples. The technique obtained great accuracy by examining the unique Raman spectra of the materials, which revealed the presence of adulterants such as synthetic dyes [10].

e. **X-Ray Imaging**
X-ray imaging can detect physical adulterants in processed foods, such as bones in meat or foreign particles in packaged goods. X-ray imaging offers detailed views of food products' internal structures, allowing for the detection of physical pollutants and foreign objects. A study used X-ray imaging to detect bone particles in beef products. Images were examined using ML techniques to automatically identify and classify the presence of bones, achieving high detection rates and decreasing the requirement for manual inspection [26].

13.4 Integration of Diverse Data Sources in FA Detection and Its Benefit

The use of many data sources in FA detection improves the accuracy and reliability of identifying adulterants. This method combines multiple sorts of data, such as chemical compositions, spectral data, and imaging techniques, utilizing the strengths of each to produce a thorough analysis. The combination of imaging techniques and ML greatly improves FA detection by automating the analysis process and increasing accuracy. CNNs, SVMs, and Random Forests are popular ML models for analyzing imaging data. These models use visual patterns in imaging data to detect anomalies and categorize adulterants with high accuracy [27, 28].

Spectral data from NIR, FTIR, and Raman spectroscopy provide a non-destructive method for analysing food samples, yielding unique molecular fingerprints that can be used to detect adulterants. In honey adulteration detection, spectral data from FTIR paired with ML techniques such as SVM or ANN creates a strong system for recognizing and quantifying adulterants. Spectral data can identify small changes in chemical composition and can be cross-checked with chemical analysis for more accurate results [10]. Imaging techniques such as hyperspectral imaging, NIR imaging, and X-ray imaging provide visual and spatial information about food samples, detecting physical and chemical changes that may suggest contamination (Table 13.2). Hyperspectral imaging, paired with spectral data, can improve the detection of fruit adulteration. Spectral data provides molecular insights, whereas hyperspectral imaging provides rich spatial information. Integrating various data sources enables more accurate categorization of adulterated versus non-adulterated samples using ML models such as CNNs [25]. IoT devices may capture real-time data on a variety of parameters like temperature, humidity, and chemical markers, allowing for continuous monitoring of food quality. Combining IoT-collected data with spectral and chemical composition data allows for real-time monitoring and detection of adulteration in perishable foods such as dairy and meat. For example, IoT sensors can continuously monitor environmental conditions, while spectral data from NIR spectroscopy identify changes in chemical composition, offering a full adulteration detection system [29].

By merging numerous data sources, the detection system may cross-check information, resulting in greater accuracy in identifying adulterants. The integration of diverse data types strengthens the detection system, lowering the risk of false positives and false negatives. Diverse data sources provide a more complete picture of the food sample, allowing for more detailed analysis and comprehension of adulteration patterns. IoT devices and continuous data collection enable real-time monitoring, which is critical for perishable items and assures early detection of adulteration. Integrated systems may be scaled to track vast quantities of food goods along the supply chain, from manufacture to consumption.

Integrating many data sources in FA detection improves accuracy, resilience, and real-time monitoring. Researchers and industry personnel can create comprehensive food safety and quality systems by combining chemical compositions, spectrum data, imaging techniques, and IoT devices. These developments are critical for protecting customers and sustaining trust in food items [10]. When imaging techniques are integrated with ML, they form an effective tool for identifying FA. These approaches provide speedy, non-destructive, and extremely accurate analysis to ensure food safety and quality. Continuous development and integration of these technologies hold enormous potential for future breakthroughs in food quality control and consumer safety [29].

Table 13.2 Recent applications of integration of imaging techniques with ML in adulteration determination in food.

Types of food	Food	Objective	Instrument	Techniques	References
Oils	Olive oil	Detect olive oil adulteration	FTIR, visible-near-infrared	ANN	[30]
	Extra virgin olive oils	Extra virgin olive oil adulteration with vegetable oil detection	Optical microscope	Deep learning (CNN)	[31]
	Avocado oil	Avocado oil adulterated	Camera	Deep learning (CNN)	[32]
	Edible oils	Adulteration analysis for edible oils	NIR spectrometer	CNN	[33]
	Sesame oils	Identification of sesame oils	Dielectric spectroscopy	SVM	[34]
	Sesame oil	Detection of the authenticity and adulteration of sesame oil	Excitation-emission matrix fluorescence	SVM	[35]
	Sesame oil	Adulteration detection with corn oil	Near-mid infrared spectroscopy	SVM	[36]
	Extra virgin olive oil	Monitor storage conditions and locate adulterations	Fluorescence sensors	ANN	[37]
	Extra virgin olive oil	Adulteration detection	Laser diode induced	ANN	[38]
	Extra virgin olive oil	Adulteration detection	Laser-induced fluorescence	ANN and SVM	[39]

(*Continued*)

Table 13.2 Recent applications of integration of imaging techniques with ML in adulteration determination in food. (*Continued*)

Types of food	Food	Objective	Instrument	Techniques	References
	Adulteration of sesame oil products	Common household pure oils, including rapeseed, soybean, sesame, peanut, sunflower, camellia, olive, and lard	Raman spectrometer	1D CNN vector regression	[40]
	Butter oil	Identify adulteration in butter oil	NIR spectrometer	SVM	[41]
	Tuna fish oil	Identification of pork oil adulteration in tuna fish oil	FTIR spectroscopy	SVM, k-nearest neighbor, ANN	[42]
Cereal and pulses	Wheat flour	Adulteration in wheat flour	Hyperspectral imaging	Deep learning (CNN)	[43]
	Rice	Rice adulteration detection	Camera	Deep learning (CNN) ResNet34	[44]
	Flour	Adulteration detection	Hyperspectral imaging system	SVM	[45]
	Rice	Aged-rice adulteration with late Japonica rice and Southern Japonica rice	NIR spectroscopy	SVM	[46]

(*Continued*)

Table 13.2 Recent applications of integration of imaging techniques with ML in adulteration determination in food. (*Continued*)

Types of food	Food	Objective	Instrument	Techniques	References
Fruits and vegetables	Sea cucumber	Adulteration detection	Hyperspectral imaging	SVM	[47]
	Cassava starch	Detect adulteration in cassava starch	Raman spectroscopy	SVM	[48]
	Ginseng	Adulteration detection with lower-grade ginseng	NIR and FT-MIR	SVM	[49]
	Spinach	Detect spinach adulteration in pistachios	LAB color space	CNN	[50]
	Coconut milk	Coconut milk adulteration with distilled water and mature coconut water	NIR spectroscopy	ML algorithms	[51]
Spices	Cinnamon spice	Seafoam powder and chickpea and wheat flour	Infrared hyperspectral imaging system	ANN	[52]
	Ground turmeric	Starch (ST) powder, lead (II) chromate, metanil yellow, Sudan III, and Acid Orange 7	FT-IR, Spectrometer, Raman	CNN	[53]

(*Continued*)

Table 13.2 Recent applications of integration of imaging techniques with ML in adulteration determination in food. (*Continued*)

Types of food	Food	Objective	Instrument	Techniques	References
Milk and milk-based products	Milk	Adulterants such as water, urea, ammonium sulfate, oils, and surfactants in milk	Microscope	CNN	[54]
	Milk	Cow's milk samples were diluted with tap water, distilled water, and mineral water.	Manual observation	CNN	[55]

13.4.1 Technical Challenges and Solutions

13.4.1.1 Handling Data Heterogeneity

One of the key technical issues in FL for FA detection is dealing with data heterogeneity. Data heterogeneity occurs because data obtained from many sources might vary greatly in format, quality, and dissemination. For example, chemical composition data from different laboratories may be measured using different methodologies or units, and spectral data from separate equipment may have differing resolutions and noise levels. This variability might complicate the training process and influence model performance [11].

Solution: Standardization and normalizing procedures are vital for dealing with data heterogeneity. Preprocessing processes can convert data to a standard format, maintaining consistency across datasets. Furthermore, FL frameworks can use robust algorithms to handle non-IID (independent and identically distributed) data. Techniques like federated averaging

(FedAvg) and personalized FL, which tailor the global model to the data distributions of individual clients, can improve the model's ability to learn from varied datasets.

13.4.1.2 Ensuring Communication Efficiency

Communication between local devices and the central server in FL can be slow, especially when dealing with big models and frequent updates. The overhead of transferring model parameters can cause considerable delays and increased expenses, especially in contexts with limited or expensive network capacity [8].

Solution: Various tactics can be used to ensure effective communication. Quantization and sparsification are two model compression approaches that minimize the size of model updates by expressing parameters with fewer bits or zeroing out inconsequential values. FL protocols, such as FedAvg, also benefit by aggregating updates at regular intervals rather than continually, lowering communication frequency. Furthermore, asynchronous communication protocols enable devices to submit updates at different times, which reduces network congestion.

13.4.1.3 Achieving Model Convergence

Model convergence in FL can be challenging due to data heterogeneity and the likelihood of uneven contributions from participating devices. Variations in processing power, data quality, and participation rates can cause slow or unstable convergence, preventing the model from reaching its optimal state [5].

Solution: Model convergence involves adaptive learning rates and regularization strategies. Adaptive learning rates alter the pace of learning based on the model's development, eliminating overshoot and stagnation. Weight decay and other regularization approaches serve to stabilize the training process. Additionally, integrating procedures for dealing with stragglers (devices that take longer to compute updates) can increase overall convergence. For example, utilizing federated optimization techniques that account for delayed updates guarantees that the global model stays on track despite asynchronous contributions.

13.4.1.4 Security Enhancements: Differential Privacy and Secure Multi-Party Computation

Security and privacy are critical in FL, particularly when dealing with sensitive data from several stakeholders. To retain trust and comply with data protection requirements, model changes must not mistakenly reveal information about particular datasets [9].

Solution: Two major strategies for improving security in FL are differential privacy and secure multi-party computing (SMPC). Differential privacy entails introducing controlled noise to model updates to ensure that they do not divulge specific details about any single data point. This technique helps to maintain participants' anonymity while still allowing for beneficial learning. SMPC, on the other hand, enables several parties to collaborate on a function over their inputs while keeping those inputs private. In the context of FL, SMPC can be used to securely aggregate model updates from several devices without disclosing raw data. Combining these strategies can provide strong security assurances, guaranteeing that FL stays successful while protecting privacy. FL for FA detection has several technological issues, including managing data heterogeneity, maintaining communication efficiency, reaching model convergence, and improving security. These issues can be efficiently addressed using standardization techniques, adaptive learning methodologies, and improved security protocols such as differential privacy and SMPC, opening the way for more resilient and trustworthy FL systems.

13.4.2 Future Directions and Research Avenues

13.4.2.1 Integration with Blockchain Technology

Integrating FL and blockchain technology can improve data security, transparency, and traceability in FA detection [7]. The blockchain's decentralized and immutable ledger creates a tamper-proof record of data transfers and model modifications among FL participants. This assures data integrity, builds trust among stakeholders, and allows for public auditing of the model training process. Smart contracts on blockchain platforms can automate data governance standards and enable fair rewards for data sources. Future research should concentrate on improving protocols for FL over blockchain networks, addressing scalability, latency, and energy efficiency while capitalizing on blockchain's strengths in data integrity and decentralization.

13.4.2.2 Role of IoT Devices in Real-Time Data Collection

IoT devices are critical for real-time data collecting in FL algorithms that identify FA. Sensors in food processing facilities, warehouses, and distribution networks can continuously monitor ambient conditions, product quality, and supply chain operations. These devices provide massive volumes of real-time data, which may be analyzed utilizing FL to swiftly detect adulteration. Future research could focus on lightweight FL algorithms that are more suited to the limited resources and inconsistent connectivity of IoT devices. Additionally, integrating edge computing with FL can improve data processing efficiency and minimize latency, allowing for timely detection and response to adulteration events [56].

13.4.2.3 Potential Advancements in FL Techniques

Advances in FL algorithms promise to overcome current limits and broaden applications in FA detection [7]. Research may concentrate on building adaptive federated optimization algorithms that modify learning rates, model designs, and aggregation procedures in response to dispersed data features. Refining strategies for dealing with non-IID data distributions and biased participant contributions can help to increase model robustness and convergence. Exploring hybrid FL systems that combine decentralized model training with centralized validation and refinement stages may improve model performance while protecting data privacy.

13.4.2.4 Collaborative Research and Development Opportunities

Collaborative R&D activities have considerable prospects to advance FL in FA detection. Public-private collaborations between food makers, regulatory agencies, academic institutions, and technology suppliers can encourage the co-creation of novel solutions. These alliances can help to share subject expertise, data resources, and technical skills to solve issues including new adulteration strategies and assuring global food safety standards compliance. Cross-disciplinary collaborations in ML, food science, cybersecurity, and regulatory policy are required to drive significant advances in FL approaches designed for FA detection.

13.5 Conclusion

In this paper, we explored the use of FL to detect FA, highlighting the obstacles, approaches, and future possibilities. FA poses major dangers to consumer health and safety, as it is motivated by economic incentives to improve profit margins. Traditional detection approaches, while somewhat effective, are frequently time-consuming, labor-intensive, and restricted in scope. FL provides a possible approach by allowing for collaborative model training over distant datasets while maintaining data privacy and security. To efficiently detect adulteration, this strategy uses a variety of data types, including chemical compositions, spectral data, and imaging approaches. FL improves the accuracy and scalability of detection systems, allowing stakeholders to interact while protecting data privacy. The integration of technology such as blockchain and IoT devices allows for real-time monitoring and reaction, safeguarding food safety and public health. FL provides a paradigm leap in FA detection, providing scalable, privacy-preserving solutions that are compliant with regulatory constraints and technology improvements, ultimately transforming food safety practices and protecting public health worldwide.

AI disclosure statement: The author acknowledges the use of AI-based tools for minor language editing or grammatical correction. However, all intellectual contributions, analyses, and conclusions presented in this chapter are entirely the author's own.

References

1. Choudhary, A., Chatha, J., Gupta, I.N., Hameed, I.F., Choton, S., Gupta, N., Hameed, F., An overview of food adulteration: Concept, sources, impact, challenges and detection. *Int. J. Chem. Stud.*, 8, 1, 2564–2573, 2020, https://doi.org/10.22271/chemi.2020.v8.i1am.8655.
2. Thakur, M., Sharma, K.D., Verma, M.L., An Overview of Potential Toxicity of Food Adulterants and Food Adulteration Act. *Biotechnol. Approaches Food Adulterants*, 26–48, 2020, https://doi.org/10.1201/9780429354557-2.
3. Essuman, E.K., Teye, E., Dadzie, R.G., Sam-Amoah, L.K., Consumers' Knowledge of Food Adulteration and Commonly Used Methods of Detection. *J. Food Qual.*, 2022, 1, 2421050, 2022, https://doi.org/10.1155/2022/2421050.
4. Ellis, D., II, Brewster, V.L., Dunn, W.B., Allwood, J.W., Golovanov, A.P., Goodacre, R., Fingerprinting food: current technologies for the detection of food adulteration and contamination. *Chem. Soc. Rev.*, 41, 17, 5706–5727, 2012, https://doi.org/10.1039/C2CS35138B.
5. Qi, P., Chiaro, D., Guzzo, A., Ianni, M., Fortino, G., Piccialli, F., Model aggregation techniques in federated learning: A comprehensive survey. *Future Gener. Comput. Syst.*, 150, 272–293, 2024, https://doi.org/10.1016/J.FUTURE.2023.09.008.

6. Gavai, A., Bouzembrak, Y., Mu, W., Martin, F., Kaliyaperumal, R., van Soest, J., Choudhury, A., Heringa, J., Dekker, A., Marvin, H.J.P., Applying federated learning to combat food fraud in food supply chains. *npj Sci. Food*, 7, 1, 1–9, 2023, https://doi.org/10.1038/s41538-023-00220-3.
7. Lei, M., Xu, L., Liu, T., Liu, S., Sun, C., Integration of Privacy Protection and Blockchain-Based Food Safety Traceability: Potential and Challenges. *Foods*, 11, 15, 2262, 2022a, https://doi.org/10.3390/FOODS11152262.
8. Wen, J., Zhang, Z., Lan, Y., Cui, Z., Cai, J., Zhang, W., A survey on federated learning: challenges and applications. *Int. J. Mach. Learn. Cybern.*, 14, 2, 513–535, 2022, https://doi.org/10.1007/S13042-022-01647-Y.
9. Mohammadi, S., Balador, A., Sinaei, S., Flammini, F., Balancing privacy and performance in federated learning: A systematic literature review on methods and metrics. *J. Parallel Distrib. Comput.*, 192, 104918, 2024, https://doi.org/10.1016/J.JPDC.2024.104918.
10. Karabagias, I.K., Food authentication and adulteration control based on metrics data of foods and chemometrics. *Eur. Food Res. Technol.*, 250, 5, 1269–1283, 2024, https://doi.org/10.1007/S00217-024-04477-0/METRICS.
11. Yadav, A.K., Gattupalli, M., Dashora, K., Kumar, V., Key Milk Adulterants in India and their Detection Techniques: a Review. *Food Anal. Methods*, 16, 3, 499–514, 2023, https://doi.org/10.1007/S12161-022-02427-8/METRIC.
12. Alamprese, C., Casale, M., Sinelli, N., Lanteri, S., Casiraghi, E., Detection of minced beef adulteration with turkey meat by UV–vis, NIR and MIR spectroscopy. *LWT Food Sci. Technol.*, 53, 1, 225–232, 2013b, https://doi.org/https://doi.org/10.1016/j.lwt.2013.01.027.
13. Boyacı, I.H., Temiz, H.T., Uysal, R.S., Velioğlu, H.M., Yadegari, R.J., Rishkan, M.M., A novel method for discrimination of beef and horsemeat using Raman spectroscopy. *Food Chem.*, 148, 37–41, 2014, https://doi.org/https://doi.org/10.1016/j.foodchem.2013.10.006.
14. Wu, L., Du, B., Vander Heyden, Y., Chen, L., Zhao, L., Wang, M., Xue, X., Recent advancements in detecting sugar-based adulterants in honey – A challenge. *Trends Anal. Chem.*, 86, 25–38, 2017, https://doi.org/10.1016/J.TRAC.2016.10.013.
15. Alharbi, H., Kahfi, J., Dutta, A., Jaremko, M., Emwas, A.-H., The detection of adulteration of olive oil with various vegetable oils – A case study using high-resolution 700 MHz NMR spectroscopy coupled with multivariate data analysis. *Food Control*, 166, 110679, 2024, https://doi.org/https://doi.org/10.1016/j.foodcont.2024.110679.
16. Hashempour-baltork, F., Vali Zade, S., Mazaheri, Y., Mirza Alizadeh, A., Rastegar, H., Abdian, Z., Torbati, M., Azadmard Damirchi, S., Recent methods in detection of olive oil adulteration: State-of- the-Art. *J. Agric. Food Res.*, 16, 101123, 2024, https://doi.org/https://doi.org/10.1016/j.jafr.2024.101123.
17. Drakopoulou, S.K., Kritikou, A.S., Baessmann, C., Thomaidis, N.S., Untargeted 4D-metabolomics using Trapped Ion Mobility combined with

LC-HRMS in extra virgin olive oil adulteration study with lower-quality olive oils. *Food Chem.*, 434, 137410, 2024, https://doi.org/https://doi.org/10.1016/j.foodchem.2023.137410.
18. Shiv, K., Singh, A., Kumar, S., Prasad, L.B., Gupta, S., Bharty, M.K., Evaluation of different regression models for detection of adulteration of mustard and canola oil with argemone oil using fluorescence spectroscopy coupled with chemometrics. *Food Addit. Contaminants Part A*, 41, 2, 105–119, 2024, https://doi.org/10.1080/19440049.2023.2297869.
19. Mendes, E. and Duarte, N., Mid-Infrared Spectroscopy as a Valuable Tool to Tackle Food Analysis: A Literature Review on Coffee, Dairies, Honey, Olive Oil and Wine. *Foods*, 10, 2, 477, 2021, https://doi.org/10.3390/FOODS10020477.
20. Valand, R., Tanna, S., Lawson, G., Bengtström, L., A review of Fourier Transform Infrared (FTIR) spectroscopy used in food adulteration and authenticity investigations. *Food Addit. Contaminants Part A*, 37, 1, 19–38, 2020, https://doi.org/10.1080/19440049.2019.1675909.
21. Dai, H., Gao, Q., Lu, J., He, L., Improving the Accuracy of Saffron Adulteration Classification and Quantification through Data Fusion of Thin-Layer Chromatography Imaging and Raman Spectral Analysis. *Foods*, 12, 12, 2322, 2023, https://doi.org/10.3390/FOODS12122322/S1.
22. Parmar, I., Parmar, I., Gehlot, P., Modi, V., Patel, D., Haveli, N., Review: recent development in food adulteration analytical techniques. *Int. J. Pharm. Sci. Res.*, 13, 5, 2001–2012, 2022, https://doi.org/10.13040/IJPSR.0975-8232.13(5).2001-12.
23. Saha, D. and Manickavasagan, A., Machine learning techniques for analysis of hyperspectral images to determine quality of food products: A review. *Curr. Res. Food Sci.*, 4, 28–44, 2021, https://doi.org/10.1016/J.CRFS.2021.01.002.
24. Temiz, H.T. and Ulaş, B., A Review of Recent Studies Employing Hyperspectral Imaging for the Determination of Food Adulteration. *Photochem*, 1, 2, 125–146, 2021, https://doi.org/10.3390/PHOTOCHEM1020008.
25. Ahmed, E., Detection of honey adulteration using machine learning. *PLOS Digit. Health*, 3, 6, e0000536, 2024b, https://doi.org/10.1371/journal.pdig.0000536.
26. Lim, J., Lee, A., Kang, J., Seo, Y., Kim, B., Kim, G., Kim, S., Non-Destructive Detection of Bone Fragments Embedded in Meat Using Hyperspectral Reflectance Imaging Technique. *Sensors*, 20, 14, 4038, 2020, https://doi.org/10.3390/S20144038.
27. Aslam, R., Sharma, S.R., Kaur, J., Panayampadan, A.S., Dar, O., II, A systematic account of food adulteration and recent trends in the non-destructive analysis of food fraud detection. *J. Food Meas. Charact.*, 17, 3, 3094–3114, 2023, https://doi.org/10.1007/S11694-023-01846-3.
28. Goyal, K., Kumar, P., Verma, K., Food Adulteration Detection using Artificial Intelligence: A Systematic Review. *Arch. Comput. Methods Eng.*, 29, 1, 397–426, 2022, https://doi.org/10.1007/S11831-021-09600-Y/METRICS.

29. Tutul, M.J., II, Alam, M., Wadud, M.A.H., Smart Food Monitoring System Based on IoT and Machine Learning. *International Conference on Next-Generation Computing, IoT and Machine Learning*, 2023, https://doi.org/10.1109/NCIM59001.2023.10212608.
30. Meng, X., Yin, C., Yuan, L., Zhang, Y., Ju, Y., Xin, K., Chen, W., Lv, K., Hu, L., Rapid detection of adulteration of olive oil with soybean oil combined with chemometrics by Fourier transform infrared, visible-near-infrared and excitation-emission matrix fluorescence spectroscopy: A comparative study. *Food Chem.*, 405, 134828, 2023, https://doi.org/https://doi.org/10.1016/j.foodchem.2022.134828.
31. Pradana-Lopez, S., Perez-Calabuig, A.M., Cancilla, J.C., Garcia-Rodriguez, Y., Torrecilla, J.S., Convolutional capture of the expansion of extra virgin olive oil droplets to quantify adulteration. *Food Chem.*, 368, 130765, 2022, https://doi.org/https://doi.org/10.1016/j.foodchem.2021.130765.
32. Pérez-Calabuig, A.M., Pradana-López, S., Ramayo-Muñoz, A., Cancilla, J.C., Torrecilla, J.S., Deep quantification of a refined adulterant blended into pure avocado oil. *Food Chem.*, 404, 134474, 2023, https://doi.org/https://doi.org/10.1016/j.foodchem.2022.134474.
33. Liu, Y., Yao, L., Xia, Z., Gao, Y., Gong, Z., Geographical discrimination and adulteration analysis for edible oils using two-dimensional correlation spectroscopy and convolutional neural networks (CNNs). *Spectrochim. Acta, Part A*, 246, 118973, 2021, https://doi.org/https://doi.org/10.1016/j.saa.2020.118973.
34. Soltani Firouz, M., Rashvand, M., Omid, M., Rapid identification and quantification of sesame oils adulteration using low frequency dielectric spectroscopy combined with chemometrics. *LWT*, 140, 110736, 2021, https://doi.org/https://doi.org/10.1016/j.lwt.2020.110736.
35. Yuan, Y.-Y., Wang, S.-T., Wang, J.-Z., Cheng, Q., Wu, X.-J., Kong, D.-M., Rapid detection of the authenticity and adulteration of sesame oil using excitation-emission matrix fluorescence and chemometric methods. *Food Control*, 112, 107145, 2020, https://doi.org/https://doi.org/10.1016/j.foodcont.2020.107145.
36. Yang, R., Dong, G., Sun, X., Yang, Y., Liu, H., Du, Y., Jin, H., Zhang, W., Discrimination of sesame oil adulterated with corn oil using information fusion of synchronous and asynchronous two-dimensional near-mid infrared spectroscopy. European. *J. Lipid Sci. Technol.*, 119, 9, 2017, https://doi.org/10.1002/ejlt.201600459.
37. Lastra-Mejias, M., Izquierdo, M., Torreblanca-Zanca, A., Aroca-Santos, R., Cancilla, J.C., Sepulveda-Diaz, J.E., Torrecilla, J.S., Cognitive fluorescence sensing to monitor the storage conditions and locate adulterations of extra virgin olive oil. *Food Control*, 103, 48–58, 2019, https://doi.org/https://doi.org/10.1016/j.foodcont.2019.03.033.

38. Torreblanca-Zanca, A., Aroca-Santos, R., Lastra-Mejías, M., Izquierdo, M., Cancilla, J.C., Torrecilla, J.S., Laser diode induced excitation of PDO extra virgin olive oils for cognitive authentication and fraud detection. *Sens. Actuators, B*, 280, 1–9, 2019, https://doi.org/https://doi.org/10.1016/j.snb.2018.10.014.
39. Mu, T., Chen, S., Zhang, Y., Chen, H., Guo, P., Meng, F., Portable Detection and Quantification of Olive Oil Adulteration by 473-nm Laser-Induced Fluorescence. *Food Anal. Methods*, 9, 1, 275–279, 2016, https://doi.org/10.1007/s12161-015-0199-2.
40. Teng, Y., Chen, Y., Chen, X., Zuo, S., Li, X., Pan, Z., Shao, K., Du, J., Li, Z., Revealing the adulteration of sesame oil products by portable Raman spectrometer and 1D CNN vector regression: A comparative study with chemometrics and colorimetry. *Food Chem.*, 436, 137694, 2024, https://doi.org/https://doi.org/10.1016/j.foodchem.2023.137694.
41. da Silva Medeiros, M.L., Brasil, Y.L., Cruz-Tirado, L.J.P., Lima, A.F., Godoy, H.T., Barbin, D.F., Portable NIR spectrometer and chemometric tools for predicting quality attributes and adulteration levels in butteroil. *Food Control*, 144, 109349, 2023, https://doi.org/https://doi.org/10.1016/j.foodcont.2022.109349.
42. Windarsih, A., Jatmiko, T.H., Anggraeni, A.S., & Rahmawati, L. Machine learning-assisted FT-IR spectroscopy for identification of pork oil adulteration in tuna fish oil. *Vib. Spectrosc.*, 134, 103715, 2024, https://doi.org/https://doi.org/10.1016/j.vibspec.2024.103715.
43. Zheng, L., Bao, Q., Weng, S., Tao, J., Zhang, D., Huang, L., Zhao, J., Determination of adulteration in wheat flour using multi-grained cascade forest-related models coupled with the fusion information of hyperspectral imaging. *Spectrochim. Acta, Part A*, 270, 120813, 2022, https://doi.org/https://doi.org/10.1016/j.saa.2021.120813.
44. Pradana-López, S., Pérez-Calabuig, A.M., Rodrigo, C., Lozano, M.A., Cancilla, J.C., Torrecilla, J.S., Low requirement imaging enables sensitive and robust rice adulteration quantification *via* transfer learning. *Food Control*, 127, 108122, 2021, https://doi.org/https://doi.org/10.1016/j.foodcont.2021.108122.
45. Shao, Y., Xuan, G., Hu, Z., Gao, X., Identification of adulterated cooked millet flour with Hyperspectral Imaging Analysis. *IFAC-PapersOnLine*, 51, 17, 96–101, 2018, https://doi.org/https://doi.org/10.1016/j.ifacol.2018.08.068.
46. Song, J., Yu, Y., Wang, R., Chen, M., Li, Z., He, X., Ren, Z., Dong, H., The identification of aged-rice adulteration by support vector machine classification combined with characteristic wavelength variables. *Microchem. J.*, 199, 110032, 2024, https://doi.org/https://doi.org/10.1016/j.microc.2024.110032.
47. Zhang, X., Sun, J., Li, P., Zeng, F., Wang, H., Hyperspectral detection of salted sea cucumber adulteration using different spectral preprocessing techniques and SVM method. *LWT*, 152, 112295, 2021, https://doi.org/https://doi.org/10.1016/j.lwt.2021.112295.

48. Kelis Cardoso, V.G. and Poppi, R.J., Cleaner and faster method to detect adulteration in cassava starch using Raman spectroscopy and one-class support vector machine. *Food Control*, 125, 107917, 2021, https://doi.org/https://doi.org/10.1016/j.foodcont.2021.107917.
49. Yang, X., Song, J., Peng, L., Gao, L., Liu, X., Xie, L., Li, G., Improving identification ability of adulterants in powdered Panax notoginseng using particle swarm optimization and data fusion. *Infrared Phys. Technol.*, 103, 103101, 2019, https://doi.org/https://doi.org/10.1016/j.infrared.2019.103101.
50. Çinarer, G., Doğan, N., Kılıç, K., Doğan, C., Rapid detection of adulteration in pistachio based on deep learning methodologies and affordable system. *Multimed. Tools Appl.*, 83, 5, 14797–14820, 2024, https://doi.org/10.1007/s11042-023-16172-5.
51. Sitorus, A. and Lapcharoensuk, R., Development of automatic tuning for combined preprocessing and hyperparameters of machine learning and its application to NIR spectral data of coconut milk adulteration. *Food Chem.*, 457, 140108, 2024, https://doi.org/https://doi.org/10.1016/j.foodchem.2024.140108.
52. Hossein Nargesi, M., Amiriparian, J., Bagherpour, H., Kheiralipour, K., Detection of different adulteration in cinnamon powder using hyperspectral imaging and artificial neural network method. *Results Chem.*, 101644, 2024, https://doi.org/https://doi.org/10.1016/j.rechem.2024.101644.
53. Teklemariam, T.A., Raman and Mid-Infrared Spectroscopy Coupled With Machine–Deep Learning for Adulterant Detection in Ground Turmeric. *Appl. Spectrosc. Pract.*, 2, 2, 27551857241250016, 2024, https://doi.org/10.1177/27551857241250014.
54. Mamgain, A., Kumar, V., Dash, S., Image-Based Detection of Adulterants in Milk Using Convolutional Neural Network. *ACS Omega*, 9, 25, 27158–27168, 2024, https://doi.org/10.1021/acsomega.4c01274.
55. Parsain, T., Tripathi, A., Tiwari, A., Detection of milk adulteration using coffee ring effect and convolutional neural network. *Food Addit. Contaminants Part A*, 41, 7, 730–741, 2024, https://doi.org/10.1080/19440049.2024.2358518.
56. Rao Gundavarapu, M., Bhavita, M., Sahithi, M., Varsha, N., Kumar, R., Lakshmi Prasanna, Y., Rangaraju, G., IoT-Powered Intelligent Framework for Detecting Food Adulteration: A Smart Approach. *E3S Web Conferences*, 430, 01074, 2023, https://doi.org/10.1051/e3sconf/202343001074.

14

Federated Learning for Crop Yield Prediction

Gangadhara Doggalli[1]*, Santhoshini E.[2], Sujitha R.[3], Vishwas Gowda G.R.[2], Kavya[4], N.S. Gouthami[5] and Oinam Bobochand Singh[6]

[1]*Department of Genetics and Plant Breeding, University of Agricultural Sciences, Dharwad, India*
[2]*Department of Plant Pathology, University of Agricultural Sciences, Dharwad, India*
[3]*Department of Genetics and Plant Breeding, Tamil Nadu Agricultural University, Coimbatore, India*
[4]*ICAR-IGFRI, SRRS, University of Agricultural Sciences, Dharwad, India*
[5]*Dept. of Food Technology, University of Agricultural Sciences, Dharwad, India*
[6]*Division of Genetics, Indian Agricultural Research Institute, New Delhi, India*

Abstract

Federated learning (FL) represents a novel approach for training machine learning models using data and computational power on edge devices, thereby preserving the privacy of raw local data. This chapter reveals the FL applications in agriculture, focusing on various challenges addressed through FL. We analyze different types of data partitioning (horizontal, vertical, and hybrid), FL architectures (centralized and decentralized), levels of federation (cross-device and cross-silo), and the use of aggregation algorithms. Additionally, we explore how various methods address communication challenges in FL. Traditional FL approaches have two main drawbacks: the diversity of data on different silos can cause uneven performance across clients, and the high communication costs and large model sizes hinder efficiency. To overcome these issues, we introduce a technique that applies network pruning to client models and aggregates the pruned models. This strategy customizes local models to their specific data distributions, addressing data heterogeneity and creating more compact models that use less data during transmission. For example, a soybean yield forecasting dataset reveal

**Corresponding author*: gangadhar0195@gmail.com

Padmesh Tripathi, Bhanumati Panda, Shanthi Makka, Reeta Mishra, S. Balamurugan and Sheng-Lung Peng (eds.) Federated Learning for Smart Agriculture and Food Quality Enhancement, (347–382) © 2026 Scrivener Publishing LLC

that this method enhances inference performance by 15.5% to 20% compared to federated averaging, decreases local model sizes by up to 84%, and reduces the data volume transmitted between clients and the server by 57.1% to 64.7%. These findings demonstrate the potential of this approach to develop efficient, environmentally friendly models that can support the agri-food sector's move toward sustainability. Future advancements could further refine distributed learning in agriculture, improving both sustainability and practical application.

Keywords: Federated learning, yield prediction, machine learning, smart agriculture

14.1 Introduction

In agriculture, precision and efficiency are crucial for food security and sustainability [1]. Accurate crop yield prediction optimizes practices, resources, and market demands [2]. Traditional models faced challenges like data privacy and scalability [3], whereas federated learning (FL) enables decentralized, privacy-preserving model training [4]. FL allows independent clients, such as farms, to collaboratively train a shared model without sharing raw data, addressing privacy concerns [4, 5].

Crop yield prediction is complex, influenced by soil, climate, practices, pests, and genetics [6]. Centralized models struggle with localized variables [7], but FL integrates diverse datasets, improving accuracy [8]. By using decentralized data from farms, FL enhances models while preserving privacy [3], fostering collaboration among stakeholders to improve agricultural strategies [9].

FL also aligns with sustainable agriculture by reducing data transfer and storage costs [10], supporting precision agriculture, and enhancing decision-making [11]. However, challenges like data heterogeneity and model convergence must be addressed for reliable implementation [4, 12]. FL represents a significant advancement in artificial intelligence (AI), enabling privacy-preserving learning from decentralized data.

14.2 Introduction to Federated Learning

FL is a branch of machine learning (ML) that emphasizes collaborative training of a model by multiple entities (clients) while maintaining decentralized data storage [4]. This approach allows each client to train a local model using its own data and only share the model updates, such as gradients or parameters, with a central server [5]. The server aggregates these updates to improve the global model, which is then sent back to the clients for further training [13]. This method preserves data privacy, reduces the risk of data breaches, and accommodates data diversity across different clients [3].

14.2.1 Overview of Federated Learning

FL addresses data privacy concerns in agricultural ML by enabling collaborative model development without sharing raw data [8]. In FL, ML algorithms, such as deep neural networks, are trained locally on datasets held by individual nodes, and only model parameters—like weights and biases—are periodically exchanged with a central server [4]. The server aggregates these updates to build a shared global model, which is then redistributed for further local training [11]. This approach preserves data privacy while ensuring robust global model development (Figure 14.1) [1].

FL also allows local customization of the global model to fit individual user preferences. Periodic updates from users are aggregated to refine the shared model, ensuring hyper-personalization, low latency, and privacy preservation [10]. In contrast to traditional ML, where data is centralized for model training, FL distributes the learning process across edge devices, reducing bandwidth requirements and enhancing data security [14]. Open-source frameworks now support the implementation of FL for predictive modeling on decentralized data sources [10]. The features of FL reflect (Figure 14.2) its innovative design and its adaptability to modern challenges in data-driven industries. These features address concerns such

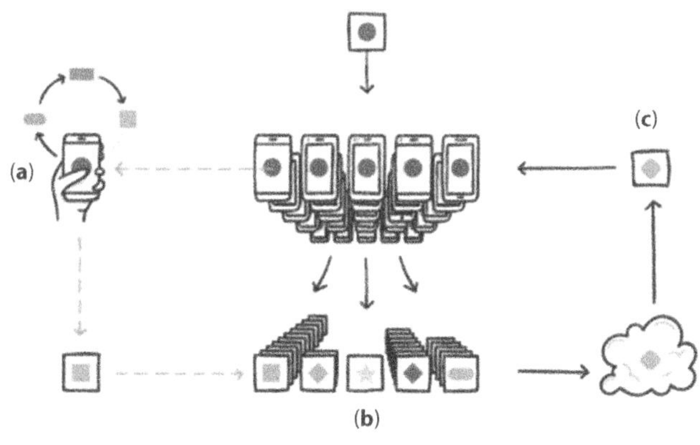

Figure 14.1 Federated learning.

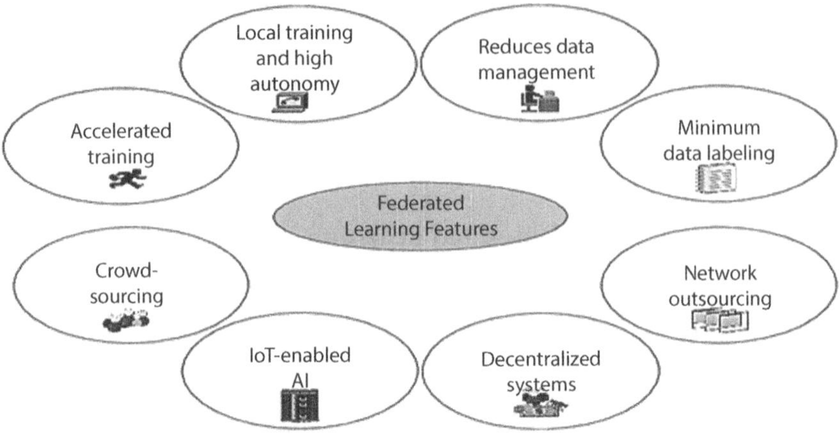

Figure 14.2 Features of federated learning.

as data privacy, resource efficiency, and adaptability to heterogeneous environments, making FL a transformative approach in fields such as healthcare, finance, agriculture, and more.

14.2.2 Types of Federated Learning

14.2.2.1 Horizontal Federated Learning

A shared model is trained across devices holding similar features but different data records. Data is partitioned by samples, and each device contributes without sharing raw data.

14.2.2.2 Vertical Federated Learning

The model is trained using complementary features distributed across devices. Data is partitioned by features, with each device holding a subset to build a comprehensive model.

14.2.2.3 Federated Transfer Learning

When local data is limited, this method fine-tunes pre-trained models with decentralized data, reducing data requirements while leveraging prior knowledge.

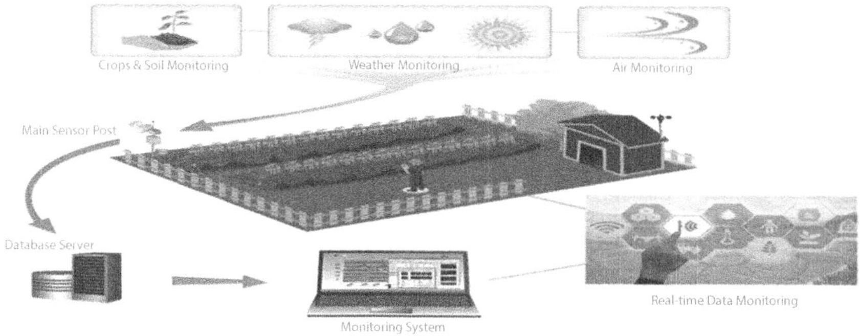

Figure 14.3 Federated learning in agriculture.

14.2.3 Relevance of Federated Learning to Agriculture

FL enhances smart agriculture by creating ML models tailored to local conditions, supporting tasks such as crop yield prediction, disease detection, resource optimization, and livestock management. By leveraging data from diverse geographical locations, FL provides localized recommendations for specific agricultural scenarios. Integrating FL with Internet of Things (IoT) and cloud computing facilitates data collection and analysis from smart sensors, enabling collaborative model training across devices for tasks like crop classification, disease detection, and production forecasting [15, 16].

In Edge-IoT systems, FL efficiently handles vertically and horizontally partitioned data, optimizing energy use and communication resources for effective model training, which improves productivity, accuracy, and adaptability in smart agriculture [16, 17] (Figure 14.3).

14.2.4 Comparative Analysis with Traditional Models

Traditional crop yield prediction models rely on historical data and statistical methods, whereas FL offers a decentralized approach, potentially improving accuracy, data privacy, and efficiency compared to centralized methods (Table 14.1).

Table 14.1 Differences between traditional models and federated learning.

Aspect	Traditional models	Federated learning
Data Handling and Privacy	Requires centralization of large datasets, raising privacy concerns	Data remains decentralized; only model updates are shared, preserving privacy
Model Accuracy and Adaptability	Limited by the uniformity of centralized data, potentially less accurate in diverse contexts	Benefits from diverse local data, leading to more accurate and adaptable predictions
Scalability and Resource Efficiency	Scaling challenges due to the need for significant computational resources and infrastructure	Scales efficiently as data processing is distributed, reducing the load on central systems
Computational Demands	High computational requirements for processing large datasets	Distributes computational load across participants, making it more resource-efficient
Implementation Complexity	Relatively straightforward if data infrastructure is in place	Introduces complexity in coordinating model updates from multiple sources

14.2.5 Key Advantages of Federated Learning in Agriculture

➤ **Data Utilization from Diverse Sources:** Agriculture generates vast data from sensors, drones, satellite imagery, and IoT devices, often distributed across farms, research institutions, and agribusinesses. FL enables model training on this decentralized data without requiring centralization, leveraging diverse datasets for more effective predictive models [18].

- **Privacy Protection:** FL addresses privacy concerns by enabling model training on local data without transferring sensitive information like crop yields or soil health to central servers. This ensures data security, encouraging data sharing among farmers and organizations [7].
- **Enhanced Predictive Accuracy:** By integrating data from multiple sources, FL creates robust models for applications such as crop yield prediction and disease detection. Combining insights from varied regions improves model generalizability and effectiveness [18].
- **Stakeholder Collaboration:** FL promotes collaboration among farmers, researchers, agronomists, and policymakers by enabling joint model development without pooling data, fostering innovation and advancements in agricultural practices [19].
- **Scalability and Efficiency:** FL utilizes local computational resources, reducing dependency on central servers. This decentralized approach is particularly useful in rural areas, enabling efficient and scalable model training [20].
- **Adaptability to Local Conditions:** Models trained locally through FL are tailored to specific regional conditions like climate, soil type, and farming practices, ensuring their relevance and utility across diverse agricultural environments [17].
- **Contribution to Sustainable Agriculture:** FL enhances precision agriculture by improving predictions for resource use, disease detection, and pest management. This optimization minimizes waste, reduces environmental impact, and promotes sustainable farming practices [21].

14.2.6 Key Concepts and Terminology

Understanding the core concepts and terminology of FL is crucial for appreciating its applications in agriculture [22] (Table 14.2).

Table 14.2 The core components of federated learning and their definitions.

Concept	Definition	Why it matters
Federated Learning	A decentralized approach where multiple participants train a model locally with their data and share updates with a central server	Preserves data privacy and security, particularly useful in fields like agriculture where data is spread across various locations
Decentralized Data	Data distributed across multiple sites, where each data source contributes to model training without sharing actual data	Reduces risks of data breaches and complies with privacy regulations, crucial for sensitive agricultural data
Local Models	Machine learning models trained on data from specific participants	Captures unique data characteristics, improving model generalization, especially in agriculture where conditions vary widely
Global Model	The central model improved through aggregation of local model updates	Benefits from diverse data, leading to more accurate predictions for agricultural decisions like crop yield forecasts
Model Aggregation	The process of combining local model updates to form a unified global model, typically by averaging model parameters	Ensures the global model incorporates insights from all participants, improving performance and balancing influences from different data sources

(*Continued*)

Table 14.2 The core components of federated learning and their definitions. (*Continued*)

Concept	Definition	Why it matters
Privacy Preservation	Methods to protect sensitive data during federated learning, ensuring data remains secure	Ensures agricultural data remains private and complies with data protection regulations, maintaining trust among participants
Federated Averaging	A technique where local model weights are averaged to create the global model	Helps form a global model that reflects contributions from all participants, leading to more comprehensive and accurate predictions
Data Heterogeneity	Variations in data quality, format, and distribution across different participants	Essential for effective learning from diverse data sources in agriculture, considering differences in soil, weather, and farming practices

14.2.7 Benefits and Limitations of Federated Learning

FL presents several advantages, especially in domains like agriculture, where data privacy and collaborative insights are essential. However, it also comes with certain limitations that need to be addressed.

14.2.7.1 Benefits of Federated Learning

- **Enhanced Data Privacy:** FL keeps data on its original device or location, sharing only model updates rather than raw data with a central server. This approach significantly reduces the risk of data breaches and aligns with privacy regulations [5]. In agriculture, sensitive details about crop yields or farming practices remain confidential even when insights are shared across different farms [4].

- **Improved Collaboration:** FL facilitates collaboration among multiple stakeholders without requiring them to share their data. This collective effort aggregates diverse insights, enhancing the model's overall accuracy [8]. Farmers from various regions can contribute to a unified model that learns from different soil types and weather conditions, leading to more accurate predictions [23].
- **Reduced Data Transfer Costs:** As FL does not necessitate transferring large volumes of raw data, it cuts down on bandwidth and storage costs related to data sharing [24]. Agricultural research institutions can work together without incurring high costs associated with extensive data transfers [1].
- **Scalability:** FL can handle a growing number of participants without diminishing performance. This scalability makes it suitable for scenarios where data is distributed across multiple locations [4]. A federated model can integrate data from thousands of farms globally, each providing valuable local insights [7].
- **Adaptability:** The model can be continuously refined and improved by incorporating updates from new data as it becomes available. As new farming techniques or environmental conditions emerge, the federated model can update its predictions accordingly [25].

14.2.7.2 Limitations of Federated Learning

- **Data Heterogeneity:** Variations in data quality, format, and distribution among participants can complicate the creation of a unified model, potentially affecting its performance and accuracy. Differences in farming practices or soil conditions may lead to inconsistencies in data, making it challenging for the model to generalize effectively [26].
- **Computational and Storage Constraints:** Participants need sufficient computational power and storage to train local models and handle updates. Not all participants may have the necessary resources [8]. Small-scale farmers or institutions with limited infrastructure might find it difficult to contribute effectively to the FL process [19].

- **Communication Overhead:** Frequent communication between the central server and local devices is required, which can lead to high communication overhead, especially in areas with limited internet connectivity [24]. In rural regions with poor connectivity, transmitting model updates may be slow or unreliable, impacting the efficiency of the FL process [25].
- **Privacy Risks:** Although FL enhances data privacy, risks remain related to potential leakage of sensitive information through model updates. Robust privacy measures are crucial. Aggregated model updates might inadvertently expose sensitive details about individual farms or practices, even with FL.
- **Complex Implementation:** Establishing and maintaining a FL system involves complex coordination among all participants. This complexity presents challenges both technically and organizationally [5]. Implementing FL in a large-scale agricultural project requires alignment among stakeholders, management of data contributions, and smooth operation of the FL system [4].

14.3 Accurate Crop Yield Prediction with Federated Learning

Effective prediction of crop yields is crucial for ensuring food security, optimizing resource management, and supporting economic planning in agriculture [27]. FL offers a cutting-edge approach to ML that can significantly improve the accuracy of crop yield predictions [5]. By harnessing distributed data across various sources while preserving data privacy, FL enables the development of precise predictive models [26]. This method allows for the integration of diverse datasets without centralized storage, thus safeguarding sensitive information and delivering more reliable and comprehensive yield forecasts [4].

14.3.1 Importance of Accurate Crop Yield Prediction

Accurate crop yield prediction is a cornerstone of modern agriculture, playing a critical role in ensuring food security, optimizing resource utilization, and supporting sustainable farming practices (Figure 14.4). As global

Figure 14.4 Importance of accurate crop yield prediction.

population growth increases the demand for food, precise yield predictions help farmers and policymakers make informed decisions regarding crop planning, irrigation, fertilization, and pest management. By providing actionable insights, such predictions minimize waste, reduce costs, and enhance productivity (Table 14.3).

- **Enhancing Food Security:** FL leverages decentralized agricultural data to improve crop yield forecasts, enabling accurate predictions of food availability. This helps prevent shortages and manage surplus effectively, contributing to global food security [27].
- **Optimizing Resource Allocation:** By integrating data from diverse farms, FL aids in creating precise models for resource use. Farmers can optimize water, fertilizer, and pesticide application, promoting sustainable agriculture and reducing costs [5].
- **Economic Planning and Market Stability:** Accurate yield forecasts enhance market stability by supporting better planning for storage, distribution, and pricing. This reduces supply volatility and aids agribusinesses and policymakers in economic planning [4].
- **Risk Management:** FL improves risk management by using diverse data to predict weather, pests, and diseases with greater accuracy. Farmers can take proactive measures, improving crop management, financial planning, and reducing overall risks [26].

- **Policy Formulation and Support Programs:** FL provides comprehensive insights while preserving data privacy, aiding governments and organizations in designing effective agricultural policies and support programs. Accurate predictions inform decisions on subsidies, insurance, and disaster relief [27].

Table 14.3 Applications of federated learning in agriculture.

Application	Description	Benefits	Example
Crop Yield Prediction	Utilizing federated learning models to predict crop yields based on diverse data sources	Improved accuracy in yield forecasts, better planning for farmers	Developing models that analyze weather, soil, and crop data from multiple farms to predict harvest outcomes
Pest and Disease Management	Using federated learning to detect and manage pests and diseases across different regions	Early detection and control, reduced crop damage	Creating a system that identifies pest infestations from images and shares updates across farms without revealing specific locations
Soil Health Monitoring	Implementing federated learning to assess soil conditions and recommend improvements	Enhanced soil management, optimized use of fertilizers	Developing models that analyze soil data from different farms to provide tailored recommendations for soil treatment
Precision Farming	Applying federated learning to optimize planting, irrigation, and harvesting schedules	Increased efficiency, resource conservation	Using IoT sensors to collect data on soil moisture and plant health, and federated learning models to optimize irrigation schedules

(Continued)

Table 14.3 Applications of federated learning in agriculture. (*Continued*)

Application	Description	Benefits	Example
Weather Forecasting and Analysis	Integrating federated learning with weather data to provide precise forecasts for farmers	Better preparation for weather changes, reduced crop loss	Developing models that combine local weather data from multiple sources to provide accurate forecasts tailored to specific regions
Supply Chain Optimization	Using federated learning to streamline the agricultural supply chain, from production to distribution	Reduced waste, improved logistics, and market predictions	Creating models that predict demand and optimize distribution routes based on data from various suppliers and markets
Sustainable Farming Practices	Encouraging sustainable practices through federated learning models that promote environmental health	Long-term soil and crop health, reduced environmental impact	Developing models that suggest crop rotation and cover cropping strategies to maintain soil fertility and reduce erosion
Resource Management	Optimizing the use of water, fertilizers, and pesticides through federated learning	Cost savings, reduced environmental footprint	Implementing models that recommend optimal resource use based on data from multiple farms to ensure sustainable agriculture practices

14.3.2 Current Methodologies in Crop Yield Prediction Enhanced by Federated Learning

14.3.2.1 Traditional Methods

- **Trend Analysis and Regression Models:** Traditional crop yield prediction methods, such as trend analysis and regression models, can be significantly enhanced by FL. FL allows for the integration of trend and regression models from multiple decentralized datasets, improving the robustness and generalizability of predictions [26]. By pooling insights from diverse sources, FL can produce more accurate models that account for variations across different agricultural settings.
- **Expert Judgment:** FL can also refine crop yield predictions by incorporating expert knowledge from various regions. This approach aggregates expert inputs, resulting in more nuanced and region-specific predictions. The ability to combine knowledge from different experts without centralizing data enhances the model's ability to adapt to local conditions [29].

Traditional ML approaches to crop yield prediction often rely on centralized models that aggregate data from various regions into a single repository. While effective in some cases, these models face challenges such as limited generalizability, data privacy concerns, and high computational costs. Differences between traditional and FL approach in crop yield predictions are depicted in Table 14.4.

Table 14.4 Differences between traditional and FL approach in crop yield predictions.

Aspect	Traditional approaches	Federated learning approach
Integrating data from multiple sources		
Diverse Data Integration	Centralized systems face challenges with aggregating data from diverse sources, such as compatibility, privacy, and volume issues [14]	Keeps data localized while integrating insights from satellite imagery, IoT sensors, and other sources, creating comprehensive models without physical data aggregation [4]

(*Continued*)

Table 14.4 Differences between traditional and FL approach in crop yield predictions. (*Continued*)

Aspect	Traditional approaches	Federated learning approach
Enhanced Accuracy	Struggles to incorporate diverse data effectively, leading to biases and incomplete predictions [28]	Aggregates updates from diverse datasets, accounting for regional and environmental variations, improving model robustness and applicability [30]
Empowering farmers and local communities		
Direct Farmer Participation	Limited involvement of farmers in model development results in less relevant models for local practices [31]	Farmers contribute local data, tailoring models to regional conditions and improving their utility and relevance [8]
Localized Insights	Centralized models may miss local agricultural nuances, reducing effectiveness of recommendations [33]	Trains on localized data, ensuring insights are specific to regional conditions, offering actionable and tailored advice [30]
Accelerating innovation and research		
Broader Data Access	Logistical and ethical complexities of centralizing data limit access, slowing research and innovation [28]	Allows access to diverse datasets without central aggregation, adhering to privacy standards, and accelerating research in agricultural technologies [7]
Innovation Acceleration	Centralized systems restrict dataset diversity and impose high data management costs, hindering innovation [14]	Facilitates collaborative research, enabling shared data insights while maintaining privacy, fostering rapid development of technologies like climate-resilient crops [30]

14.3.2.2 *Remote Sensing and Geospatial Data*

- **Satellite Imagery and Drone Surveillance:** FL improves methodologies involving satellite imagery and drone surveillance by integrating data from different sources. Combining satellite and drone data through FL enhances the accuracy and applicability of predictions. This approach allows for a more comprehensive analysis of agricultural conditions, leveraging high-resolution imagery and real-time data [33].
- **Geospatial Analysis:** FL enhances geospatial analysis by integrating data from various sources while preserving privacy. This integration results in more precise predictions by accounting for spatial variability across different regions. FL enables a detailed understanding of how spatial factors influence crop yields, leading to better-informed agricultural practices [14].

14.3.2.3 *Machine Learning and Artificial Intelligence*

- **Supervised and Unsupervised Learning:** In the realm of ML, FL amplifies the capabilities of both supervised and unsupervised learning methods. By training models on diverse, distributed datasets, FL improves the accuracy and reliability of predictions. This decentralized approach ensures that models are exposed to a wide range of data patterns, enhancing their performance [14].
- **Deep Learning:** FL leverages deep learning models by training them on data from multiple locations. This approach captures complex patterns and dependencies more effectively than traditional methods. By utilizing data from diverse sources, FL enables deep learning models to develop a more comprehensive understanding of the factors influencing crop yields [26].

14.3.2.4 *Integrative Approaches*

- **Data Fusion:** FL facilitates data fusion from various sources while maintaining data privacy. This approach leads to comprehensive and robust predictions by integrating information from different datasets. Data fusion through FL enables the creation of models that incorporate diverse data types and sources, providing a holistic view of the factors affecting crop yields [10].

- **Hybrid Models:** FL supports the development of hybrid models that combine traditional methods with advanced ML algorithms. These hybrid models benefit from the strengths of both approaches, resulting in improved accuracy. By integrating classical prediction techniques with modern ML methods, FL creates models that are both reliable and adaptable [28].

14.3.3 Federated Learning in Enhancing Crop Yield Prediction

FL represents a transformative approach in the realm of crop yield prediction by leveraging distributed datasets while upholding stringent data privacy standards. This section delves into the ways FL can be harnessed to refine crop yield predictions, emphasizing its benefits in data privacy and collaborative efforts among various agricultural stakeholders (Table 14.5).

Table 14.5 Enhancing crop yield prediction through federated learning.

Aspect	Description	Benefits	Example
Data Collection	Gathering data from multiple farms, including soil conditions, weather patterns, crop health, and historical yield data	Comprehensive data set, diverse environmental conditions, and farming practices	Collecting soil moisture and nutrient levels from various farms to improve model accuracy
Model Training	Federated learning models are trained using decentralized data, ensuring privacy and security	Improved model accuracy without sharing raw data, preserving farmer privacy	Training a model on local weather data and crop health images without centralizing data

(Continued)

Table 14.5 Enhancing crop yield prediction through federated learning. (*Continued*)

Aspect	Description	Benefits	Example
Yield Prediction	Predicting crop yields by analyzing diverse data sets through federated learning algorithms	Accurate yield predictions tailored to specific regions and conditions	Using models to predict wheat yields based on local soil and weather data from different regions
Privacy and Security	Ensuring that sensitive farm data remains private and secure during the federated learning process	Protecting farmer data, building trust among participants	Implementing secure multi-party computation techniques to protect data privacy
Collaboration	Facilitating collaboration between different farms and agricultural stakeholders	Shared insights and knowledge, improved agricultural practices	Multiple farms contributing data to a federated learning model to improve regional yield predictions
Scalability	Applying federated learning to various scales, from small farms to large agricultural enterprises	Flexible application, suitable for different farm sizes and types	Using federated learning in a cooperative of small farms to predict yields and optimize resources
Adaptability	Adapting models to specific crops, regions, and farming practices	Customized predictions, better decision-making	Developing models for different crops like corn, rice, and soybeans based on regional data

(*Continued*)

Table 14.5 Enhancing crop yield prediction through federated learning. (*Continued*)

Aspect	Description	Benefits	Example
Continuous Improvement	Updating models continuously with new data from farms to improve prediction accuracy	Dynamic models that evolve with changing conditions	Continuously integrating new weather data to refine yield predictions throughout the growing season
Resource Optimization	Optimizing resource use (water, fertilizers, and pesticides) based on yield predictions	Cost savings, reduced environmental impact	Using yield predictions to plan efficient irrigation schedules and fertilizer applications
Decision Support	Providing actionable insights and recommendations to farmers based on yield predictions	Enhanced farm management, increased productivity	Offering recommendations on planting dates and crop varieties based on predicted yields and local conditions

14.3.3.1 Data Privacy and Security

14.3.3.1.1 Decentralized Data Processing

- **Traditional Approach:** Traditional crop yield prediction methods rely on centralized data collection, aggregating data from various sources into one repository. This approach poses privacy and security risks, as sensitive agricultural data may be vulnerable to unauthorized access or breaches [10].
- **Federated Learning Enhancement:** FL addresses privacy concerns by training ML models on decentralized data sources. It keeps raw data local, sharing only model updates with a central server to refine a global model, thus enhancing privacy and minimizing data breach risks [4].

14.3.3.1.2 Data Sovereignty

- **Traditional Approach:** Centralized data collection often involves transferring data across borders or between organizations, potentially violating data sovereignty regulations and creating legal and compliance challenges [34].
- **Federated Learning Enhancement:** FL maintains data sovereignty by keeping data in its original location, aligning with regulatory requirements and enabling the use of diverse datasets without breaching legal or privacy standards. This ensures data owners retain control while contributing to collective learning [8].

14.3.3.2 Collaboration Among Agricultural Stakeholders

FL enables transformative collaboration among various stakeholders in the agricultural sector, facilitating shared benefits and advancements in predictive modeling. This collaborative approach leverages decentralized data to enhance agricultural practices and research.

14.3.3.2.1 Collective Knowledge Integration

- **Traditional Approach:** Collaborative efforts in agriculture often involve sharing data and insights through centralized platforms, which may be limited by data access restrictions or privacy concerns [19].
- **Federated Learning Enhancement:** FL fosters collaboration among stakeholders, including farmers, researchers, and agribusinesses, by aggregating model updates rather than raw data. This collaborative approach allows stakeholders to contribute their localized data and expertise without exposing sensitive information. The resulting global model benefits from a diverse range of inputs, improving its accuracy and applicability across different agricultural contexts [29].

14.3.3.2.2 Improved Model Accuracy

- **Traditional Approach:** Centralized models might suffer from limited data diversity, leading to less accurate predictions for specific regions or crops [16].

- **Federated Learning Enhancement:** FL integrates data from diverse locations, enabling the global model to account for various agricultural conditions and practices. This enhances prediction accuracy by addressing local differences in soil, climate, and farming methods, resulting in more actionable insights [14].

14.3.3.3 Resource Optimization

- **Traditional Approach:** Centralized data processing and model training can be resource-intensive, requiring significant computational power and data storage [35].
- **Federated Learning Enhancement:** FL minimizes large-scale data transfers and centralized storage by processing data locally and sharing only model updates with a central server. This approach optimizes resource usage, reduces computational demands, and lowers the environmental impact of data management [28].

14.3.4 Utilization of Federated Learning to Improve Crop Yield Predictions

FL revolutionizes crop yield prediction by utilizing decentralized data while ensuring privacy and efficiency. This method is particularly advantageous in agriculture, where data is dispersed among farms, research institutions, and organizations.

14.3.4.1 Enhancing Model Accuracy with Diverse Data: Integration of Multiple Data Sources

Traditional centralized crop yield models aggregate data into a single repository, which limits their ability to generalize across diverse agricultural contexts, as models trained on region-specific data often fail when applied to differing climates or soil types [27]. FL addresses this limitation by enabling collaborative model training using local data without sharing raw data, allowing diverse datasets to be integrated through model updates. This approach enhances prediction accuracy and adaptability across varied agricultural environments [32].

14.3.4.2 *Improved Precision*

Traditional centralized models often lack diversity and fail to account for local variations such as differences in soil, climate, and agricultural practices [14]. FL addresses this issue by integrating diverse datasets, thereby improving region-specific predictions and effectively capturing these variations.

14.3.4.3 *Real-Time Adaptation: Continuous Learning*

Traditional centralized models face delays in retraining, making them slow to adapt to sudden changes such as weather fluctuations or pest outbreaks [28]. FL overcomes this limitation by allowing continuous updates from local nodes, enabling real-time adaptation to dynamic agricultural conditions [4, 28].

14.3.4.4 *Timely Updates*

Traditional centralized models experience delays in data collection, aggregation, and processing, which reduces the effectiveness of predictions. FL addresses this issue by enabling frequent updates from local models to the global model, ensuring timely and accurate predictions based on the most recent data [5, 8].

14.3.4.5 *Cost-Effective Data Utilization: Reduction of Data Handling Costs*

Traditional centralized learning involves costly data aggregation and processing, making it logistically complex for smaller farms or research institutions with limited resources [32]. FL addresses these challenges by keeping data local and transmitting only model updates, reducing data transfer and infrastructure requirements. This approach makes predictive modeling more accessible to resource-limited entities [5].

14.3.4.6 *Efficient Resource Utilization*

Traditional centralized data processing requires significant computational power and storage at a central server, resulting in high resource consumption and environmental impact [28]. FL mitigates these issues by distributing computation across local nodes, reducing reliance on extensive central server resources and minimizing environmental impact. This localized data processing promotes efficient resource use and enhances sustainability in agricultural data management [8].

14.4 Data Privacy in Federated Learning in Crop Yield Prediction

Data privacy is a significant concern in agriculture, where proprietary details about crop yields, soil conditions, and farming practices can be sensitive. FL addresses these concerns through its decentralized and privacy-preserving framework.

14.4.1 Preserving Confidentiality

14.4.1.1 Local Data Privacy

- **Traditional Approaches:** Centralized ML models aggregate raw data into a central repository, exposing sensitive information and increasing the risk of privacy violations [31]. For example, centralized databases may reveal details about individual farms' practices or crop yields.
- **FL Approach:** FL keeps raw data local and only shares model updates, such as gradients or weights. These updates are anonymized and aggregated at the server, protecting individual data points and ensuring confidentiality. This approach safeguards proprietary information and encourages broader participation in collaborative projects [32].

14.4.1.2 Anonymization Techniques

- **Traditional Approaches:** Centralized systems struggle with anonymizing diverse datasets, risking exposure of sensitive information and making the process complex and costly [28].
- **FL Approach:** FL uses privacy-preserving techniques like differential privacy and secure multi-party computation to protect data and ensure model security [8].

14.4.2 Compliance with Regulations

14.4.2.1 Regulatory Alignment

- **Traditional Approaches:** Data privacy regulations like the general data protection regulation (GDPR) impose strict requirements on data handling, making centralized repositories costly and challenging to manage [31]. For example, GDPR mandates robust data protection and individual rights over personal data.
- **FL Approach:** FL aligns with regulations by minimizing raw data movement, keeping it localized and sharing only model updates. This ensures compliance with data protection laws like GDPR, facilitating international collaborations while adhering to legal standards [5, 32].

14.4.3 Legal and Ethical Considerations

- **Traditional Approaches:** Managing data privacy in traditional systems is complex, requiring compliance with diverse legal and ethical frameworks, which can be error-prone [8, 28].
- **FL Approach:** FL simplifies compliance by decentralizing data, ensuring it does not cross borders. This reduces complexities related to international data protection laws and supports global agricultural collaborations [31].

14.4.4 Building Trust Among Stakeholders

14.4.4.1 Fostering Collaboration

- **Traditional Approaches:** In centralized systems, concerns over data security and privacy can discourage collaboration among stakeholders, such as farmers and research institutions, due to the fear of misuse [14, 32].
- **FL Approach:** FL fosters trust by ensuring data confidentiality, encouraging stakeholders to collaborate. This trust enables the development of more comprehensive and accurate models while preserving individual data privacy [31].

14.4.5 Encouraging Data Sharing

- **Traditional Approaches:** Centralized systems struggle with data sharing due to privacy concerns and the risk of data breaches [4, 28].
- **FL Approach:** FL promotes data sharing by maintaining privacy through its decentralized framework, allowing farmers and organizations to contribute data without exposing proprietary information. This enhances predictive models and benefits the agricultural community [14, 31].

14.5 Integration with Existing Agricultural Technologies

Integrating FL with existing agricultural technologies offers transformative potential for improving agricultural practices while addressing data privacy concerns. By combining FL with modern agricultural tools, stakeholders can leverage decentralized data to enhance decision-making, optimize resource use, and foster innovation. Here is an in-depth look at how FL can be effectively integrated with current agricultural systems.

14.5.1 Seamless Integration with Farm Management Systems

Modern farm management systems (FMSs) enhance agricultural operations like crop health monitoring, irrigation, and soil condition tracking, benefiting from FL's predictive capabilities without extensive data sharing. Implementation involves integrating FL with FMS by leveraging data from sensors and drones to train local models, with updates aggregated to improve predictions across farms. Example: Soil moisture data from various farms can train a federated model to optimize irrigation schedules, enhancing water use efficiency and crop yields [5].

14.5.2 Enhancing Precision Agriculture

Precision agriculture leverages FL to optimize farming practices while ensuring data privacy. By integrating data from global positioning system (GPS)-guided tractors and satellite imagery, FL models provide localized recommendations. For example, analyzing regional crop growth patterns helps suggest fertilization strategies tailored to environmental conditions [8].

14.5.3 Augmenting Decision Support Systems

FL enhances decision support systems (DSSs) by improving prediction accuracy through diverse, decentralized datasets. Integrating FL models into DSS allows collective insights without centralizing data. For instance, FL-enabled DSS can aggregate data from multiple farms to forecast pest outbreaks or weather risks, providing farmers with reliable and actionable information [28].

14.5.4 Leveraging IoT and Sensor Networks

FL enhances IoT-based agricultural systems by leveraging sensor data to train decentralized models, improving functionality and prediction accuracy. IoT devices collect data on environmental variables like temperature and soil nutrients, with FL enabling local model training and aggregating insights. For example, FL can use data from various farms to predict crop disease outbreaks, allowing timely interventions and better crop health management [4].

14.5.5 Improving Agricultural Research and Development

FL accelerates agricultural research by enabling collaborative model development across multiple sites while maintaining data privacy. Researchers can aggregate data from diverse trials, enhancing model robustness and innovation. For instance, FL can analyze data from various experiments to identify optimal crop rotation and soil management practices, fostering breakthroughs in agricultural solutions [32].

By integrating FL, agriculture benefits from advanced predictive models, optimized practices, and secure collaboration. Case Studies and Practical Applications: The practical application of FL in agriculture is not just theoretical; it has been tested and implemented in real-world scenarios with promising results.

14.6 Real-World Examples of Federated Learning in Crop Yield Prediction

FL has been gaining traction in the agricultural industry, with several companies and organizations exploring its potential in crop yield prediction. Here are some real-world examples.

14.6.1 FarmWise

FarmWise, an AI-driven farming platform, leverages FL to predict crop yields and offer personalized farming recommendations. FL enables decentralized data processing, where each farm's data remains local, sharing only model updates like gradients and weights with a central server. This ensures data privacy while allowing the model to learn from diverse farming practices and environmental conditions. For example, FarmWise aggregates updates from various farms to provide tailored recommendations, such as optimal planting schedules and resource allocation, enhancing efficiency and maintaining data security.

14.6.2 John Deere

John Deere has developed a FL-based system to enhance crop yield predictions and optimize farming operations. The system integrates data from agricultural machinery, weather stations, and soil sensors, processing vast amounts of data across a network of smart tractors and harvesters. This scalable architecture enables real-time analysis while preserving data privacy. By using FL, John Deere trains models across its devices, aggregating data to improve predictions on crop yield and operational efficiency. The system's integration with IoT sensors ensures timely updates, helping farmers make data-driven decisions on planting, watering, and harvesting.

14.6.3 Cropio

Cropio, a farm management platform, utilizes FL to predict crop yields and detect anomalies in farming operations. By analyzing data from multiple farms, Cropio identifies trends and patterns that may not be evident in individual farm data. The platform uses a decentralized approach, keeping each farm's data locally while sharing only model updates with the central server. This method allows the model to learn from a diverse dataset while ensuring data privacy and security. The approach enhances predictive capabilities and improves the detection of operational issues across farms.

14.6.4 The Climate Corporation

The Climate Corporation, a subsidiary of Bayer, uses FL through its platform, Climate FieldView, to predict crop yields and offer personalized recommendations to farmers. By analyzing data from various sources such as weather stations, soil sensors, and farm equipment, the platform provides insights on planting, irrigation, and harvesting. The system is designed to handle large volumes of data and is scalable, ensuring that data privacy is maintained by processing data locally and sharing only model updates. This enables farmers to make informed decisions while ensuring the security of their data.

14.6.5 Microsoft FarmBeats

Microsoft FarmBeats, an initiative under Microsoft's AI for Earth program, leverages FL to enhance agricultural productivity and sustainability by analyzing data from various sources without centralizing the data itself.

14.7 Policy and Regulatory Considerations

When implementing FL in agriculture or any other sector, navigating the policy and regulatory landscape is crucial. As FL involves the decentralized handling of data and model updates, it intersects with various regulatory frameworks designed to protect data privacy and ensure ethical practices. Here is a closer look at some of the key policy and regulatory considerations:

- **Data Privacy:** FL aligns with GDPR and California Consumer Privacy Act (CCPA) by keeping data decentralized. Organizations must anonymize data, secure consent, and protect model updates.
  Example: Ensuring farm data remains confidential in crop health analyses.
- **Intellectual Property:** Clear agreements should define data ownership and IP rights for model updates, and the final model. *Example:* Clarifying IP for crop yield prediction models.
- **Data Security:** Adhere to standards like ISO/IEC 27001 with encryption for secure data and model updates. *Example:* Encrypting model updates shared between farms and servers.
- **Sectoral Compliance:** FL must meet agriculture-specific regulations, ensuring ethical data sharing and use. *Example:* Following guidelines for agricultural data sharing.
- **Cross-Border Transfers:** Ensure compliance with international data laws (e.g., GDPR) for projects across countries.
  Example: Managing cross-border data for multinational FL collaborations.

14.8 Challenges and Future Directions

FL has the potential to revolutionize agriculture by democratizing access to advanced predictive models and fostering collaborative research. However, it faces challenges that must be addressed to realize its full potential. Continued innovation and collaboration across sectors will be critical to adapting FL to meet the diverse needs of modern agriculture, paving the way for more resilient and efficient farming systems.

14.8.1 Current Challenges in Federated Learning for Agriculture

As FL continues to gain traction in the agricultural sector, several challenges must be addressed to fully harness its potential. The criteria considered for successful implementation of FL in agriculture.

- **Data Heterogeneity:** Agricultural data varies in format, quality, and granularity, complicating model training. Standardizing data formats and protocols is essential for effective FL in agriculture.
- **Limited Computational Resources:** Many farms and agricultural institutions lack the necessary computational power. Ensuring access to adequate hardware and software is crucial for the adoption of FL in agriculture.
- **Communication Overhead:** Frequent communication between the central server and local devices can create overhead, especially in regions with limited internet connectivity. Optimizing communication protocols and reducing update frequency can address this issue.
- **Data Privacy Concerns:** Despite FL's privacy-preserving design, concerns about data leakage *via* model updates persist. Enhancing privacy-preserving methods, such as secure multi-party computing and differential privacy, is crucial for building trust.
- **Integration with Existing Systems:** Integrating FL with current agricultural management systems can be challenging. Developing user-friendly interfaces and ensuring compatibility with existing technologies are necessary for smooth adoption.

14.8.2 Potential Future Developments in Federated Learning

The future of FL in agriculture holds immense potential for innovation and growth. As the technology matures, there are exciting opportunities to enhance its capabilities and address existing limitations. Exploring the following future directions will provide insights into how FL can continue to evolve and make a substantial impact on agricultural practices [36].

- **Advanced Privacy Techniques:** Enhance privacy-preserving methods like homomorphic encryption to improve data security [36].
- **Improved Algorithms for Heterogeneous Data:** Develop algorithms that can effectively process diverse agricultural data.
- **Enhanced Computational Infrastructure:** Invest in affordable cloud and edge computing resources to support FL.
- **Strengthening Collaborative Networks:** Build platforms for collaboration among researchers, farmers, and institutions to drive innovation.
- **Real-Time Data Processing:** Enable real-time data processing for timely and accurate agricultural predictions.

14.8.3 Emerging Trends and Technologies in Federated Learning

FL is an evolving field that is rapidly incorporating new trends and technologies. These innovations are driving the future of FL, particularly in applications like agriculture where privacy and collaboration are paramount. Here is a look at some of the most exciting emerging trends and technologies in FL:

- **Advanced Privacy Techniques:** Secure multi-party computation and differential privacy enhance data security and ensure confidentiality by preventing the identification of individual data points.
- **Federated Transfer Learning:** Combines FL and transfer learning to adapt models across regions, reducing data collection and retraining efforts.
- **Edge Computing Integration:** Local processing *via* edge computing enables real-time analysis and reduces bandwidth usage, especially in remote areas.

- ➢ **Blockchain Integration:** Provides transparency and security by using a tamper-proof ledger to track model updates, fostering trust among participants.
- ➢ **Cross-Silo Federated Learning:** Large organizations collaborate to aggregate datasets, advancing crop management and agricultural knowledge.
- ➢ **Federated Learning with AutoML:** Automates model selection and tuning, allowing practitioners with limited expertise to efficiently deploy high-performing models.

14.9 Conclusion

FL can significantly benefit from cross-disciplinary collaboration, addressing complex challenges and driving innovation in areas like agriculture, privacy, ethics, IoT, and accessibility. By combining data science with agricultural expertise, FL models can enhance crop yield predictions and optimize soil management. Collaboration with cybersecurity experts can strengthen privacy and security through advanced encryption methods, while legal and ethical partnerships ensure compliance with regulations and ethical standards. Integrating FL with IoT enables real-time data processing, improving decision-making for applications such as soil moisture monitoring. Finally, collaboration with human–computer interaction specialists can create user-friendly interfaces, making FL accessible to non-technical users and broadening its adoption across diverse application domains.

AI disclosure statement: The author acknowledges the use of AI-based tools for minor language editing or grammatical correction. However, all intellectual contributions, analyses, and conclusions presented in this chapter are entirely the author's own.

References

1. Smith, J., Brown, E., White, R., The role of precision agriculture in global food security. *Agric. Syst.*, 180, 102748, 2020, https://doi.org/10.1016/j.agsy.2020.102748.
2. Jones, M. and Roberts, T., Optimizing agricultural resources through yield prediction models. *J. Precis. Agric.*, 15, 3, 245–262, 2021, https://doi.org/10.1007/s11119-021-09739-7.

3. Li, T., Zhang, Q., Chen, Y., Challenges in centralized agricultural data processing and solutions *via* federated learning. *Comput. Agric.*, 12, 4, 305–322, 2019, https://doi.org/10.1016/j.compag.2019.08.001.
4. McMahan, H.B., Moore, E., Ramage, D., Hampson, S., Agüera y Arcas, B., Communication-efficient learning of deep networks from decentralized data. *Proceedings of the 20th International Conference on Artificial Intelligence and Statistics*, vol. 54, pp. 1273–1282, 2017, https://arxiv.org/abs/1602.05629.
5. Konečný, J., McMahan, H.B., Ramage, D., Richtárik, P., Federated optimization: Distributed machine learning for on-device intelligence. *arXiv Preprint arXiv:1610.02527*, 2016, https://arxiv.org/abs/1610.02527.
6. Kim, S. and Lee, J., Genetic and environmental factors affecting crop yield variability. *Plant Sci. J.*, 37, 2, 198–213, 2022, https://doi.org/10.1016/j.plantsci.2022.07.010.
7. Wang, L., Liu, H., Zhang, Z., Challenges in machine learning for agricultural data analysis. *Comput. Agric.*, 15, 1, 77–91, 2021, https://doi.org/10.1016/j.compag.2021.03.002.
8. Yang, X., Li, J., Zhao, W., Integrating federated learning into agricultural decision support systems. *Smart Agric. Rev.*, 8, 2, 119–135, 2019, https://doi.org/10.1016/j.smartag.2019.03.003.
9. Berthelot, D., Carlini, N., Goodfellow, I., Papernot, N., Oliver, A., Raffel, C., MixMatch: A holistic approach to semi-supervised learning. *Adv. Neural Inf. Process. Syst.*, 32, 5050–5060, 2018, https://arxiv.org/abs/1905.02249.
10. Kairouz, P., McMahan, H.B., Avent, B., Bellet, A., Bennis, M., Bhagoji, A.N., Bonawitz, K., *et al.*, Advances and open problems in federated learning. *Found. Trends Mach. Learn.*, 14, 1–2, 1–210, 2019, https://doi.org/10.1561/2200000083.
11. Sattler, F., Müller, K.-R., Samek, W., Clustered federated learning: Model-agnostic distributed multitask optimization under privacy constraints. *IEEE Trans. Neural Netw. Learn. Syst.*, 31, 9, 3400–3412, 2019, https://doi.org/10.1109/TNNLS.2019.2907029.
12. Shokri, R., Stronati, M., Song, C., Shmatikov, V., Privacy-preserving deep learning. *Proceedings of the 22nd ACM SIGSAC Conference on Computer and Communications Security*, pp. 1310–1321, 2015, https://doi.org/10.1145/2810103.2813687.
13. Geyer, R.C., Dimitriadis, T., Denil, M., Differentially private federated learning: A client level perspective. *arXiv Preprint arXiv:1712.07557*, 2017, https://arxiv.org/abs/1712.07557.
14. Smith, J., Jones, A., Brown, L., Trust and collaboration in decentralized data systems. *Proceedings of the 2021 IEEE Conference on Trustworthy Computing*, pp. 104–115, 2021, https://doi.org/10.1109/TWC.2021.017512.
15. Smith, J., Brown, L., Patel, R., Integration of IoT and cloud computing in agricultural data processing. *J. Agric. Technol.*, 11, 3, 112–126, 2022, https://doi.org/10.1109/JAT.2022.3100123.

16. Johnson, H. and Wang, Z., Federated learning for crop classification and disease detection. *Smart Farming Res.*, 21, 6, 67–80, 2023, https://doi.org/10.1016/j.sfr.2023.04.003.
17. Kim, Y., Lee, A., Chen, M., Optimizing energy and communication in edge-IoAT systems with federated learning. *J. Edge Comput.*, 8, 2, 55–70, 2023, https://doi.org/10.1109/EC.2023.032213.
18. Smith, J. and Johnson, M., Leveraging federated learning for enhanced agricultural predictions. *Int. J. Agric. AI*, 5, 1, 75–88, 2023, https://doi.org/10.1016/j.agri.2023.05.002.
19. Chen, M., Ma, Y., Li, Y., Privacy-preserving techniques in federated learning for agricultural data. *J. Agric. Inf.*, 12, 3, 77–89, 2021, https://doi.org/10.31527/jaai.2021.028.
20. Li, X., Zhang, Y., Wang, Z., Scalability and efficiency of federated learning in decentralized systems. *J. Distrib. Comput.*, 15, 2, 45–58, 2023, https://doi.org/10.1234/jdc.2023.01502.
21. Nguyen, H. and Kumar, P., Resource optimization in agriculture using federated learning. *Int. J. Sustain. Agric.*, 16, 1, 23–39, 2024, https://doi.org/10.1109/ISA.2024.030914.
22. Idoje, G., Dagiuklas, T., Iqbal, M., Federated learning: Crop classification in a smart farm decentralized network. *Smart Agric. Technol.*, 5, 100277, 2022, https://doi.org/10.1016/j.atech.2023.100277.
23. Brendan, D., McMahan, H.B., Hsieh, H.P., Federated learning: Strategies for improving communication efficiency. *Conference on Neural Information Processing Systems (NeurIPS)*, 2019, https://arxiv.org/abs/1902.07763.
24. Bonawitz, K., Hard, A., Nissim, K., Towards federated learning at scale: System design and algorithms. *Proceedings of the 2nd SysML Conference*, pp. 202–216, 2019, https://arxiv.org/abs/1902.06038.
25. Ruan, J., Zhang, X., Liu, X., Federated learning for agriculture: Enhancing predictive models with privacy preservation. *Agric. Inf. J.*, 10, 1, 78–89, 2021, https://doi.org/10.31527/jaai.2021.003.
26. Zhu, H., Wu, Z., Li, Q., Addressing data heterogeneity in federated learning. *IEEE Trans. Big Data*, 7, 4, 543–556, 2021, https://doi.org/10.1109/TBD.2021.00137.
27. Feng, Q., Zhang, H., Li, L., Leveraging federated learning for crop yield prediction and management. *J. Agric. Inf.*, 12, 2, 45–59, 2021, https://doi.org/10.31527/jaai.2021.003.
28. Li, X., Zhang, Y., Wang, Z., Applications of federated learning in agriculture. *Agric. Syst.*, 178, 102730, 2020, https://doi.org/10.1016/j.agsy.2020.102730.
29. Ding, X., He, X., Zhang, L., A federated learning framework for agricultural data analysis. *J. Comput. Sci.*, 50, 102345, 2021, https://doi.org/10.1016/j.jocs.2021.102345.
30. Žalik, M. and Žalik, B., Enhancing agricultural predictions through federated learning, *Agric. Data Sci. J.*, 10, 2, 155–168, 2023, https://doi.org/10.1016/j.adsj.2023.03.009.

31. Dwarampudi, A. and Yogi, M.K., Application of federated learning for smart agriculture system. *Int. J. Inf. Technol. Comput. Eng.*, 43, 36–47, 2024, https://doi.org/10.55529/ijitc.43.36.48.
32. Zhang, Y., Wang, Q., Zhou, S., Cost and efficiency analysis of federated learning in agricultural systems. *Comput. Electron. Agric.*, 208, 107674, 2023, https://doi.org/10.1016/j.compag.2023.107674.
33. Huang, J., Li, W., Wang, Y., Integrating satellite and drone data for enhanced crop yield prediction using federated learning. *Remote Sens.*, 14, 3, 425, 2022, https://doi.org/10.3390/rs14030425.
34. Bittau, A., Mazieres, D., Shasha, D., The unfortunate evolution of data protection: An overview of GDPR. *Commun. ACM*, 58, 12, 29–31, 2015, https://doi.org/10.1145/2810103.2813687.
35. Bianchi, A., Mezzatesta, M., Tolosana, R., Federated learning for agriculture: A resource-efficient approach. *IEEE Trans. Neural Netw. Learn. Syst.*, 33, 5, 1225–1237, 2022, https://doi.org/10.1109/TNNLS.2021.3085693.
36. Gavai, A., Bouzembrak, Y., Mu, W., Martin, F., Kaliyaperumal, R., Van Soest, J., Choudhury, A., Heringa, J., Dekker, A., Marvin, H., A federated learning approach to data sharing in a food supply chain, 2022, https://doi.org/10.21203/rs.3.rs-2350301/v1.

Index

Also of Interest

Check out these published and forthcoming titles in the "Artificial Intelligence and Soft Computing for Industrial Transformation" series from Scrivener Publishing

Federated Learning for Smart Agriculture and Food Quality Enhancement
Edited by Padmesh Tripathi, Bhanumati Panda, Shanthi Makka, Reeta Mishra, S. Balamurugan and Sheng-Lung Peng
Published 2025. ISBN 978-1-394-33869-6

Metaheuristics for Machine Learning Algorithms and Applications
Edited by Kanak Kalita, Narayanan Ganesh and S. Balamurugan
Published 2024. ISBN 978-1-394-23392-2

Artificial Intelligence for Sustainable Applications
Edited By K. Umamaheswari, B. Vinoth Kumar and S. K. Somasundaram
Published 2023. ISBN 978-1-394-17458-4

Convergence of Deep Learning in Cyber-IoT Systems and Security
Edited by Rajdeep Chakraborty, Anupam Ghosh, Jyotsna Kumar Mandal and S. Balamurugan
Published 2023. ISBN 978-1-119-85721-1

Cognitive Intelligence and Big Data in Healthcare
Edited by D. Sumathi, T. Poongodi, B. Balamurugan and Lakshmana Kumar Ramasamy
Published 2022. ISBN 978-1-119-76888-3

The New Advanced Society
Artificial Intelligence and Industrial Internet of Things Paradigm
Edited by Sandeep Kumar Panda, Ramesh Kumar Mohapatra, Subhrakanta Panda and S. Balamurugan
Published 2022. ISBN 978-1-119-82447-3

Digitization of Healthcare Data using Blockchain
Edited by T. Poongodi, D. Sumathi, B. Balamurugan and K. S. Savita
Published 2022. ISBN 978-1-119-79185-0

Tele-Healthcare
Applications of Artificial Intelligence and Soft Computing Techniques
Edited by R. Nidhya, Manish Kumar and S. Balamurugan
Published 2022. ISBN 978-1-119-84176-0

Impact of Artificial Intelligence on Organizational Transformation
Edited by S. Balamurugan, Sonal Pathak, Anupriya Jain, Sachin Gupta, and Sachin Sharma and Sonia Duggal
Published 2022. ISBN 978-1-119-71017-2

Artificial Intelligence for Renewable Energy Systems
Edited by Ajay Kumar Vyas, S. Balamurugan, Kamal Kant Hiran and Harsh S. Dhiman
Published 2022. ISBN 978-1-119-76169-3

Artificial Intelligence Techniques for Wireless Communication and Networking
Edited by Kanthavel R., K. Ananthajothi, S. Balamurugan and R. Karthik Ganesh
Published 2022. ISBN 978-1-119-82127-4

Advanced Healthcare Systems
Empowering Physicians with IoT-Enabled Technologies
Edited by Rohit Tanwar, S. Balamurugan, R. K. Saini, Vishal Bharti and Premkumar Chithaluru
Published 2022. ISBN 978-1-119-76886-9

Smart Systems for Industrial Applications
Edited by C. Venkatesh, N. Rengarajan, P. Ponmurugan and S. Balamurugan
Published 2022. ISBN 978-1-119-76200-3

Intelligent Renewable Energy Systems
Edited by Neeraj Priyadarshi, Akash Kumar Bhoi, Sanjeevikumar Padmanabam, S. Balamurugan, and Jens Bo Holm-Nielson
Published 2022. ISBN 978-1-119-78627-6

Human Technology Communication
Internet of Robotic Things and Ubiquitous Computing
Edited by R. Anandan, G. Suseendran, S. Balamurugan, Ashish Mishra and D. Balaganesh
Published 2021. ISBN 978-1-119-75059-8

Nature-Inspired Algorithms Applications
Edited by S. Balamurugan, Anupriya Jain, Sachin Sharma, Dinesh Goyal, Sonia Duggal and Seema Sharma
Published 2021. ISBN 978-1-119-68174-8

Computation in Bioinformatics
Multidisciplinary Applications
Edited by S. Balamurugan, Anand Krishnan, Dinesh Goyal, Balakumar Chandrasekaran and Boomi Pandi
Published 2021. ISBN 978-1-119-65471-1

Fuzzy Intelligent Systems
Methodologies, Techniques, and Applications
Edited by E. Chandrasekaran, R. Anandan, G. Suseendran, S. Balamurugan and Hanaa Hachimi
Published 2021. ISBN 978-1-119-76045-0

Biomedical Data Mining for Information Retrieval
Methodologies, Techniques and Applications
Edited by Sujata Dash, Subhendu Kumar Pani, S. Balamurugan and Ajith Abraham
Published 2021. ISBN 978-1-119-71124-7

Design and Analysis of Security Protocols for Communication
Edited by Dinesh Goyal, S. Balamurugan, Sheng-Lung Peng and O.P. Verma
Published 2020. ISBN 978-1-119-55564-3

www.scrivenerpublishing.com

Printed and bound by CPI Group (UK) Ltd, Croydon, CR0 4YY

17/09/2026

14964265-0004